AF651654

PERSPECTIVES IN GEOMORPHOLOGY

VOLUME ONE

By the Same Author

- Climate Change: Bio-Diversity and Green Economy with *S. Padmaja* and *Ganesh Sharma*
- Indian Geomorphology
- Mathematical Modelling in Geographical Information System Global Positioning System and Digital Cartography with *D.R. Ram, Rama Prasad* and *P.R. Binda*
- Modeling in Resource Management and Environment Through Geomatics with *P.R. Binda*
- Ranthambhore Sanctuary: Dilemma of Eco-Development
- Ravine Erosion in India
- Sustainable Development: Issues and Case Studies with *Sri K. Chattopadhyay*
- Tropical Geomorphology: A Morphogenetic Study of Rajasthan

About the Editor

Prof. H. S. Sharma (*b.* 1945) holds post-graduate and doctoral degrees in Geography from the University of Saugar, Madhya Pradesh. Prof. Sharma is a former Professor and Head, Department of Geography, University of Rajasthan, Jaipur. He is also a former Director of the Indira Gandhi Centre for Human Ecology, Environment and Population Studies (HEEPS), Academic Staff College and Dean, Faculty of Science, University of Rajasthan.

Prof. Sharma has over 100 research papers in National and International journals like *Nature* and 16 books to his credit. He has supervised more than three dozen Ph.D.s. Prof. Sharma has travelled widely across U.K., U.S.A., Germany, Holland, Canada, Norway, Seychelles and Malaysia to attend and chair technical sessions in International Conferences and Exchange Programmes.

Prof. Sharma has successfully conducted several research projects sponsored by the U.G.C., Ministry of Environment and Forests, DRDO, New Delhi, NORAD, Norway and ISRO (SAC) Ahmedabad.

Prof. Sharma has been President of IIG, IGI and NAGI. He is also a President of the Commonwealth Human Ecology Council (India Chapter). Recently, he has been awarded Senior Fellowship by Indian Council of Social Science Research, New Delhi to work on Climate Change and its Impact on Land and Water Resources of Rajasthan.

Concept's International Series in Geography No. 2

PERSPECTIVES IN GEOMORPHOLOGY

Volume One

Recent Trends

Edited by
H.S. Sharma

CONCEPT PUBLISHING COMPANY PVT. LTD.
NEW DELHI-110059

© 1981 H.S. Sharma

All rights reserved. No part of this work may be reproduced, stored, adapted, or transmitted in any form or by any means, electronic, mechanical, photocopying, micro-filming recording or otherwise, or translated in any language, without the prior written permission of the copyright owner and the publisher. The book is sold subject to the condition that it shall not, by way of trade or otherwise, be lent, resold, hired out, or otherwise circulated without the prior publisher's written consent in any form of binding or cover other than that in which it is published.

The views and opinions expressed in this book are author(s) own and the facts reported by them/her have been verified to the extent possible, and the publishers are not in any way liable for the same.

ISBN-13: 978-81-7022-123-4 (Set)

Edition 2016

Published and Printed by

Concept Since 1974

Concept Publishing Company Pvt. Ltd.
A/15-16, Commercial Block, Mohan Garden
New Delhi-110059 (India)
T: 25351460, 25351794, *F*: 091-11-25357109
E: publishing@conceptpub.com, W: www.conceptpub.com
Editorial Office: H-13, Bali Nagar, New Delhi-110 015, India.

Cataloging in Publication Data--Courtesy: D.K. Agencies (P) Ltd. <docinfo@dkagencies.com>

Perspectives in geomorphology / edited by H.S. Sharma.
volumes cm. -- (Concept's international series in geography ; no. 2)
Reprint. Originally published: 1981-1982.
Includes bibliographical references and index.
Contents: volume 1. Recent trends -- volume 2. Quantitative fluvial geomorphology -- volume 3. Applied geomorphology volume 4. Essays on Indian geomorphology.
ISBN 9788170221234 (set)

1. Geomorphology. 2. Geomorphology--India. I. Sharma, H. S. (Hari Shanker), 1945- editor. II. Series: Concept's international series in geography ; no. 2.

GB401.5.P47 2016 DDC 551.4 23

PUBLISHER'S NOTE

THIS IS the second set of volumes in our International Geography Series. It follows close upon the heels of the five-volume set on *Perspectives in Agricultural Geography* edited by Dr. Noor Mohammad. We take this opportunity at the very outset to acknowledge the co-operation we have received from the Editor Dr. Noor Mohammad and the numerous contributors from all over the world who have made Series One possible. The wide acclaim and the heartening response from our readers—researchers and students in the field of agricultural geography—have encouraged us to plan the publication of this second set entitled *Perspectives in Geomorphology* edited by Dr. Hari Shanker Sharma. Here again we would like to thank the Editor and the contributors from various countries of the world who have given a truly international perspective to these volumes. We are confident that this series will also be equally well-received all over the world.

The Concept Series in Geography is now an on-going one and we are very happy to announce that plans are afoot to make available a third set of volumes in the very contemporary subject of Urban Geography. This series is expected to run into fifty volumes and the publication programme, hopefully, will commence early in 1982. The response to our brochure has been very encouraging and more than 150 articles have already been received including several from overseas contributors.

With the publication of the initial volumes of this new series we hope to announce the subject of the fourth series so that both students and prospective contributors will receive information in advance.

In the next few years we hope in this manner, to provide geographers with a complete compendium of the latest advances in major areas in geography. Here, we might emphasize again, that the benefits that the students in this country as well as in other developing countries will obtain will be immeasurable.

Finally, we crave once again, the indulgence of our readers for any shortcomings in these volumes. We shall learn with experience.

PREFACE

In the last three decades there has been considerable progress in all branches of science. The science of geomorphology has also achieved a great deal during this time, and geographers, geologists, hydrologists, pedologists and specialists from other fields who have contributed to the study of geomorphology, have become aware of the comparative suddenness of the change within the context of land form studies. As a matter of fact, it is becoming virtually impossible for a single scholar to command all aspects of geomorphology. However, there has been increasing specialization at all levels on the one hand and a need for synthesising the results of these researches, to benefit the scholars working in the field of geo-science.

The present project which has been sponsored by the Concept Publishing Company, New Delhi is an humble attempt to disseminate the knowledge of concepts, theories, models and recent trends in geomorphology with reference to the present conceptual revolution in this field. With this ambition in view original research papers dealing with modern quantitative techniques were invited from geo-scientists from all over the world. The basic themes along with the tentative titles of the papers were sent to the contributors in various branches of geo-science. The contributors have responded well with original research papers.

Later on, these papers were thoroughly edited and the papers found worthy for publication were classified and arranged into four volumes.

The first volume of the series contains the papers dealing with the modern trends in geomorphology. This volume has been divided into five sections. Section one deals with the stream lines of geomorphic research, climatic and climato-genetic geomorphology, planation surfaces, granitic inselbergs and duri-crusts. Section two discusses Karst processes and land forms. Papers dealing with glacial and peri-glacial land-forms have been included in Section three. Section four contains papers dealing with the coastal processes and land forms in tropical region. Section five contains the papers dealing with remote sensing and geomorphology.

As considerable amount of work has been done in fluvial geomorphology, therefore, volume two has been devoted to the papers dealing with the various aspects of quantitative fluvial geomorphology.

Volume three contains papers on applied geomorphology. Papers dealing with land classification, geomorphological mapping, terrain evaluation, and application of photo-interpretation and remote sensing in geomorphological mapping have been included in this volume.

Geo-scientists in India have also become aware and realised the significance of land form studies in their researches. As a matter of fact considerable amount of work has been done in various aspects of geomorphology. The fourth volume of the series contains papers on Indian geomorphology. It is earnestly hoped that these volumes will serve the purpose of enlightening the readers regarding the kind of research which is being taken up by modern geomorphologists today. As the editor of these volumes and, in addition a contributor, I count myself highly fortunate in my fellow contributors. All those whose papers are included in these volumes have prepared and supplied their material in the face of manifest difficulties—mainly the difficulties of time which pressed hard upon every busy academician. Most of my fellow contributors have patiently dealt with manuscripts and with queries relating to their papers. Those who sent their materials early have been tolerant of the subsequent delay, as late comers have tolerantly borne pressure from myself to complete their papers.

It would be ungenerous to conclude this introduction without acknowledging the guidance, encouragement and help of a number of senior scientists.

I am indebted to Professor Moonis Raza for his suggestions in making the outline of the project during my visit to the Centre for Regional Development, Jawaharlal Nehru University, New Delhi from January to April, 1980. I am highly thankful to Prof. Vaidyanadhan of Waltair University and Prof. Pramila

Kumar of Saugar University for their suggestions and encouragement. I am also thankful to Shri P.S. Dhinwa and Shri Ram Avatar Sharma for their assistance throughout the preparation of these volumes. Staff members of the Geography Department, University of Rajasthan, Jaipur, also deserve thanks for their co-operation and kindness in the preparation of this project. Special thanks are due to the publisher who has calmly suffered the repeated redistribution of authors throughout the world and who maintained effective contacts with myself. In addition I may suitably take this occasion to make a block statement of gratitude, on behalf of all the contributors, to the numerous friends and colleagues who have at various stages read part or all of the manuscripts and have supplied comments, criticisms and suggestions.

Department of Geography
Rajasthan University
Jaipur (India)

H.S. SHARMA

Independence Day
15 August 1981

CONTENTS

Publisher's Note v
Preface vii
List of Contributors xiv
List of Plates xvi
List of Figures xx
List of Tables xxiv

SECTION ONE

RECENT TRENDS, CLIMATE AND PLANATION SURFACES

Introduction 3

1. Modern Trends in Geomorphology
—*André Cailleux* and *Lucien Romani* 9

2. Climatic Influence on Geomorphology
—*J. Tricart* 21

3. Climatogenetic Geomorphology
—*Cuchlaine A.M. King* 37

4. Planation Surfaces : Theory and Reality
—*C.R. Twidale* 63

5. Peneplanation Planes in an Alpine Country : Georgia as an Example
—*L.I. Maruashvili* 83

6. Has Deccan Foreland Rejuvenated ?
—*Pramila Kumar* and *R.K. Rai* 101

7. Polygenetic Erosion Surfaces of the Barakar Basin
—*Nageshwar Prasad* and *Md. Naseemuddin* 115

8. Granitic Inselbergs
—*C.R. Twidale* 127

9. Geomorphology of Jawaharlal Nehru University, New Campus, New Delhi : A Study of Quartzitic Tors
—H.S. Sharma 169

10. Geomorphological Significance of Duricrusts, with Special Reference to Australia
—C.R. Twidale 187

SECTION TWO

KARST—PROCESSES AND LAND FORMS

INTRODUCTION 207

11. Karst—Karst Denudation—Chemical Erosion—Climate
—Ivan Gams 211

12. The Problem of Cavern Formation
—J.N. Jennings 223

13. Features of Mountainous Karst of the Southern Part of the USSR : Georgia as an Example
—L.I. Maruashvili 255

SECTION THREE

GLACIAL AND PERI-GLACIAL PROCESSES AND LAND FORMS

INTRODUCTION 279

14. Ancient Glaciation of the Caucasus (The USSR)
—L.I. Maruashvili and *R.D. Khazaradze* 281

15. Periglacial Geomorphology : Present Problems and Future Prospects
—Jaromir Demek 301

SECTION FOUR

COASTAL LAND FORMS

INTRODUCTION 325

16. Coastal Landforms
—*C.A.M. King* 327

17. Coastal Landforms in the Humid Tropics
—*E.C.F. Bird* 351

18. A New Concept—Submarine Terrain Evaluation
—*Jag Mohan Agrawal* 375

19. Cartographic Practices Used in Representation of Submarine Terrain
—*Satya Prakash* and *J.M. Agrawal* 381

SECTION FIVE

REMOTE SENSING AND GEOMORPHOLOGY

INTRODUCTION 399

20. Remote Sensing for Geomorphology
—*K.R. Rao* and *Naresh Bedi* 401

INDEX 419

LIST OF CONTRIBUTORS

Jag Mohan Agrawal
Principal, Civil Hydrographic Officer, Naval Hydrographic Office, Dehra Dun, India

Naresh Bedi
Scientist, National Remote Sensing Agency, No. 4, Sardar Patel Road, Secunderabad, India

E.C.F. Bird
Department of Geography, University of Melbourne, Parkville, 3052

Andre Cailleux
Laboratory De Geomorphologie, 9, Avenue de la Tremouille, 94100 ST. Maur-des-Fosses, France

Jaromir Demek
Ceskoslovenska Geograficka Spolecnost,
Pr. Ceskoslovenske Akademii Ved, Sekretariat 66282 Brno, Mendlovo nam la

Ivan Gams
Department of Geography, Askerceva 12/II, 61000, Lujubljana—Yugoslavia

J.N. Jennings
Department of Biography and Geomorphology, Australian National University, Post Box No. 4, Canberra ACT 2600

R.D. Khazaradze
Geomorphologic Department of Vakhushti Institute of Geography, the Academy of Sciences of the Georgian SSR, Tbilisi, USSR

Cuchlaine A.M. King
Department of Geography, University of Nottingham, University Park, Nottingham, N67 2RD

L. I. Maruashvili
Chief of the Geomorphologic Department of Vakhushti Institute of Geography, the Academy of Sciences of the Georgian SSR, Tbilisi, USSR

Md. Naseemuddin
Lecturer in the Post-Graduate Department of Geography, Bhagalpur University, Bhagalpur, India

Pramila Kumar
Professor of Geography, University of Saugar, Sagar, India

Nageshwar Prasad
Department of Geography, Bhagalpur University, Bhagalpur, India

R.K. Rai
Reader in Geography, North Eastern Hill University, Shillong, India

K.R. Rao
Director, National Remote Sensing Agency, No. 4, Sardar Patel Road, Secunderabad, India

Lucien Romani
Physicist, Director of the Laboratorie Eiffel, France

Satya Prakash
Principal, Civil Hydrographic Officer, Naval Hydrographic Office, Dehra Dun, India

H.S. Sharma
Career Scientist, (U.G.C.) Department of Geography, University of Rajasthan, Jaipur, India

J. Tricart
Director du C.G.A., 43 rue Goethe, 67083 Strasbony Cedex—France

C.R. Twidale
Department of Geography, University of Adelaide, Box 498, G.P.O. Adelaide, South Australia-5001

LIST OF PLATES

Chapter Four

1. Deep narrow gutters cut in grus (weathered granite) near Paarl, Cape Province, South Africa. The dry cemented crust acts as a caprock and retards widening of the gutters

2. (a) Pediments cut in schist, Lucerne Valley, Southern California

2. (b) Pediments cut in siltstone and with coarse debris mantle fringing the Flinders Ranges near Brachina, South Australia

3. The granite platform bordering Corrobinnie Hill, on Northern Eyre Peninsula, South Australia, is dumped and grooved and extends 700-800 m from the base of the residual. Patches of grus stand on the platform and there are also small blocks and boulders—erstwhile corestones

4. (a) Isolated inselbergs surrounded by low angle pediment cones near Broken Hill, Western New South Wales. The inselbergs are formed of discrete masses of quartz and the plains are cut in schist

4. (b) Pediment aprons cut in granite and surrounded by inselbergs of the same material in Lebowa, northern Transvaal, South Africa

4. (c) Pediment and inselberg landscape developed in crystalline rocks in Arizona (W.C. Bradley)

5. (a) The remarkably flat plains bordering the Kuiseb Canyon in central Namibia are cut in schists. The residual hills are granitic or gneissic

5 (b) The plains around Meekatharra in central western Australia are cut in granitic rocks

Chapter Eight

1. Granite bornhardts standing abruptly above pediplain near Uwerus, central Namibia
2a. Castle koppie a few kilometres northeast of Salisbury, Zimbabwe/Rhodesia
2b. Haytor West, a tor or castellated residual on eastern Dartmoor, southwestern England
3. Block strewn granite nubbin near Naraku, northwest Queensland
4. Remarkable Rocks, Kirkpatrick Point, southwestern coast of Kangaroo Island, South Australia, showing angular blocks standing on bare dome protruding from aeolianite (wind-blown calcarenite)
5. Stock of gneissic granite emplaced in schists, Namib Game Park SSW of Karibib, central Namibia
6. Isolated domes of porphyritic foliated granite near Nanutarra, western Pilbara region, Western Australia
7. Domes exposed in escarpment below African Land Surface (X) near Witrivier, eastern Transvaal. The Bushman's Kop, the prominent residual at right clearly stood above the general level of the African plain.
8. The Armchair and other granite domes in the northern Flinders Ranges, South Australia, occur below the level of a pronounced summit surface which is of exhumed type and of Lower Cretaceous age
9. A reservoir the northwestern edge of Ucontitchie Hill northwestern Eyre Peninsula, South Australia, showing the contrast in joint density implied by the small corestones exposed in the excavation and the massive rock of the inselberg
10. Quarry face at Ebaka, southern Cameroun, showing domical compartment of massive granite, overlain and bordered laterally by weathered granite
11. The Leeukop, in western Transvaal, with, in the foreground, the crest of an incipient dome (X) of which no sign existed in the natural landscape but which has been exposed in a shallow reservoir excavation
12. Rille or gutters extent beneath the natural soil level, and indeed, here at Dumonte Rock, near Wudinna,

northwestern Eyre Peninsula, covered in the sub-surface. X-X is the natural soil level, and the gutters are exposed in an excavation intended for water storage

13. Flared slope and platform, side of Ucontitchie Hill, northwestern Eyre Peninsula
14. Yarwondutta Rock and reservoir, northwestern Eyre Peninsula, showing flared marginal slope of the residual, and the concave weathering front exposed in the artificial excavation. The shoulders (S) mark the present natural soil level and hill-plain junction and the former piedmont
15. Part of northwestern slope of Yarwondutta Rock showing the stepped morphology and the shoulders (S) marking former hill-plain junctions
16. The foliation of the rock is expressed in well developed dipping cleavage which is exploited by weathering. Below the shoulder (right) which is a remnant of a palaeosurface above which stand the major domed forms, the weathering is more pronounced presumably because it took place in the subsurface
17. The Humps, in the southwest of Western Australia is a group of granite domes surrounded by a plain on which laterite is well developed
18. Subangular to subrounded corestones are being exposed on Tcharkulda Hill, northwest Eyre Peninsula, as the grus matrix is washed away. Some (X), *in situ* are seen sitting on the platform (foreground) and below the large residual boulders
19. Part of the Enchanted Rock complex in the Llano of central Texas, with, on the lower slope, a thick sheet of rock that has broken up into large polygonal slabs. A few remnants of the sheet stand in isolation on the upper slope face
20. Eastern side of Ucontitche Hill, Eyre Peninsula, showing multiple flares and exposed end of sheeting joint
21. Castle Rock, east of Albany in the southwest of Western Australia is an angular blocky residual standing on the crest of a broad granite dome. The flares on its sidewalls indicate that it has been exposed

in at least two phases, each comprising subsurface piedmont weathering followed by lowering of the plain and exposure of the erstwhile weathering front

22. Large residual blocks and boulders resting on large radius dome at the Devil's Marbles, N.T.
23. Angular residual or castle koppie in the summit surface (Tosa de Gargantillar), altitude circa 2640 m, in the Andorran Pyrenees
24. Angular, frost-riven castellated granite residual located on the high plain surface of the Rocky Mountains in Colorado
25. Angular granite residual in the Agulhas Negras at an altitude of 2787 m in the Serra da Mantiquiera, Rio Province, Brazil

Chapter Ten

1. (a) Silcrete mesa and plateau, central Australia (CSIRO)
   (b) Butte eroded in granite capped by laterite south of Cloncurry, northwest Queensland (CSIRO)
2. Ridges capped by folded silcrete
   (a) Southwest Queensland
   (b) Oodnadatta area of northern south Australia
3. Calcrete mesa in the Ashburton valley, northwest of Western Australia
4. Western fault-scarp of Lake Eyre salina showing coastal cliffs capped by gyprete
5. Silcrete gibber or hamada near Kamaran Downs, southwestern Queensland
6. The Humps, a complex of granite domes or bornhardts in the southwest of Western Australia, with laterite exposed in a shallow quarry (foreground) at its base

Chapter Thirteen

1. The goffers of calcite in the cave of Mtiskalta
2. The swallow holes on the Bzipi range
3. The macaroni type stalactites in the Abrskil cave
4. The Arsen Okrojanashvili's cave
5. The karst landscape of the Okhachkue massif

Chapter Fourteen

1. Chalaati glacier
2. Peak Ushba
3. Sakeni erratic block in Apkhazeti
4. View of one part of central Caucasus from a slope of Mt. Elbrus
5. Discharging of Irtishi glacier into Triberi glacier
6. Adishi glacier

Chapter Twenty

1. Multispectral data Analysis M Das—A Computer system for analysing landsat data in CCT form
2. Multispectral Additive colour viewer
3. Landsat RBV picture of a Part of Andhra Pradesh, India, dt. 15-3-80
4. Metsat (AVHRR) picture of India, Middle-East and Arabian Sea, dt. 13-11-79, shows generation of cyclone
5. Landsat MSS image, Band 7, of a Part of Karnataka, India, dt. 6-1-73

LIST OF FIGURES

1.1 Inclinaison Moyenne de l'orbite du Satellite Sur l' Equateur de la Planete en Degres *13*
2.1 The Structure of the Ecotope *30*
4.1 Diagram showing stages of youth, maturity and old age in peneplanation (after Holmes, Principle of Physical Geology) *66*
4.2 Extract from South Australian Lands Department. Topographic map series 1:500000 sheet Pordia 5921-V, showing granite inselberg surrounded by lo angle pediment cores that merge downslope with a rolling peneplain *70*
4.3. Diagrammatic section illustrating constrast between backwearing (left) and downwearing (right) *71*

5.1 Scheme of the main orographical parts of Transcaucasian (scale, 1:500 000) *86*
5.2 Geomorphological scheme of Dzirula Massif *88*
5.3 Location of Akhuni Mountain and Mountain Range (stairs of Erosional Surfaces) *91*
5.4 Longitudinal profile of Akhuni Range *92*
6.1 The Deccan Foreland *103*
6.2 The Sonar-Bearma Basin and adjoining area superimposed profiles *106*
6.3 The Sonar-Bearma Basin and adjoining area—altimetric frequency curve *107*
6.4 The Sonar-Bearma Basin : Erosion surfaces *108*
7.1 The Barakar Basin—Physiognomy *116*
7.2 The Barakar Basin—Erosion surfaces *117*
8.1 The Everard ranges, northern south Australia, showing large fracture delineated blocks *128*
8.2 Field sketch of the Swakop valley, south of Karibile, Central Namibia, showing palaeosurface (X) and domes and nubbins exposed in the valley floor *129*
8.3(a) Stages in the baselevelling of land mass by scarp retreat and pedimentation and (iii) continued erosion following stream rejuvenation *131*
8.3(b) Two stages development of inselbergs showing (i) deep fracture controlled subsurface compartment weathering and (ii) differential erosion of the compartments *131*
8.4 North western Eyre Peninsula showing distribution of granite outcrops, major structural lines and (inset) regional lineaments *133*
8.5(a) Association of inselbergs and palaeodrainage in part of the wheat belt of southwestern Australia *135*
8.5(b) Shashe river and Castle Koppie, Zimbabwe *136*
8.6 Fracture controlled margins of some granite bornhardts *137*
8.7 Distribution of stress in an antiform, t-tension, np-neutral plane, c-compression *138*
8.8 Development of flared slopes by subsurface moisture attack in the pediment, followed by erosion and exposure of the concave weathering front *142*

8.9 Contour plan of Yarwondutta rock with X the site from which plate 15 was taken, looking northeast, and S the shoulders of the flared slopes *143*

8.10a,b,c,d Plans of Kokerbin, (W.A.) Carappee, (S.A.). Mt. Stirling bornhardts, (W.A.), and the Humps, (W.A.), showing stepped form *146*, *147*, *148*, *149*

8.11 Diagrammatic section through northern Eyre Peninsula showing relationship between duricrusts and flared zones on granitic residuals *151*

8.12 Inselberg near Gokomere, Zimbabwe, showing marginal disintegration of dome to form boulder-strewn nobbin areas (drawn from air photographs) *155*

8.13 Granite outcrops in the Mount Wuddinna area of north western Eyre Peninsula, showing small areal extent of the Cottage Loaf (drawn from air photographs) *158*

8.14 Morphological Map of the Devil's Marbles area N.T. *159*

9.1 Delhi Metropolitan *170*

9.2 Delhi and Environs—Superimposed Profile *175*

9.3 The two stage development of Quartzitic Tors *178*

9 4 Slope Histogram J.N.U. Campus, New Delhi *182*

9.5 Geomorphological Map : Jawaharlal Nehru University *183*

10.1 Diagrammatic cross-section through the Beda Valley and Camp Hill, a remnant of the quartzite-capped Arcoona Plateau *190*

10.2 Diagrammatic section through the folded silcrete involved in the Betoota and Curalle anticlines, south-west Queensland *193*

10.3 Folded silcrete in the northeastern piedmont of the Flinders Ranges, south Australia *194*

10.4(a) Laterite plateaux and high plains in the gulfs regions of south Australia *196*

10.4(b) Laterite and bedrock geology of part of northern Kangaroo Island *197*

10.5 Distribution of laterite and silcrete in Australia *200*

11.1 Relation between the discharge of the proper karst river in Provence of South France *213*

11.2 Solutional deepening of the depressions *217*
11.3 Climate conditions *218*
11.4 Corrosion as non-form-decisive process stands behind the proper karst *219*
12.1 Mixing corrosion *227*
12.2 Offloading of valleyside by erosion to cause sheeting joints (after Renault) distribution of stresses indicated by ellipsoids *230*
12.3 Putrid Pit, Bungonia Caves, New South Wales open shaft cave in joint planes due to Vadose seepage solution *234*
12.4 Cathedral cave, Wellington, New South Wales *236*
12.5 Barber Cave *239*
12.6 Four states of dynamic phreatic cake *240*
12.7 Development of an Alpine a karst cave of the Holloch type *246*
13.1 Scheme of location of karst Massifs of Western Georgia *259*
13.2 The schematic plan of Akhali Atone Cave *260*
13.3 Vertical section of Tooliani Precipice of the Bzibi range *266*
13.4 Plan of the multi-storeyed cave of Tsutskhvati *269*
14.1 Maximum pleistocene glaciation of the Caucasus *283*
15.1 Schematic representation of alas development as a result of permafrost degradation *307*
15.2 Main cryogenic textures caused by syngenetic freezing of unconsolidated deposits *210*
15.3 Syngenetic freezing of lake-floor deposits *310*
15.4 Diagram to show freezing of loose deposits *311*
17.1 Humid tropical coasts *354*
19.1 Portion of Hydrographic Survey *383*
19.2 Portion of a Navigational Chart *384*
19.3 Portion of an Oceanic Plotting Sheet *385*
19.4 Portion of a General Bathymetric Chart of Ocean *386*
19.5 Portion of a Shaded Relief Chart *388*
19.6 From Canadian Hydrographic Department *390*
19.6(a) Portion of Shaded Relief Chart (from National Geographic Magazine) *395*

20.1 Regional Geological map of parts of Karnataka and Kerala *411*
20.2 Regional Geomorphological map of Karnataka and Kerala *413*

LIST OF TABLES

1.1 The natural satellites of Mors *14*
1.2 Density of Phobos *14*
1.3 Ravonis Mons *17*
3.1 Climatic zones after Budel 1957 *43*
3.2 Rates of regional erosion after Corbel (1964) in m^3/km^2/-year *48*
3.3 Climate related to sediment yield according to Fournier 1960 *49*
3.4 Relationship between mean slope and drainage density after Peltier 1962 *51*
3.5 Climate morphological zones of the earth after Budel 1948 *53*
3.6 Morphogenetic regions according to Peltier 1950 *57*
5.1 Erosion steps of the Akhuni Ridge *93*
8.1 Many stepped Inselbergs have been noted in southern-western and central Australia *145*
14.1 Distribution of the number and areas of the contemporary glaciers at the Caucasus *284*
14.2 Dimensions of the largest contemporary glaciers of the great Caucasus *284*
17.1 Humid Tropical Sectors *385*
20.1 Geomorphic units obserable on Aerospace Imagery *409*

SECTION ONE

RECENT TRENDS, CLIMATE AND PLANATION SURFACES

INTRODUCTION

GEOMORPHOLOGY is concerned with the study of landforms, i.e., with the explanation of the present relief on the earth and its development. In recent years there has been rapid advancement in geomorphological knowledge with the use of sophisticated tools and techniques and current trends of research have not only put geomorphology on sound footing but have proved its utility beyond in solving human problems. The study of landforms can be achieved through a variety of methods which corresponds to alternative concepts in geomorphological research, e.g., the descriptive (morphography) or the historic-genetic approach. In recent years as a third approach the functional or dynamic concept has found increasing acceptance by students of geomorphology (Ahnert 1973 and 1978, Joly, 1977).

As compared with the historic-genetic approach which continues to be a strong and necessary part in the study of landforms (Semmel, 1979) the dynamic geomorphology is mainly concerned with processes and relation (including feedback) between form, material and processes. General Systems theory (Chorley 1962, Chorley and Kennedy, 1971), model construction, applied physics and applied chemistry are fundamentals. The growing research in this field is represented by journals such as "Catena" or "Earth Surface Processes". During the last years under the guideline of the dynamic or functional concept in geomorphology quite an amount of new data and information on geomorphological processes has been found. All parts of geomorphology have been involved. The current status of research was well represented at the First Benelux Colloquium on geomorphological processes (April 1976 in Leuven), Belgium; papers published in "Earth Surface Processes" 2, 2/3, 1977 or by the summary of F. Ahnert (1978, 1954-1962).

In spite of a larger amount of field work, field or laboratory experiments and of model construction more questions are open than answered.

The subsequent papers in the present volume reflects the diverse trends in geomorphology. Andre Cailleux and Lucien Romani in their paper on 'Modern Trends in geomorphology' have stressed the major tendencies of research with examples of Phobos and Diemos. The authors analysing the morphology of Phobos and Diemos have supported the Romani's hypothesis according to which Phobos and Diemos would be volcanic plugs of Pavones satellized during two paroxymal eruptions. Besides this, like other sciences geomorphology tends towards quantification, appeal to neighbouring sciences such as astronomy and vulcanology, importance of masses, energy and velocities, importance of time in the study and the solution of problems.

Climate exerts a significant influence on the development of landforms. The concept of climatic geomorphology has merged rather recently. The papers by Tricart and C.A.M. King have reviewed thc significance of climate in geomorphological studies. Tricart has made it clear that climatic geomorphology is somewhat dangerous expression. Some binary minds can be inclined to take it literally. In some papers climatic geomorphology has been opposed to structural geomorphology or to cyclic landforms. According to him this is not justified and they do not exist *per se*. They are not independent geomorphologies. They just focus upon certain aspects of landform shaping. Isolating them is an obstraction. Climatic geomorphology has no intrinsic reality, it is just a point of view, an aspect of geomorphology as a whole. He has recommended the systems approach as a logical tool for the study of climatic geomorphology. This approach enables us to be conscious of the complexities of our environment.

C.A.M. King has distinguished climato-genetic geomorphology from climatic geomorphology. In her paper she has discussed the nature and scope of climato-genetic geomorphology. According to her climato-genetic geomorphology is a complex field of study that covers a wide range of possible approaches. Its essential aim is to elucidate the processes that have shaped the landscape overtime by studying the present landforms.

The landforms should also provide evidence of present and past climate through a knowledge of the relationship between processes and climate. She has reviewed that much of the work on climato-genetic geomorphology has been done in continental Europe. Tricart (1961-65), Tricart and Cailleux (1965) and Birot (1968) deserve more discussion than space has allowed. She concludes that in view of the recent large scale fluctuations of climate both present and past processes must be considered. The study is, therefore, a historical one involving consideration of changes over time as well as space.

The problem of erosion surfaces has been most controversial and much has been written pro and cons regarding this than any other aspect of geomorphology. C.R. Twidale in his paper discusses the theory and reality of planation surfaces. He has reviewed various types of surfaces confused with the peneplain. The author has stressed that planation surfaces both ancient and modern are outcome of running water. Given time, rivers can reduce large land masses to base level. The most common form of regional plain is the peneplain. Both valley side slopes and fringing pediments develop and as valleys extend and widen divides are lowered in due course an extensive rolling and undulating surface of low relief, a peneplain, is formed. Further degradation produces a flat erosional planation surface or altiplane. Conversely, slight rejuvenation can cause a return of the peneplain form.

L.I. Marushvili in his paper "Peneplanation planes in an Alpine country" has discussed the chronological history of the Georgian region. He has concluded that the concept of W. Penck of "piedmont steps is totally abstract scheme worked out on the examples of Mid-European Hercynian and other mountains of non-Alpine type. The concept of planation of crests by glacial and chronological process mentioned by some geomorphologists is also fully groundless.

Geomorphological studies of the different parts of the Deccan Foreland and references in geological publications should make it possible to synthesise certain facts which seem to be universally present in this region. The paper by Pramila Kumar and Rai has presented a systematic account of various evidences of rejuvenation of the Deccan Foreland. Unlike an old peneplain, the Deccan Foreland has the characteristic of a rejuvenated land with interrupted cycles of erosion. These are reflected in the landforms, the slopes and the processes and features associated with rivers and may lead once to certain logical conclusions.

The paper by Nageshwar Prasad and M. Naseemuddin also discusses the polygenetic erosion surfaces of the Barakar Basin. Various evidences of this region further support the thesis presented by Pramila Kumar and Rai.

Inselbergs are the remnants of circumdenudation remaining when scarped retreat and pedimentation have all but base levelled the landmass. C.R. Twidale has elaborately discussed granitic inselbergs. He has distinguished three types of inselbergs, i.e., Bornhardts, Castle Koppies, and Nubbins. He has explained the development of these three forms of inselbergs with specific examples of Australia. Bornhardts are dominated by sheet structures. The residuals are formed in 2 stages—The first involved differential fractural controlled subsurface weathering, resulting in the development of some zone of deep weathering and distributed amongst them as incipient inselbergs. Relief compartments of massive rock that have been but little affected by weathering attack. In the second of the two stages erosional agents most commonly rivers and streams erode the landmass, the weathered compartments being lowered to base level leaving the fresh rock compartment in relief. Bornhardts are modified by weathering to give Nubbins and Castle Koppies. Nubbins are developed in humid, partially humid tropical conditions and are due to the blocky disintegration of the outer shell on shells of domical forms. Castellated features on the other hand result from the pronounced marginal attack of Bornhardts in conditions of specially aggressive weathering or as a consequence of long continued weathering.

The Aravallis of Rajasthan are ancient mountains system which are also characterized by subdued topography and extensive development of inselbergs and tors. The study by H.S. Sharma, of the north-eastern termination of the Aravallis discusses the development of tors on quartzitic rocks. The author has advocated the development 2 stages model of quartzitic tors like Twidale. General agreement has now been reached that tors are truly the result cf semi-arid morphogenic process. With year round high temperatures and heavy seasonal rainfall chemical weathering extends to a great depth. Thus he has concluded that structural control and climate combine to produce one or more striking features like tors.

Presence of duricrusts and laterites in tropical and sub-tropical regions are considered as evidence of erosion surfaces. They are very significant from geomorphological point of view as they form cape rocks. C.R. Twidale in his paper on duricrusts has emphasized the significance of duricrusts. He has considered these features as stabilized landmasses, morpho-stratigraphic makers in order to determine the age of the land-surface. He has concluded that duricrusts have a direct influence on landforms development in the tropics and sub-tropics but also provide information of the environment in which landscape evolved and the age or age-ranges of their development.

REFERENCES

Ahnert, F., 1973 : Inhalt and Stellung der funktionalen Methods in der Geomorphologie. Erdkundliches Wissen (Festochrift Plave) 33, 105-113.

——1978 : Gegenwartige Forschungstendenzen der physischen Geographie. Die Erde 109, 49-80.

Birot (1968) : The cycle of erosion in different climates. Translated by C.I. Jackson and K.M. Clayton Bastsford, London, pp. 144.

Chorley R.J., (1962). : Geomorphology and general systems theory. US Geog. Survey Prof. Paper 500-B, pp. 10.

Chorley, R.J. and B.A. Kennedy, (1971): Physical Geography—a systems approach. London. pp. 370.

Cooke, R.U. and J.C. Doornkamp, (1974): Geomorphology in Environmental management, Oxford, pp. 413.

Leser W., (1976): Landschaftsokologie, UTB 521, Stuttgar, pp. 432.

Joly, F., (1977): Point de vue sur la geomorphologie. Annales de Geographie, 86, pp. 522-541.

Semmel, A., (1979): Geomorphologie als geowissenschaftliche Disziplin-praklische Erfahrungen theoretische Moglichkeiten. In: Stuttgarter Geographische Studies 93, (Festschrift Meckelein), pp. 23-32.

Tricart, J., (1961): Les carateristiques fondamantales du systeme morphogenetique des pays tropicaux humides. Information geographique 25, pp. 155-69.

Tricart, J., (1965) : Principles et methodes de la geomorphologie. Paris, Masson, pp. 496.

Tricart, J. and Cailleux, A., (1965): Introduction a la geomorphologie climatique Paris, S.E.D.E.S. pp. 106.

ONE

André Cailleux and *Lucien Romani*

MODERN TRENDS IN GEOMORPHOLOGY

As IN MOST other sciences, to describe every detail of modern trends on geomorphology would need an entire book. In this chapter, we intend to stress the major tendencies, and, as occasion arises, to point out some possible ways of research or, on the reverse, a few dangerous snags. After this general view, a very recent, concrete example will be given and finally conclusions will be drawn. Except for the example, no author's name will be quoted, for which we apologize : there would be too many of them.

QUANTITATIVE TENDENCY

Like every science, geomorphology trends towards quantification through countings, measurements, computations, and other kinds of mathematical operations. So strong is this trend, that some would be tempted to drown themselves into it recklessly. Such people forget that one must first thoroughly and for a long time observe the terrane, aerial or satellite photographs and . . . maps, and ask oneself questions. Only afterwards countings, measures, experiments and models may be called for. A model is an hypothesis with numerical values and (or) equations, thereby admitting variants, which are to be studied, even if finally rejected. The computer will be very useful, provided the number of data to be treated is sufficient, and the results that can be hoped are of real interest : using computers needs time and money, which should be spared.

The domain of geomathematics is so varied that an author qualified in one field may not be so in another; if you consult him outside his own field do not overlook that he may be mistaken.

TENDENCY TOWARDS DIVERSE EXTENSIONS

APPEALING TO NEIGHBOURING SCIENCES

The sciences appealed to are not only mathematics, but also theoretical and applied physics (soil mechanics, energetics, teledetection . . .), chemistry (weathering, dissolution . . .), biology (role of vegetative covering, relief-building or destroying plants and animals . . .), even astronomy as we will see later. Basic knowledge in each of these diverse domains shall be acquired by the geomorphologist : a burden, but what a strength !

COMPLEXIFICATION

It necessarily results from the above-mentioned interconnections. But it has a counterpart :

HELP GIVEN BY APPLIED GEOMORPHOLOGY

In our fully developing world, the geomorphologist, with his experience and established contacts, is more and more consulted, either for a dam, a highway, the asset of a building, the land planning, the national development or any other far-reaching project. More and more geomorphologists are engaged by private or public organizations, which is a good thing in itself, and specially for young people in search of a job.

DIVERSIFICATION OF SUBJECTS, METHODS AND CONCEPTS

Primarily concerned with the landscape, geomorphology, as physics or biology previously did, takes advantage of extending the scale of its interests on both sides. On the one hand, the very small forms, the elementary processes—either physical or chemical, biological, human giving way to weathering, erosion and deposition. On the other hand, on the global scale, continents and oceans and their reciprocal connections.

As far as the matter and processes are concerned, the changes of states—solid, liquid, gaseous—are of prime importance : without them, we would have no volcanoes, no rivers, no glaciers. More global balances have to be done, and geomorphologists can contribute to their setting up : this is one of their most important tasks. The case is the same for energy. The role of internal sources of energy (heat flux, tectonics, stress . . .) and external ones (solar radiation, tides . . .) must be better understood. From such studies also important material benefits and profits may arise.

Some landforms may result from catastrophes, anastrophes or retroactions (either positive or negative). The bibliography, at the end of this chapter, contains a few references about these interesting concepts.

IMPORTANCE OF TIME

In many domains—resources, labour, future of the young—time is more and more important. As well known, this is also the case for geomorphology : present-day landforms cannot be explained, in many regions, without calling for inheritances, paleoclimates or other conditions dating from bygone times.

On the other hand, facing the huge tasks to be done, the future belongs to those who will have succeeded in organizing their own time. Overburdening scientific workers, as it often happens, with fruitless administrative tasks, overwhelming them with progress reports to write hinders what man properly wanted to foster : their scientific activity.

WIDENING IN SPACE

Until about 1950, geomorphology has been primarily concerned with land above sea-level, including sea-shores. But since that time, oceanographic researches have made considerable progress, the landforms of the sea-bottom are better and better described and mapped, and ingenious or imposing explanations are proposed.

By the way let us disapprove the title (The sixth continent) for a book treating of the Ocean !

But let us come back to serious things. Men have put the foot on the moon, rocks have been brought back from there, photographs of distant planets and moons have been taken at close range by spacecrafts, remarkable landforms have been discovered : Mars volcanoes twice higher and twenty times more voluminous than those of the earth; at least six volcanoes simultaneously active on Io, the satellite of the Jupiter; elsewhere, regions wholly pitted with craters due to impact of meteorites, or larger cosmic bodies. All these forms, all these processes largely broaden our views and open new horizons before our eyes.

Such are, briefly summarized, the two major trends of modern geomorphology : towards quantification and towards most varied extensions. Those trends will be now concretized by an example. We will take it in the domain which was most recently opened to us : among our neighbours in the Cosmos, the planetary bodies.

AN EXAMPLE : VOLCANOES AND SATELLITES OF THE PLANET MARS

MORPHOLOGY OF PHOBOS AND DEIMOS—THE SATELLITES OF MARS

The greater, Phobos, is only 27 km long (Table 1.1 and Fig. 1.1). Both are far from spherical, rather ellipsoidal, with 3 axes clearly unequal : for Phobos, 27.0 - 21.6 - and 18.8 km. A same kind of inequality is observed for the smaller Deimos. The ratio of an axis of Deimos to the corresponding axis of Phobos is 0.556 - 0.565 - 0.580, these three values being astonishingly similar, with a mean of 0.567 : both satellites have the same general form, they are almost homothetic as Veverka (1978) noted it. This strongly suggests a common kind of origin.

Moreover, if for each planet of the solar system, we establish the ratio of the diameter of the largest satellite, to the diameter of the smallest, we get for Uranus 8, Saturn 9, Neptune 17, Jupiter 560, but for Mars less than 1, 8; and this places Phobos and Deimos far away from all the others.

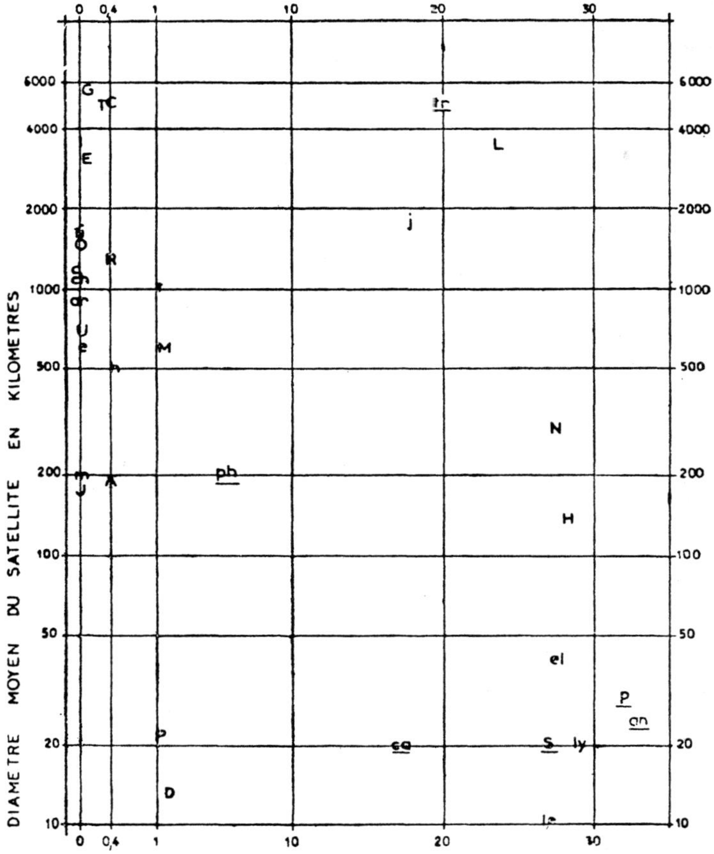

FIG. 1.1 : *Inclinaison Moyenne de l orbite du Satellite Sur l'Equateur de la Planete en Degrés*

Légend : A Amalthea; *an ananke*; ar Ariel; C Callisto; *Ca Carme*; Ch Charon; D Deimos; d dione; E Europe; e Encelade; el Elara; G Ganymede; H Himalia; h Hyperion; I Io; J Janus; j Iaetus: L Luna; le leda; ly Lysithea; M Mimas; m Miranda, N Nereis; O obron; P Phobos; *p Pasiphae*; *ph Phoebe*; R Rhex; *S. Sinope,* T Titan; t Tethys; ti Titania; *tr Triton*; U Umbriel.

Les symboles et les noms en *italiques* sont Ceux des satellites *rétrograde*.

TABLE 1.1 : **The natural satellites of Mars**

Numerical Data	Phobos	Deimos
Diameters of the representative ellipsoid, km :		
2a, towards Mars	27.0	15.0
2b, in orbital plane	21.6	12.2
2c, orthogonal	18.8	11.0
Estimated volume, km^3	5300 –5400	1200 –1500
Mass in g, ±15%	1.1×10^{19}	?
Density (water=1)	2.1 ±20%	?
Geometric albedo, %	5–6	5–6
Diameter of the largest crater, km	10	3
Half great axis of orbit (Mars radius=1)	2.76	6.91
Eccentricity	0.0150	0.0008
Inclination on Equator of Mars	1°1′	0°9′ –2°7′
Orbital period	7 h 39 mn	30 h 18 mn

The albedo—the percentage of light reflected to light received is among the smallest presently known for the planetary bodies : 5 to 6 per cent. In other words, they are very dark, almost black. Such an albedo, jointly with the rather small density of Phobos, excludes any known composition other than a type-I carbonaceous chondrite, or a basaltic rock with 30 to 40 per cent bubbles, in volume (Table 1.2).

TABLE 1.2

	Density	Diffuse reflectance*			
Wavelength μm		230	400	560	650
1. Diverse (Maxim.)	3.3	5	23	30	31
2. Basalts (Minim.)	2.7	3.5	9	12.5	12.5
3. Meteorite Orgueil	2.2	2.5	4.5	8	8
4. Phobos	2.1	2.0	4.9	5.0	5.0
5. Highly vesiculous basaltic glass	1.2	2.5	3.5	3.5	3.5

* 1, 2, 3, 5 according to Dollfus and alii (1980) ;
4 according to Pang and alii (1980).

Like the Moon and Mercury, Phobos and Deimos have no atmosphere, no liquid water, and even no trace of its previous existence, no fluvial valley, no sea and no folded mountain range. The most common landform is impact craters, formed when extraneous bodies, encountered in space the satellite. The surface is entirely pitted with such craters, saturated with them. If all of them were due to meteorites, as on the Moon and Mercury, this would mean no less age than 3900 Ma (megaannus, million years).

The surface of Phobos is partly covered with bundles of shallow, parallel, 150 to 200 m broad grooves. The bundles have diverse orientations. Such forms are unknown in any other planetary body. The grooves are resected by younger craters. Many grooves, in oblique light, show a series of small, slightly elongated ($a/b = 1.10$ to 1.25) shallow depressions which are quite different from impact craters.

Deimos shows no grooves. It is almost entirely covered with dust, which blunts most landforms. In addition to craters, it is marked, on its southern hemisphere by a rather irregular, 10 km broad and 2 km deep excavation which greatly differs from impact craters and results from breaking or tearing away.

ASTRONOMICAL DATA—DIFFICULTIES

The 38 well studied satellite of the solar system have very diverse diameters, drawn up from bottom to top on the Fig. 1.1. The angles between the planes of their orbits and the equatorial plane of the planet also are very diverse, going from zero to more than 30°, as may be seen from left to right on the same figure. The majority (32 of them) revolve in the same direction as the planets, and only 6 revolve in the opposite, retrograde, direction.

With the exception of Phobos and Deimos, all satellites may be grouped in two well defined categories :

(a) Upper part of Fig. 1.1, above 180 km in diameter, 28 satellites. Some of them are very large, others have almost circular orbits at a very low angle with the equator of the planet, others (23) simultaneously present both characteristics.

(b) In the lower right corner, the 8 remote satellites of Jupiter, which are small, with orbits at a high angle and very eccentric, in other words having the form of elongated ellipses.

At the first glance on the same Fig. 1.1, one sees that Phobos and Deimos are located on the lower left corner, far apart from those two categories. This fact raises a first question : why are they so exceptional ? Are they previous small planets, subsequently captured by Mars ? This seems unlikely for several reasons. For instance orbits with low eccentricity and low inclination on the equator (Table 1.1) make a capture at random very unlikely.

Other difficulties occur. Phobos describes around Mars a circle with a radius equal to 2.76 times the radius of the planet, hence smaller than 3.02, value which is for it the Roche limit, beneath which a satellite is not stable, and is in the long run doomed to disruption. This contributes to explain us that, unlike the much more distant Deimos, Phobos could not keep the dust it encountered in the space : this dust is gone away towards the more attractive Mars. But meanwhile a question is raised : Phobos being unstable cannot have revolved around Mars for a long time, more than about 200 Ma. Then how can we explain the saturation with impact craters, since if they were due to meteorites, as on the Moon and Mercury, the saturation would have required 3900 Ma, that is 20 times more ?

Finally, the strange grooves of Phobos until now defied every explanation.

For all these reasons, a new hypothesis must be found. Geomorphology offers us one, proposed by Lucien Romani. Let us now look into it.

PHOBOS AND DEIMOS EJECTED BY A MARTIAN VOLCANO

Each of them would result from an enormous subterranean explosion which, not very long ago, would have provoked a blocked-vent eruption, as those known on the Earth (Rittmann, 1963) but in this case happening under a big volcano located on the equator of Mars. Such an ejection would automatically explain the quasi-coincidence of the orbits with the equatorial plane, and the grooves of Phobos (by jerky friction and rotation in the volcanic pipe). An enormous mass of smaller clasts and gases was ejected during and after the eruptions, causing friction on the satellite, hence the quasi-circularity of its orbit. The impacts on the satellites were mostly due, not to meteorites, but to the volcanically ejected clasts; the saturation became effective in a rather short time : the difficulty raised by the young age is removed.

Concerning this hypothesis, which facts for or against are provided by the comparative geomorphology of Mars and the Earth ?

There are volcanoes on Mars. The five greatest are, in the well established relative chronology, among the most recent features : probably no more than 200 Ma, possibly less, which is short compared to the age of the planets (4600 Ma). Among those 5, one—Pavonis Mons—almost exactly straddles the equator (Table 1.3). It presents two nested caldeiras. The ratio of their diameters 40.5/75 = 0.54, is very near the ratio of homothety between Phobos and Deimos, 0.56. The similarity is striking. One may guess that a first paroxysmal explosion is responsible for the ejection of Phobos, followed by a subsidence giving the great caldeira; another explosion would have given Deimos and the small caldeira. The ratios of dimensions (a+b) caldeira/(b+c) satellite are : For Phobos : 75/20.2 = 3.71, for Deimos: 40.5/11.6 = 3.49 and this again is satisfactory.

TABLE 1.3 : **Pavonis Mons**

Latitude of centre	great caldeira	0°7′ N
	small caldeira	0°4′ N
Dimensions* kilometres	great caldeira	78×73
	small caldeira	40.5×40.5
Depth, kilometre		>1
Mean altitude of rim above surrounding plains, km		17
Diameter of the base, km		450

* According to Mutch (1976), p. 187, fig. c. According to the map (Anonymous, 1976) those values should be raised by about 10%.

At first sight, the amount of energy required to satellize such an enormous mass as Phobos seems gigantic. Calculation gives 1.4×10^{23} joule. On the Earth, the greatest volcanic explosion known is that the Tambora (Indonesia, 1815); its energy has been estimated to 8.4×10^{20} joule (Rittmann, 1963). Under similar conditions, energies are proportional to cross-sections. Let us consider the sections of the caldeiras. For the Tambora, the area is 29 km^2. For Pavonis, 4460, that is 154 times more. One can guess, for Pavonis, an energy of $154 \times 8.4 \times 10^{20} =$ $1.29.10^{23}$ joule, which is in good agreement with the above quoted evaluation of 1.4×10^{23} joule. It is true that part of the energy dissipates as warmth, and this slightly vitiates the comparison (total energy for the Tambora, mechanical energy for Pavonis). But there is a counterpart : the two paroxysmal eruptions of Pavonis are maxima which occurred during millions or billions of years (otherwise, there should be other satellites around Mars, which is not the case). Whereas the eruption of the Tambora is only the maximum in the annals of quantitative volcanology, no more than two centuries.

On the other hand, the same eruption of the Tambora ejected 150 km^3 of pyroclastic ejecta. Phobos, with its volume of 5350 km^3, is only 36 times greater; this is small compared to the above quoted ratio of caldeira areas, 154, and raises no difficulty.

Moreover, Pavonis and the Tambora also have similar forms, the external slope having, for the Tambora, a mean value of 8°, with a maximum of 11°, and for Pavonis, respectively 5°5′ and 15°, which bracket very well the Tambora values.

Finally, the morphology of Phobos and Deimos themselves affords a new evidence. Veverka and Duxburry (1977, p. 4217) mention chains and clusters of craters, due to low velocity impacts. Such is not the case of meteoritic impacts (about 20 km/s), but of the many bodies ejected by Pavonis, whose velocity may have been very small, much less than the escape velocity on Mars (5 km/s).

Summing up, in the present state of knowledge, Romani's hypothesis, according to which Phobos and Deimos would be volcanic plugs of Pavonis, satellized during two paroxysmal eruptions, gives us at least a tenable explanation.

CONCLUSION

As the Earth, with the Tambora, can help us to understand Mars and its satellites, in the same way Mars and the other planetary bodies help us to acquire a better understanding of our Earth and appreciation of its specific characters. In the solar system, the Earth is the only body on which liquid water permanently exists and forms oceans; and also the only one where there are folded mountains : and may be there is a link between these two features (Cailleux, 1979).

In any case, the example we gave sets forth very clear trends of research : quantification from beginning to end; appeal to neighbouring sciences, such as astronomy and volcanology; importance of masses, of energy, of velocities; importance of time in the study and the solution of problems; extension of our views towards objects more and more distant and vast. Such are the fundamental trends in modern geomorphology.

SUGGESTED READINGS

Cailleux, A., 1979 : Tibet, Andes : catastrophes, plaques et rétroaction. *Annal de géogr.*, 88, pp. 419-431, Paris.

Dollfus, A., *et. al.*, 1980 : Reflectance spectrophotometry extended to u.v. for terrestrial, lunar and meteoritic samples. *Geoch. et Cosmoch.* Acta, V. 44, pp. 1293-1310, Pergamon Press, Oxford.

Locquin, M.V., 1979 : Catastrophes et Anastrophes, essai d'épistémologie générale. *Ark'all commun.*, V. 6, f. 1, pp. 17-35.

Mutch, T.A., *et. al.*, 1976 : *The Geology of Mars*, pp. 400, Princeton Univ. Press.

Pang, K.D., *et al.*, 1978 : *The Composition of Phobos: Evidence for Carbonaceous Chondrite Surface from Spectral Analysis. Sci.*, t. 199, pp. 64-66.

Rittmann, A., 1963 : *Les volcans et leur activité*, p. 461 Ed. Masson, Paris.

Scheidegger, A.E., 1975 : *Physical aspects of natural catastrophes*, p. 13+ 289 Elsevier, Amsterdam.

Thom, R., 1972 : *Stabilité structurelle et Morphogénése*. Ediscience, Paris.

Veverka, J., *et al.* 1977 : Viking Observations of Phobos and Deimos : Preliminary Results. *J. Geophys. Res.* t.82, pp. 4213-4223.

Anonymous, 1976 : Topographic map of Mars 1 : 5,000000, Phoenicis Lacus quadrangle. map I-984, MC-17; Tharsis quadrangle, Map I-977, MC-9. USGS, Denver.

Anonymous, (sans date) : Tactical Pilot Chart 1:500000, S. TPC, Sheet N12BG, Indonesia. Ed. 1.

TWO

J. Tricart

CLIMATIC INFLUENCE ON GEOMORPHOLOGY

THE CONCEPT of "climatic geomorphology" has emerged rather recently in the scientific literature written in english. Its first expression is the title "Climatic accidents" of C.A. Cotton (1947), followed by L.C. Peltier (1950). But this first approach remains entangled in the obsolete frame of the Davisian cycle of erosion. This delay which has affected the conceptual development of current literature in english language is turned quite evident when we compare it to continental Europe. The influence of climate on the landforms is included in the German concept of landscape, which is background of modern soil science, as founded by the Russian Dokuchaev, whose main publications have been published at the same time as those of W.M. Davis. A Pole, Loczinki, has created the word "periglacial" just before World War-I in order to synthesize the influence of the cold climate around Quaternary glaciers on the shaping of landforms in South Poland, some 40 years before L.C. Peltier in U.S.A.

This historical difference can be explained, at least partially, by a difference in approach and in type of mind. The German concept of landscape is a quite concrete one. It involves a perception of all the visual aspects of a stretch of land: topography, rocks and surficial deposits, vegetation. Dokuchaev has included soil in this list. The explanation of the landscape requires the input of climate. A. von Richthoffen, before the end of last century, has given very sound advices to the "scientific travellers" in order to guide their observations in the field. Such a global evaluation of the environment has favoured the understanding of the interactions between the different elements of the landscape, and more recently, the adoption of a systems approach. The modern landscape science, as developed mainly in Eastern Germany, has become "Landschaftsekologie", an ecology of the landscape. On the other side, Davisian geomorphology is a very theoretical model, where "imagination", as recommended by the Prophet himself, is essential, not accurate descriptions of the environment, and even less, surveys in the field, as geomorphic mapping. The type of mind is the same as in geometry: abstraction. It is the opposite of the naturalistic approach of the geographers of Central Europe at the end of the past century. Under the pressure of facts, Davis has been obliged to admit some influence of climate upon landforms development, but he tried to fit it in his type of geometrical model. The same with C.A. Cotton, who considered it just as "accidents", and with L.C. Peltier.

More recently, the English speaking geomorphologists have-discovered the computor. This has strengthened their inclinan tion towards morphometry and statistics. The hortoniais analysis of drainage networks has been the first basis of thn new orientation. New? More or less, and better "less" thay "more" . . . In fact, it is the continuation of the same tendenct toward a geometrical conception of the World. Measuremen of complex objects like landforms requires a lot of simplification in order to assimilate them to algorithms. For instance, a very current landform, meander, has been the object of more than 50 different procedures of measurement. Is any one of them acceptable? The authors do not agree about what is fundamental and what is unimportant. The necessary simplification can affect important aspects and be unrealistic. Before quantifying, it is necessary to know what are the relationships between the aspect which shall be measured and others: these relationships determine the actual meaning, the significance of the measurement. If this first step fails, the measurements is meaningless, and consequently, useless. Unfortunately, examples of such measurements are not unfrequent. . .

A second problem involved in "quantitative" geomorphology, is the use of statistics. It is well known, but frequently forgotten, that a statistical correlation is not necessarily the expression of causal relationship. It is called a "statistical" law, in order to eliminate any confusion with a "physical" law. The adjective is more important than the substantive! Many of the statistical laws established by Horton are, in fact, merely probability laws. They are devoid of any geomorphological interest. What is significative is not the conformity of any actual drainage network with these laws, but the unconformity, the departure of some of them from these laws. Such unconformities mean, effectively, that the drainage network has been influenced by very strong constraints, whose nature has to be investigated. Changes in lithology, originating high and low permeability, or vigorous lineations in rigid rocks can be the explanation, among others. The system approach teach us that many statistical corrrelations are, in fact, the expression of undirect relationships, the causal effect passing through one or more inter-related parameters. This results in a dispersion of the individual points which are plotted on the graphs used to calculate the regression curves. The statistical practice is to use a greater population of data, in order to have "safer" regressions. But this is a quite theoretical view. In fact, the probability is usually very high that the greater the population, the more heterogeneous it is. We increase the "noise" which results from the influence of unknown parameters. We have personally experienced this dilemma during field measurements of boulder size in order to determine the centile or with boulder orientation. The solution adopted has consisted (1) use the more simple type of measurement possible and (2) use statistical classes sufficiently large in order to be able to treat only small populations. For instance, for the orientation of boulders, we have adopted classes of 30°, compatible with the uncertain determination of the main axis of the boulder on one side and of the axis of the river, glacier, slope etc . . . on the other. The results have been good, because we were able to concentrate the measurements on limited areas, where, when deposited, the direction of the current has not suffered important changes in the case of braided channels.

This appraisal makes us understand why there is no wonder if the tentatives made, principally in England, to establish correlations between some climatic parameters and some characteristics of landforms have failed to show any evidence. The contrary would have been remarkable and would have turned necessary a complete revision of our knowledge!

Recently, D. Dunkerley (1980) has focussed, in a very substantial review of methodologies about the evolution of slopes, two basic facts: their theoretical studies lack from a good *descriptive* basis and from a sufficient knowledge of process, and above all, of their interacting.

"Climatic geomorphology" is a somewhat dangerous expression, just for this reason. Some binary minds can be inclined to take it literally. In some papers, "climatic geomorphology" has been opposed to "structural geomorphology" or to "cyclic landforms". This is not justified. Structural geomorphology or climatic geomorphology do not exist *per se*. They are not independent geomorphologies. They just focus upon a *certain aspect* of landform shaping, a certain aspect which is, in fact, very inter-dependent of the others. Isolating it is an abstraction. We can do it, of course, but this operation is justified only if we do not forget that we did it . . . We must keep it in a dialectic of unity-diversity. Climatic geomorphology has no intrinsic reality, it is just a point of view, an aspect of geomorphology as a whole : the study of landforms.

The same landform is elaborated under the influence of various parameters which interact together. The systems approach is the best logical tool we can use presently for its study. Using its own terminology, we can say that there is a group of intrants which are climate, structure and time. They interact upon each other when shaping the form. Any process reflects their mutual inter-action. As a result, there is *certain* common features between landforms shaped in limestones under different climates, like caves, sinkholes, blind valleys etc. . . , but they are not active, when they actually exist, under certain climates, like very cold ones with permafrost and very dry ones, with just sporadic rainfalls without organized drainage. The existence of these forms is "usual" in limestone, not all "compulsory". This is the result of climate. These karstic forms are a result of limestone "solution" (in fact, formation of more soluble bicarbonate by reaction of CO_2 on Ca of the limestone). For it to be possible, water is necessary, of course.

Ice cannot yield dissolved carbon dioxide for this reaction in permafrost. In permafrost, just as in very arid countries, no water circulates in the rock, which turns the process impossible. There is, clearly, a double command of the process lithological and climatic. It is not one of them or the other, but both of them together which are necessary for elaborating karstic forms. We could take, as example, any other group of rocks : granite, sandstone or schists. In each one, the relative importance of the two climatic and lithological inputs is different. For instance, in quartzite, where both chemical and mechanical resistance are high, the influence of lithology prevails strongly. Rocky outcrops, turning evident in the forms the influence of dip for instance, are frequent under all climates. The expression of climatic influence is at minimum. On the contrary, it is strong in granite.

Time is another very important input. We perfectly know that climate has undergone frequent and drastic changes in the past. At the end of the XVIII century, de Saussure already arrived to the conviction that glaciers have been very much bigger in the Mont Blanc massive during the past. . . The study of the Quaternary climates is gaining popularity and accuracy. Nobody can ignore its results. The morphogenic efficiency of the different climates of the world varies widely. For appreciating it, we must consider, too, the different rocks. For example, dry and cold climates are very poorly efficient on limestone, as remembered above. Nevertheless, some karstic forms do exist in regions that are, presently, extreme deserts, or occupied by glaciers and permafrost. These forms are inherited. They have been shaped during the past, under quite different climates. When some scientists try to establish regression curves between some characteristics of the landforms and the present climate, this inheritances are quite confusing. They introduce a "noise" which can be sufficiently high to obscure completely the influence of the present climate. Before tempting to establish such regressions, it is absolutely necessary to know what is the structure of the system (=landform). But, even that is not sufficient. Regressions are bidimensional : they express just the relationship between *two* parameters, and no more than two. Shaping of landforms depends on much more than two parameters. Correlation analysis is unable to investigate these various relationships and these mutual inter-actions. At best, a ratio of weight of a factor A in the explanation of a factor B is given. But this is a purely statistical object, not a natural one. A system is something more than a multiple correlation. The statistical tool is poor, unsuitable, for studying it.

The problem of the ability of inherited landforms to persist is a difficult one. Only few investigations have been devoted to it. Roughly, we can assimilate it to a response problem : the time-lag of response to climatic change of a landform. But, actually, it is more complex because we have to take into consideration a threshold effect. What intensity and duration must have the climatic change to be able to modify the shaping of the land-forms? Very little is known about that. One of the reasons for this sad situation is the short duration since the last drastic climatic change, at the beginning of the Holocene. Another one is the existence, in many regions of the world, of appreciable climatic fluctuations during the Holocene itself (Little Ice Age of middle and high latitudes, changes to drier or moister climates in the arid regions etc. . .). We need to know very much more before elaborating an acceptable model, as stressed by Dunkerley (1980). Presently, we just can identify the main inputs of the system "landform", but any quantification of them should be a brilliant demonstration of bold ignorance.

Let us focus now upon the influence of climate upon land-form shaping. Above all, we must not forget, as a result of excessive abstraction, that this influence acts always upon a certain structure (lithology, disposition of the rocks, recent tectonic movements), at a certain moment of the geomorphic evolution with inherited elements whose evaluation is, usually, far from easy. The influence of climate is just a sub-system of the system "landform shaping".

Considered under this angle, the influence of climate on the shaping of forms is complex in itself. Actually, some direct influence of climate on landform shaping exist, but the main influence is indirect and takes place as a component of the functioning of the ecosystem.

The existence of glaciers depends directly on climate. But the relationship is far from simple. Mean annual temperature, cumulated thickness of the successive snowfalls of the year are not satisfactorily correlated with the existence of absence of glaciers. The glacial balance is a complex function of abundance of snow, thermal regime, seasonal occurrence of snowfalls and of temperature variations. Yakoutia is much colder than Spitzberg, but there is no glaciers in Yakoutia when they well developed in Spitzberg. The low thermal conductivity of ice explains the existence, at a depth of 80-100 m below the surface, of temperatures well below the mean annual temperature of the surface. These abnormally low temperatures are inherited from the last cold period . . . The problem of soil temperature is even more complex, as a result of the great differences which exist between the thermal characteristics of the different rocks. Permafrost cannot be correlated accurately with a certain minimum annual mean of the air temperature. Its limit is, very roughly, an annual mean temperature of –2°. The parameters which turn this limit very rough are the following :

— Snow cover, acting as an insulator : where snow is scarce during winter soil freezes more deeply. But the snow cover itself depends not only on climate : local exposure to wind and topography are important. The distribution of snow, in a certain area, is a kind of patchwork at the detail level !

— Water content of the soil, acting through the laws of thermodynamics : a great quantity of energy is necessary for melting ice. . . A soil with important ice segregations does not melt easily during summer. On the contrary, lakes and streams, when sufficiently deep, can prevent the formation of a permafrost beneath. The bottom layer of the water freezes very late, and when frozen, its ice acts as an insulant.

— Vegetation and soil are important too. Under cold climates, vegetation is rich in celulosis and is not easily decomposed, a result of both its particular nature and type of micro-organisms. Soils are frequently peaty. Peat a very good insulator and retains much water. In palsas, a core of permafrost persists, protected by the surficial peaty layer, when there is no permafrost around these palsas.

But, with this example of palsas, we are already discussing some indirect, ecological aspects of the influence of climate on geomorphic processes. . . Direct influence of climate on the geomorphic processes is truely restricted. If we analyse more carefully the example of glaciers, it becomes even more evident. What we examined above is the influence of climate on the *existence* of glaciers, not on glacial action. This last one cannot be correlated directly with atmospheric climate. Actually, what imports is the temperature of ice at the rock contact, below the ice. Ahlmann has classified the glaciers as cold, temperate and warm according to this criteria. But, compressive and extensive stresses, resulting from the proper movement of ice, modify the pressure at the rock contact. Their changes can result, in certain definite temperature conditions of the ice, in alternate freezing and melting. Percolation of water down to the bottom of the glacier through crevices bring calories during the melting period. During winter, these cracks allow penetration of cold air and counteract the insulating effect of ice. They contribute to modify the thermal regime at the interface ice-rock, where processes are operating. Smoothing resulting in roches moutonnèes or quarrying are two opposite processes depending on temperature conditions and changes at the interface. They generate quite different minor landforms and even, when time and lithology are favourable, quite different mesoforms.

One of the major difficulties of climatic geomorphology is the inadequacy of climatic data. They have been collected for the purpose a world-wide comparison, and for this, have been standardized, what means as independent as possible from the local environment. It is thus no wonder at all if they are not suitable for the study of this very environment, either geomorphological, or ecological—which is, for us, exactly the same. Temperature are measured with thermometers, it is to say using the direct contact of the air with an object able to dilatation with a temperature increase. Radiation is completely eliminated, using a shelter. Two hundred years ago, radiation was ignored and the thermometer was acceptable according to the level of knowledge in physics. But, nowadays, this is no more the case. Remote sensing is based upon the measurement of radiations, of radiometric temperatures. Photosynthesis, the basic process of primary production, is the transformation of radiated energy into chemical reactions. Transpiration is as much a result of radiated energy as of contact with hot air. The thermometer is not a suitable instrument to measure energy of radiation. At best, it measures a result of the absorption of a part of the radiated energy by a natural body, air water or rock. Geomorphic process need energy and solar radiation is one of their sources of energy, the main other one being gravity. Glacial processes are a function of the energy balance at the contact of ice and rock, influencing the effects of gravity energy, which provokes the movement of the glacier. The changes of this energy balance, which can be very frequent in a space spot, are the command of geomorphic processes producing roches moutonnèes, quarrying and plucking, shaping cirques U-valleys or not.

The analysis of soil freezing and of the various mechanisms in which it results would arrive exactly to the same type of conclusions. What matters for the understanding of geomorphic processes is not temperature under a shelter but temperature on the soil surface or in the soil and in the rock, below the ice, below vegetation. This type of data is not included in climatic studies. It is unfortunate, but it is. A correlation of climatic parameters with geomorphic processes is completely meaningless. The logical mistake is even greater when the correlation intends to link climatic parameters with landforms, because there is no mechanical and automatic production of a specific landform by a unique and specific process. Most landforms result of a combination of processes, of a *morphogenic system*. Once again, we meet the inadequacy of the search of correlations for studying a system.

Morphogenic systems are a part of ecosystems, modelling the ecotopes.

The structure of the ecotope has been analyzed by J. Tricart (1978). Figure 2.1 has been reproduced from this publication. Influence of climate on vegetation is universally admitted. Continental specialists of geomorphology have insisted since many years on the influence of vegetation itself on geomorphic processes (J. Tricart & A. Cailleux, 1972 and former French editions). Its principal aspects must be remembered :

Interception modifies completely the quantity of water reaching soil surface. Generally, it diminishes it, but, when mist and fog are common, it can increase it seriously, as in the mist forest of tropical mountains or in misty deserts which lay along

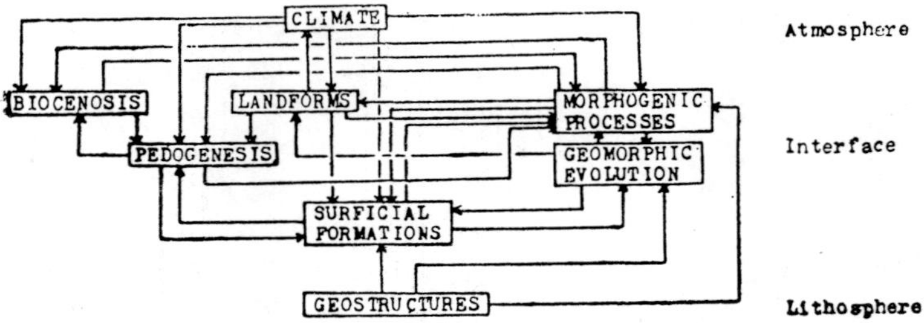

FIG. 2.1 : *The Structure of the Ecotope*

some coasts, as in Peru and North Chile. But the energetic aspect of rain is modified too. Kinetic energy of rain depends on drop size (mass) and velocity. The greatest drops commonly observed in nature reach their limit-speed after a fall of 9-10 m only. Such a height is common in forest with tall trees without dense ground cover of small trees, bushes and grass. It is frequently the case in the cultivated forests, but, too, in rain forest of the tropics. Like drops resulting of the concentration of rain on the canopies of tall trees are frequently bigger than the drops falling on this canopy, their kinetic energy is increased when they reach the ground at their limit-speed. But the energy balance must include, too, the loss of rain resulting from interception. The problem is complex and can be solved, in each particular case only with measurements. Unfortunately, they are very few. . .In mist forest, in Colombia, for example, cautious measurements have shown that the kinetic energy transmitted by drops to the ground surface is higher under the trees than on open space. But it is only an intermediary step : under the trees, a thick mulch is produced by vegetal debris, whose decomposition is slow as a result of altitude and wetness. This mulch intercepts completely the kinetic energy of drops. This energy is just a potential one and yields no work.

Another aspect of interception is the radiative one. Aerial organs of plants receive solar radiation. A very low proportion of it serves to photosynthesis : 1-6 per cent in the marshes of Silver Spring, Florida, U.S. A., according to measurements by Odum, this ratio being very near of the maximum in spontaneous ecosystems. But thermal infra-red radiation tends to increase plant tissues temperature. A regulation is achieved through transpiration, using the calories that evaporation require. This mechanism transforms a part of incident energy before it reaches ground surface. In the best ecological conditions (no stress resulting of water deficiency), the phytocoenosis can transform nearly 25 per cent of this incident energy. It is a part of it which is not available for the nourishment of other processes, weathering in particular. But, on the other side, this energy can be used to evaporate surficial water and dessication inhibits many aspects of weathering. Once more, we are in front of a complex set of interactions, requiring the use of a systems approach and of measurements.

Vegetation plays, too, a pure and simple mechanical effect: an increase of rugosity, which results in a dispersion of a part of the kinetic energy of flowing water by increasing turbulence. This rugosity is an obstacle to the concentration of run-off and to the incision of small gullies. It tends to increase the portion of the upper slopes where braided run-off results, with time, in a generalized exportation of fine debris, without incision. Or, if one prefers, the threshold surface needed for the beginning of incision, a very important geomorphic feature, with a drastic importance for solving conservation problems. The rugosity effect of vegetation applies to wind action, too, for exactly the same mechanical reasons. It is assumed empirically that wind breaks are efficient, on their leeside, on a strip of land whose broadness is approximately 15 times the height of the wind-break. This example demonstrate the importance of some purely physionomical aspects of vegetation for geomorphic processes, just as for hydrological and energetical aspects of rain interception. A dense low strata of vegetation, herbaceous in most cases, is highly efficient for energy dispersion of the kinetic energy of drops, of run-off and wind. Grass is the most suitable protection, commonly used for bare ground along embankments, road cuts, etc . . . Occurrence of trees and grass, in Nature, is partly under climatic command.

A last direct influence of vegetation on geomorphic processes is the role played by plants in weathering. But, in this particular aspect we must include, too, micro-organisms the importance of which is increasingly recognized. Penetration of roots is a source of corrosion on rocks and open cracks. It is a part of weathering and acts indirectly too, turning easier water circulation. Micro-organisms achieve fixation in their tissues of many different minerals, increasing their mobility. Frequently, it is the first step of the formation of the solute load of rivers. Mosses and lichens play a dramatic role in the disaggregation of bare rock.

Of course, we do not assume that climate is the only one command of the distribution of living beings, plants, animals and micro-organisms. Many other ones are acting too and a system analysis is necessary when studying biogeography. But, a part of this command is climatic.

Indirect influence of biocenosis is, perhaps, even more important. As shown of Fig. 2.1, it passes through soils and surficial formations. Both of them have a decisive importance for the exportation of solutes. Water percolating through soils is enriched in carbon dioxide produced by the decomposition of vegetal tissues and the respiration of roots. This water gains a certain amount of acidity which turns possible many chemical reactions with the different minerals of the rocks. Organic products are elaborated (complexes, chelates, etc. . .), which permit a mobilization of many minerals. But, most of these organic combinations are unstable, a great part of them being destroyed by micro-organisms. It results in some precipitates in the soil itself, in surficial formations and in solid rock cracks. But these precipitates are frequently formed very far from the area they come from. For example, in the Niger River basin, iron mobilized in the Fouta Djalon (Guinea), has been precipitated under the form of iron duricrust formed at the top of the ground water table contained in coarse alluvium in the region of Segou and San, more than 1,000 km downstream. Water modifies its composition and, obviously, its chemical potential when percolating through the various soil horizons which commands the chemical weathering of the solid rock beneath. Autochtonous surficial formations result, in part, from an interaction between solid rock characteristics (first, water circulation, and second mineralogy) and modifications of percolation water properties provoked by the decomposition of vegetal tissues on the soil surface. This decomposition has a very great importance. It depends of the type of the vegetal tissues falling on the ground, from one side, and on the other, from the decomposers them-

—selves, which partially under climatic command—temperature, moisture principally—but at the ground level, which is very different from the climate of the climatologists. . . Termites, ants, worms, Coleoptera, bacteria, fungi play a different role in decomposition and produce different weathering potentials. They have their own biogeographical distribution, partially influenced by climate.

Mechanical properties of ground surface determine the conditions in which geomorphic processes can act, can produce a certain amount of work. Still using the vocabulary and concepts of mechanics, we can write that they determine the efficiency of these processes. Unfortunately, lack of proper measurements does not allow to pass from a qualitative assessment to a quantification. In the example quoted above of the mist-forest of Columbian Andes, the quantity of energy transmitted by drops to the ground surface is higher under forest than at the open air. But, this energy has a null efficiency and does not produce any geomorphological work. It is completely absorbed by the mulch formed by vegetal detritus. This aspect of dynamics results directly from the characteristics of the biocenosis of the ecosystem : plants and decomposers. Splash erosion, a very important process is under the command of ground surface properties. Mechanical resistance of soil surface interacts with the cumulative kinetic energy transmitted by rain. Kinetic energy of rain depends on its intensity and duration, two parameters which are very seldom measured by climatologists, which ignore the pluviograph . . . Their importance has been recognized for conservation, as demonstrated by the Universal Soil Loss Equation of Wischmeier. This is a direct climatic command only where soil is bare, i.e. on ploughed land and in absolute deserts. Everywhere else, climatic command is indirect and modified by interception by vegetation. The other parameter, mechanical resistance of soil, is only partly under climatic command. The so-called (in a useless complicated and approximative manner) "aggregates stability" (in fact, mechanical resistance of soil), depends from the granulometry (texture) of the mineral fraction of soil. Silt and loam, and very fine sand (roughly, between silt and loam favour splash erosion. On the contrary, coarse sand, granules and stones from side, acting as a protective pavement, and clay from the other, for its tenacity, are not highly susceptible to it.

Nevertheless, organic matter and clay display a very important role as aggregate cement. Clay can be floculated or dispersed. It depends, in part, of the water pH conditions in soil. This pH is under the influence of mineralogy (calcium carbonate maintains clay floculated and exerts an influence unfavourable to splash erosion) and of the decomposition of organic matter, which tends to produce low pH. For example, upon quartzous sands, in Amazonia, under tropical rain forest, river water pH of 4 is common. This low pH of acid soils favour clay leaching and result in low resistance to splash erosion. But, on the other side, they turn the decomposition of vegetal debris more difficult and have thus the tendency to permit the persistence of a cover of half decomposed peaty debris on the ground. This is the case in sites where drainage is deficient. Leaching predominates, with an acceptable or good decomposition of organic matter, on better drained sites, resulting in a podzolisation favourable to splash erosion.

These few examples, voluntarily restricted, are intended to demonstrate how geomorphic processes take place in the functioning of the ecosystem.

It is through this dynamic of the ecosystem (ecodynamics) that a certain climatic command on geomorphic processes and shaping of landforms is functioning. It is mostly complex and undirect. A purely analytic approach, using statistics and regressions is an unsuitable method. Unfortunately, too much energy has been lost by English speaking geomorphologists in the exploration of this blind valley. Arriving at its end, they can just, at best, disappear in a dark cave.

To take the opportunity of applying to geomorphology what the ecologists can teach us is a very much more promising avenue. We see in it two very important aspects :

It oblige us to be conscious of the concrete aspects of the "Face of the Earth". Landforms are a part of the landscape. Primitive and limited measurements of drainage or slopes resulting in the isolated use of computer to produce regressions is just turning the heels to this concrete aspect, whose study is the object of geography. It is not at all a "new" geography, it is a useless one, an eroneous one. Like the unrealistic models of W.M. Davies, it can only result in isolating geomorphology both from human sciences and public questioning and from the other natural sciences. Geomorphology would be *useless.* Everybody knows the biologic law according to which a useless organ always disappears. . .

—Secondly, ecology has mastered very much better than geography the system approach. It is a chance for us from the methodological aspect, because the system approach is presently, the more suitable for understanding the complex interactions in which geomorphic processes are included. This approach can be used qualitatively and is not questioned when quantitative data become available. This is a great advantage in geomorphology, where we lack quantitative data, partly as a result of the specific difficulties of our investigations, partly for historical misunderstanding in the methodological field. The system approach, helping us to gain a better qualitative knowledge of the interactions of the different parameters gives a sound ground for tempting some measurements, which are very urgently needed.

But we ought not to demonstrate a kind of selfishness as scientists and we must not confine ourselves in "pure" science preoccupations. We have the moral obligation to participate in a better management of limited natural resources. For this, too, the systems approach is necessary. It enables us to be conscious of the complexity of our environment. It enables us to evaluate correctly what role can be played by geomorphology side by side with other scientists. Really, very few practical problems of environmental management can be solved in the limited frame of an academic speciality. Nearly all the problems have a much broader scope and require the intervention of a team of different specialists. The systems approach enable them to work together and be efficient : it is a common methodological denominator. The cooperation of these different specialists is, in itself, a svstem. One of the most important properties of a system is to have its proper characteristics, different and much more valuable than the simple addition of the characteristics of its components.

SUGGESTED READINGS

Cotton, C.A., 1947 : *Climatic accidents in landscape-making.* Whitcombe, Christchurch etc. . . 354 p., 149 fig., LVIII plates.

Dunkerley, D,, 1980 : The study of the evolution of slope form over lost periods of time : A review of methodologies and some observational data from Papua-New Guinea. *Z. für Geom.*, NS 24 (1), pp. 52-67.

Peltier L.C., 1950 : The geographical cycle in periglacial regions as it is related to climatic geomorphology. Assoc. *Amer. Geogr., Annals*, 40, pp. 214-36.

Tricart, J., Cailleux, A., 1972 : *Introduction to climatic geomorphology.* Translated by C. Kiewiet de Jonge. Longmann, London, p. 295, fig.

Tricart, J., 1978 : Le sol dans l'environment écologique. Soils and ecological environment. *Rev. de Géomorphologie Dynamique* (Paris and Strasbourg XXVII (4), p. 113-128.

Tricart. J., 1979 : Paysage et ècologie. Landscape and ecology.*Rev. de Géomorphologie Dynamique*, XXVIII (3), pp. 81-94.

THREE

Cuchlaine A.M. King

CLIMATOGENETIC GEOMORPHOLOGY

INTRODUCTION

CLIMATOGENETIC geomorphology must be distinguished from climatic geomorphology. Büdel in 1948 commented upon the interpretation of these terms. He suggested that three stages of study were necessary in considering the relationship between climate and geomorphology. The first stage he called dynamic geomorphology, which consisted of the study of single processes. The second stage was climatic geomorphology, which he described as the study of climatic variation. The third stage, which could not be accomplished without the two earlier stages, was climatogenetic geomorphology. This study involved, according to Büdel, an analysis of the relief itself. If this approach to the problem is justified, then it is clear that this type of study in geomorphology is both complex and difficult. Indeed in the present state of knowledge it may well be premature. There is also the possibility that some of the evidence may have been lost permanently through erosion.

Problems concerning both the temporal and spatial aspects of the study must be considered. From the temporal point of view the most obvious effects arise through climatic fluctuations over time. These fluctuations have been particularly great in the Pliocene and Pleistocene periods, resulting in very rapid swings from large scale glaciation to interglacial conditions. These changes affected a very large part of the earth's surface.

Climatic geomorphology, which seeks to establish the effects of climate on the present landforms, must necessarily be a historic study. Few, if any, landforms are now in equilibrium with present-day climatic conditions or unaffected by past conditions. It is not surprising, therefore, that climatogenetic geomorphology grew out of the classic denudation chronological study, based on the concept of the cycle of erosion as proposed by W.M. Davis (1899). In fact Davis (1900, 1904, 1905 and 1906) himself was one of the first to attempt to define cycles of erosion for climates other than his normal or temperate cycle. He put forward ideas concerning both glacial and arid and semi-arid cycles of erosion.

The historic approach to the study of landforms and climates can be of two types. Firstly, the cycle can be studied as it theoretically runs its course in one specific type of climate from youth to old age. Secondly, the effect of changing climates can be sought in the landforms. Davis adopted the first approach, while the second was the aim of Büdel's work. The second is the more satisfactory in view of the rapid changes of climate in the recent past, which makes the first almost entirely theoretical.

The spatial aspect of the study involves the effect of scale. Several authors, including Stoddart (1969) have drawn attention to this matter. The work of Schumm and Lichty (1965) has been much quoted in this connection. They specified three scales on which geomorphological analysis can rest. The smallest and shortest scale is called steady time and is time independent; it is associated with dynamic equilibrium. It refers to a single hill slope or similar sized feature, which is studied over a short time span. The next scale is called graded time, and again it can be considered time independent. The largest scale of all is called cyclic time and it involves a large area and is time dependent. It is at this largest scale of space and time that long-term denudation chronology and cyclic development is studied. Therefore, climatogenetic geomorphology is most likely to produce reasonable results at this scale.

On the other hand, the study of process in dynamic geomorphology is usually undertaken at the smallest scale of both time and space, owing to observational difficulties and the nature of the enquiry, which involves detailed measurements over relatively short periods of time and in small areas. The problem, therefore, involves extrapolation from the small area, dynamic studies to the large area, climatic studies.

The structural element also enters into the study at various scales in different ways. It is, however, fundamental to an analysis of the landforms. At the small scale in the local area, the type of lithology and its arrangement including dip, bedding and other characteristics, are of great importance. In the large area study the structural aspect involves larger features, such as mountain ranges, plains and plateaux. It is on these structural elements, both small and large scale, that the influence of the climatic factors is superimposed. Indeed in the large structural units the climatic elements form minor modification of the basic landform pattern, and are on a smaller scale than the primary lineaments of the landscape.

In order to consider some of these problems and the stage to which the study has attained this contribution is divided into several sections. First the range of climate over space will be briefly mentioned; and secondly, the range of climate over time will be reviewed. These short sections lead to a study of the relationships between climate and process. Next the relationships between process and form are considered, leading to the discussion of landform classification in climatic terms. Some of the work has been empirical in character, while some authors have attempted a synthetic approach.

Present-day Climatic Range

The earth is at present in the waning stages of an interglacial period, according to many authorities. There is still a substantial amount of ice on the earth, although the major northern hemisphere land-based ice sheets are temporarily in retreat. They are represented only by small relics on the highlands of central Norway and Baffin Island in Arctic Canada. There are, however, still major ice sheets on Greenland and Antarctica. The latter ice sheet is probably a semi-permanent feature of the climatic scene, at least since the middle of the Tertiary, and it shows no sign of disappearance.

At the present time, therefore, a very considerable range of climatic conditions exist on the earth. At one extreme is the intense cold of the Antarctic continent. The ice sheet over Antarctica exerts a cooling influence on the whole southern hemisphere, so that latitude for latitude temperatures are lower in the southern than the northern hemisphere. The Arctic areas are by contrast less cold, owing to the generally higher temperatures and the moderating effects of warmer waters. Nevertheless very low temperatures are recorded in winter in the interior of the northern continents, particularly in Central Asia. These northern high latitude areas have a very strong seasonal contrast. It is greater in the north because the large land masses heat up during the summer period, resulting in very great temperature ranges in central Eurasia and central and northeast North America.

The strong seasonal effect is enhanced by the long periods of winter dark and summer light in the high latitudes. This is in strong contrast to the extremes which are also characteristic of high altitude areas of low latitude, where the contrast is largely diurnal rather than seasonal in character.

The seasonal element is strong in both glacial and periglacial areas. The periglacial areas are most extensive in the zones that are too dry for glaciers to form, in northern Eurasia and northern North America, where Tundra is widespread. It is not, however, a dry landscape owing to the low temperatures and evaporation, and the presence of permafrost, which renders the ground impermeable. Part of the precipitation, which is low, falls as snow in winter, when temperatures are very low.

The seasonal element is also a marked feature of the temperate latitudes, giving the contrast between the cool, wet winters, and mild, often damp summers. Precipitation is more uniform throughout the year, but evaporation varies much more with the seasons, causing contrasts between winter and summer run-off. This contrast is further increased in the Mediterranean climate, with its hot, dry summers, and damper, warm winters. The contrast is further developed in the adjoining semi-arid areas, where the rainfall becomes more unreliable and sporadic. It is usually, however, intense when it does fall.

The most arid areas are nearer the equator. Rainfall is very rare in the true deserts, and wind locally becomes one of the major climatic elements. Temperature ranges are great on a diurnal pattern, rather than a seasonal one. On the equatorial side of the great deserts the seasonal element again becomes important, but in these areas it is the rainfall pattern, rather than the temperature, that is seasonal in its occurrence. This is the savanna region with its wet and dry seasons. The monsoon areas also experience wet and dry seasons of marked contrast.

The least seasonal type of climate is that of the hot, wet equatorial zone, where the temperature range is very small and the rainfall fairly uniformly distributed, although in some areas there are two wet and two drier seasons in the year. Thus the contrast of climate throughout the earth is great at present. It is bound to influence the geomorphological processes, and the landforms to which they give rise, in a complex way.

Climatic Range Over Time

Over the very long term there are probably two states in which the earth's climate can be; either there is ice present or there is not. The glacial periods appear to have a very approximate cycle of about 150 million years, although recent work has tended to find new evidence of glacial action on the earth in the past.

There are probably still some periods during which ice is not present. At these stages the contrasts between the equatorial areas and the polar areas are very much reduced. As the temperatures are generally more equable so the atmosphere and ocean circulations are slowed down and restricted. Belts of depressions in high latitudes are less vigorous and oceans currents are more sluggish. Thus, the precipitation is reduced and desert conditions are more likely to be widespread. Temperatures are generally much warmer in high latitudes, so that weathering conditions are different from those of the present climatic regime.

In the long term it is also necessary to take into account the movement of the continents relative to the poles, a movement that has been shown to be very great. These movements may well account for evidence of changing climatic conditions in the long term. The movements of the continents in turn, causes climatic changes, or is at least an important contributary element in these changes.

Flohn (1964) has made a theoretical climatic reconstruction of the climate of Central Europe during the late Mesozoic and early Tertiary, when the latitude would have been similar to the present, but when the earth climate was non-cryogenic. He considered that a subtropical arid zone with mean temperatures from 20 to 26°C extended up to 50°N, apart from areas with oceanic exposure, such as West Europe. Between 50 and 65°N he thought that a subtropical zone with winter rains and a mean temperature between 15 and 22°C would have existed. Such a climate is similar to the present Mediterranean climate, but with much higher temperatures. The winter rains could have been very heavy, giving extreme event or catastrophic erosion.

The earth is at present within a glacial phase of climate, so that there is a wide range of climatic conditions over space. There are, however, superimposed upon the long term glacial and non-glacial climatic epochs, shorter variations within the present glacial state. These variations within the present glacial phase make the development of a climatogenetic geomorphology difficult. They are long enough for the two large land-based ice sheets of the northern hemisphere to extend over large areas of middle and high northern latitudes, and to cause major movements of the climatic zones of almost the whole earth.

They are not, however, long enough for the processes associated with the different climatic zones to reach an equilibrium with the forms that they create. It has been estimated that there have been at least twenty major switches of climate from full glacial to full interglacial during the present glaciation. These changes cover a period that is probably several million years duration, as there is evidence of ice in the northern hemisphere about five million years ago.

The changes are relatively regular and have a period of approximately 100,000 years, according to Emiliani (1966). They are probably related in part at least to the regular movement of the earth relative to the sun as discussed by Milankovich (1930), giving regular variations in solar radiation receipt over the earth's surface.

There is a very rapid switch of conditions from glacial to interglacial, but the built-up of the ice sheets, once a critical threshold has been passed, is not quite so fast. Associated with the rapid growth and decay of the ice sheets are the equally rapid fluctuations of sea level. These rapid changes of level again complicate the development of landforms, especially in those areas that are affected by changes of base level, owing to their proximity to the coast. Thus all landscapes with few exceptions are now both polygenetic and polycyclic. It is this factor that makes climatogenetic geomorphology essential, rather than the simpler climatic geomorphology, which is of necessity a purely theoretical study.

According to Büdel (1957) the latitudes of the different climatic belts were very different in the Würm, or last glaciation, to their present positions. He gives the latitudes for the different zones as indicated in Table 3.1, which shows a general extension in the high latitudes with a compression of the zones of the lower latitudes. One of the very few zones to remain consistent as the ice waxes and wanes is the inner zone of the humid tropical forest area.

TABLE 3.1 : **Climatic zones after Büdel (1957)**

Zone	Holocene	Wurm in degrees
Glacier zone	77	54
Frost rubble zone	74	51
Tundra zone	69	45
Non-tropical forest	45	38
Mediterranean forest	37	34
Mediterranean steppe	34	28
Desert steppe	32	20 Desert and desert savanna
Desert	18	15 Dry savanna
Desert savanna	16	0 Humid savanna and tropical forest
Dry savanna	12	
Humid savanna and tropical forest	0	

The climate is continually varying, and even in the Holocene, during the last 10,000 years, there have been quite substantial changes in climate, and these have been reflected in changes of vegetation. The Little Ice Age, for example, was a colder phase lasting from about 1550 to 1850, during which time Alpine glaciers expanded to their historic maxima.

The changes of climate are associated with changes of

vegetation, as suggested in Table 3.1 given by Büdel. The changes of vegetation are as important as the climatic changes that cause them. They influence the landforming processes and their effects. The vegetation and climate also exert an important influence on the soil. Thus in attempting to establish the relationships between climate and landforms, it is essential to consider the part played by both vegetation and soil. In these aspects man can play a vitally important role, by modifying the vegetation and soil.

The climate cannot be seriously modified by man at present. It is, however, a possibility for the future as human interference becomes more ubiquitous and massive in character. The increasing consumption of fossil fuels, by altering the carbon dioxide in the atmosphere, and land clearance for agriculture, are examples of important human activities. It is very unlikely that an extrapolation from the present back into the past will provide reliable results for these reasons. This point further complicates the study of climatogenetic geomorphology, which is necessarily a long-term study with an historical basis, but which must rely on short term process measurements made under present day conditions. These may be quite unlike past conditions because of human interference, even if the climate is similar.

Nevertheless, at the extremes at least, it is possible to relate climate to process on a quantitative basis, as has been attempted in various ways by different workers.

Relationship between Climate and Process

At the extremes of the climatic range it is clear that climate is related to process. In the glacial climates the role of glacier ice is obviously important in modifying the landscape, and in the periglacial areas the various forms of ground ice are equally significant in the processes typical of the Tundra areas. At the other extreme, in the hot deserts or tropical rain forests, the processes depend on very different climatic phenomena. In between, in the many landscapes that depend largely on the action of fluvial and slope processes not related to ice or wind action, a wide range of forms occur. These forms depend on much more subtle variations in processes that are by no means easy to recognise or quantify in climatic terms.

One of the best known attempts to relate climate to geomorphological process is that of Peltier (1950). It is based on an empirical relationship between climatic values and the intensity of action of various processes. He uses the mean annual temperature plotted against the mean annual precipitation to delimit the boundaries between the zones of varying intensity of action of various processes.

Frost action is the dominant form of mechanical weathering in areas of low temperature and precipitation, the Tundra and frost rubble zones. Chemical weathering, on the other hand, is strong in the hot, wet areas, and weak in the dry ones, regardless of temperature. Weathering regions are delimited in terms of the relative importance of chemical and mechanical weathering.

Mass movement has two optimum climatic types, the hot and wet, and the cold and relatively dry, with the minimum occurring in the warm and hot, dry zones. Wind action reaches its maximum efficacy in the dry zones, both hot and cold. Pluvial erosion has its optimum in the warm and moderately wet areas, with minima at all the extremes, hot and dry, cold and dry, and hot and wet. The relationships can also be illustrated in terms of the dominant processes under different climatic conditions.

Peltier goes on to define morphogenetic regions in relation to the climatic elements. These will be mentioned in the next section, as they are more concerned with form in relation to process.

More recently Wilson (1969) has considered the relationship between processes and the present climate, as a method of studying palaeoclimate. It is necessary first to establish the relationships between climate and process, which he refers to as the climate-process system, CPS. Processes are related to temperature and precipitation in a way very similar to that of Peltier, but rather more generalised.

Mechanical and chemical weathering tend to occur in opposite climates. Mass movement of the solifluction type is important in cold, damp regions, while landsliding is more common in hot, wet ones. These two climates produce the maximum weathering of one or other type, and, therefore, have the maximum amount of loose material available.

Fluvial erosion is particularly important in morphogenetic

analysis, but is not simply related to temperature and precipitation. As precipitation increases run-off tends to increase, but vegetation plays an important part and inhibits erosion as it becomes denger with increasing precipitation. Semi-arid areas have a peak of fluvial erosion

Wilson used 1500 rivers in his attempt to relate sediment yield with rainfall. In this study he followed the work of Langbein and Schumm (1958), who, from a more limited sample of rivers, suggested that the maximum sediment yield relative to precipitation occurred at a precipitation of between 25 and 50 cm. Wilson, from his larger sample, gives a more complex curve with two maxima of sediment yield at precipitation values of about 75 cm and 175 cm.

The highest rates of erosion occurred where the climate had a strong seasonal element, with the climate being arid at one season. Wilson's results show that the mean annual values of temperature and precipitation are too general to provide a really satisfactory means of delimiting process zones. He, therefore, used the mean monthly values to define different climate-process systems in terms of climatic regions.

When the monthly values are used, the resulting plots extend across several regions, so that several climate-process systems are involved where seasonality is marked. He relates geomorphological process to air mass regimes, based on Strahler's (1969) climatic regimes, which give seven zones: (1) Polar (cA), (2) Desert (cT), (3) Equatorial (mT), (4) Mid-latitude equable (mP), (5) Tropical wet-dry (cT and mT), (6) Mediterranean (cT and mT and/or mP), (7) Continental (mT and cP). Wilson's six climate-process systems are (1) glacial, (2) periglacial, (3) humid, (4) temperate, (5) semi-arid, and (6) selva, and they relate to the seven Strahler climatic regimes.

In referring his results to palaeoclimatic conditions, Wilson considers the greater seasonality, rather than aridity, occurred in South America. This resulted in greater slope erosion, planation and valley alluviation. When the climate becomes more equable, this could lead to upland stability and lowland river incision as seasonality declines.

Flohn (1964) has used somewhat similar arguments in his assessment of the early Tertiary climate of Europe. He suggests that the early Tertiary climate could have resembled the periodically humid, semi-humid and semi-arid climates. In these climates morphological processes favour the development of pediments and isolated hills of inselberg type. The inselbergs of the present tropics are sharper than those of early Tertiary times in Central Europe, because the alternation of wet and dry seasons was less intense in the higher latitudes, where solar radiation is less. Lithological differences could also partly account for the lower inselbergs in parts of Central Europe. Glacial modification of inselbergs is not thought to be great and cannot account for differences between them and modern ones.

Wilson followed Langbein and Schumm, and various other workers, in using the sediment yield of rivers as a measure to relate geomorphological processes to climate. The sediments yield of rivers is a function of the relationship between slope processes, the material of the slope and its gradient, length and surface character. The latter includes the nature of the soil and vegetation which is of special importance. The use to which the soil and vegetation are put is also very significant. The slope processes themselves depend to a considerable extent on the climate. Sediment yield does, therefore, give a reasonable measure of the effectiveness of the processes. It cannot, however, give much useful information concerning the methods by which the processes operate.

Douglas (1967) has made detailed observations of the relationships between climate and sediment yield in various small catchments in Australia. Some were in the tropical northeast of Queensland and others in the temperate southeast of New South Wales. He chose catchments that had been disturbed as little as possible by human interference. The results showed that there was a highly significant difference between the two areas in terms of the total sediment yield. He found that there was a good correlation of 0.4999 between run-off and suspended sediment yield when all the basins were plotted together. The correlation was, however, much improved, to 0.6871, when the suspended sediment yield was plotted against the parameter p^2/P, as used by Fournier (1960), p is the amount of rainfall in the wettest month and P is the annual rainfall, thus bringing in the seasonality of the climate.

Douglas' conclusion is that the seasonality and total

amount of rainfall are more important than the temperature. Although his results were below those of Fournier by almost an order of magnitude, he nevertheless came to the conclusion that present-day sediment loads must be much in excess of those that have operated over considerable periods in the geological past. This conclusion makes extrapolation back in time very doubtful. He considered that human interference was responsible for the high rates recorded at present.

Other workers who have attempted to define world erosion rates in terms of climate include Corbel (1964), Fournier (1960) and Strakov (1967). Corbel (1964) attempted to relate different climatic zones to erosion rates in terms of precipitation. His results are given in Table 3.2. He also distinguished between mountain and plain erosion, showing that steep slopes produce considerably more erosion than flatter ones.

TABLE 3.2 : **Rates of regional erosion after Corbel (1964, p. 407) in $m^3/km^2/year$**

Climate	Arid under 200 mm		Normal 200—1500 mm		Humid over 1500 mm	
	Mountain	Plain	Mountain	Plain	Mountain	Plain
Hot 15°N–15°S	1.0	0.5	25	10	30	15
Tropical 15°–23½° N and S	1.0	0.5	30	15	40	20
Extra-tropical Temperature over 15°C	4	1	100	20	100	30
Temperate, Temperature 0—13°C	50	10	100	30	150	40
Cold Temperate, Under 0°C	50	15	100	30	180	—
Polar	50	15	100	30	150	—
Glaciated polar	50		1000		2000	
Glaciated, non-polar			—		2000	

His analysis shows that glaciated lands have by far the greatest rate of erosion, while the more humid areas have higher rates than the normal and much more than the arid areas. His results are based on relatively few drainage basins and the tropics, in particular, are poorly represented. His figures, therefore, are no more than possible orders of magnitude. It seems possible that typical tropical basins may have been omitted from the sample, while temperate upland basins were unrepresented.

Fournier (1960) produced very different results. He used 78 basins of very different dimensions to relate suspended sediment yield to climate. The values he used for climate were P (rainfall), P′ (run-off), P_n (rainfall frequency) and several others, including p^2/P, and a relief index $C_m = \overline{H} \tan \Phi$, $\overline{H}$ is the mean height of the basin and Φ the mean slope. His sample basins fell into three main groups : (1a) basins with low relief and temperate climate, (1b) basins with low relief and tropical, subtropical and semi-arid climate, (2) basins with high relief and a humid climate, and (3) basins with high relief and a semi-arid climate. Points on the first line had a C_m value of less than 6, while those on the second and third had values over 6. Each group can be represented as straight lines when p^2/P is plotted against the sediment yield in tons/km^2/year. The regression equations are given in Table 3.3. The results are all highly significant.

TABLE 3.3 : **Climate related to sediment yield according to Fournier (1960)**

Line	Regression equation	t	Degrees of freedom	P less than
1a	$Y = 6.14\,X - 49.78$	9.16	46	0.001
1b	$Y = 27.12\,X - 475.4$	22.07	38	0.001
2	$Y = 52.49\,X - 513.21$	18.10	29	0.001
3	$Y = 91.78\,X - 737.62$	28.06	21	0.001

1a low relief, temperate climate

1b low relief, tropical subtropical and semi-arid climate

2 high relief, humid climate

3 high relief, semi-arid climate

The two variables, precipitation and relief, can be linked in the equation: $\log E\ (\text{tons/km}^2/\text{yr}) = 2.65\ p^2/P(\text{mm}) + 0.46 \log \overline{H} \cdot \tan \Phi - 1.56$. Fournier has provided a world map of his regions 1 to 3 based on p^2/P. The highest rates of erosion occur in the seasonally wet tropics, with less erosion in the continuously wet equatorial regions, and the continuously dry desert areas. Values rise again in the seasonally wet Mediterranean regions, but are less in the temperate and cold regions, except in mountainous areas.

Corbel and Fournier produce very different patterns. Strakov's values are more similar to those of Fournier, but the absolute amounts are very different. One weakness of these values is the variety of size of drainage basin used, which must affect the results.

Langbein and Schumm (1958) refined the results by calculating the effective precipitation, which is defined as that required to produce a known run-off at 10°C. They were able to show that as temperature rises, so a higher effective precipitation is needed to produce a peak of sediment yield. Semi-arid areas produce high yields. The situation is most complex in the areas of higher precipitation and run-off, in which vegetation plays an important part.

Relationship between Process and Form

The discussion in the last section will have indicated that as yet it is not possible to give detailed assessments of the effect of climate on different processes in quantitative terms. The very general results obtained from the study of sediment yield in rivers is only a measure of total erosion effectiveness. It does not relate to morphological modification of the landscape. There have, however, been some attempts to assess the effect of climate on various landscape parameters, such as drainage density.

One example of this type of analysis is that of Chorley (1957) who compared mean stream length, basin area, and drainage density in three areas of mature dissection, in southwest England, Pennsylvania and Alabama. He used a measure, I_c, to combine the effects of climate and vegetation. $I_c = I$. Amount of vegetation/P . Q_s . I is Thornthwaite's (1948) precipitation effectiveness index, P is the amount of precipitation and Q_s is its intensity. I_c was 0.4 for Alabama, 1.0 for Pennsylvania and 1.3 for Exmoor in southwest England. The results showed that Exmoor, with the highest I_c value, had the greatest mean stream length and drainage area, but the lowest drainage density. Alabama, with the lowest I_c value, had the reverse. This confirmed the theoretical expectations. The results are not, however, interpretable in terms of morphology, apart from the generalisations concerning stream morphometry.

Peltier (1962) attempted to test his empirical morphogenetic regions by world-wide sampling of morphological characteristics of the landscape. One of the difficulties that he encountered was the lack of suitable large scale maps for many areas. The largest scale map from which he could derive suitable measurements was 1 : 1,000,000 for many areas. From these maps he obtained a general relief factor by measuring the maximum difference in height within a 100 mile² area. From these values the average slope may be obtained after first determining a local relief estimate from the equation $\log \bar{R}_l = 0.08355 + 0.79606 \log \bar{R}_g$, $\bar{R}_l$ is the local relief in ft/mile² and $\bar{R}_g$ is the general relief in ft/100 mile². The mean slope, $\bar{S}$, is derived from $\bar{R}_l$ by the equation $\bar{S} = 0.024\ \bar{R}_l - 0.66$. He also recorded the drainage density $\bar{G}$, which is the average number of drainage ways per mile of transect.

Although the measurements are somewhat crude they did provide consistent results when different climatic zones were plotted on separate curves. The equations for the curves were as follows in Table 3.4. The curves show a systematic displacement for the desert, moderate and semi-arid climates. These curves have the same slope, with the moderate climate lying between the other two. Thus for any given mean slope the semi-arid climate has the greatest number of drainage ways, and the desert least. This supports the view that the semi-arid areas have the most effective run-off, owing to sparse vegetation cover and considerable rainfall of a spasmodic nature.

TABLE 3.4 : **Relationship between mean slope and drainage density after Peltier (1962)**

Moderate or mesothermal climate	$\log \bar{S} = 4.23 \log \bar{G} - 1.22$
Semi-desert climate	$\log \bar{S} = 6.00 \log \bar{G} - 2.86$
Desert climate	$\log \bar{S} = 2.89 \log \bar{G} + 0.04$
Tropical climate	$\log \bar{S} = 2.04 \log \bar{G} - 0.57$

$\bar{S}$ is the mean slope and $\bar{G}$ is the drainage density.

The curve for the tropical areas lies at a different angle, suggesting that different controls operate in these areas. The number of drainage ways are considerably greater than for the other climatic types, when the slopes are steep. Extreme dissection in this type of climate is indicated, therefore, Peltier suggests that this is due to increased landslide activity on the steeper slopes, in areas where rainfall is heavy and continuous.

There is need for considerably more work of this type if a quantitative assessment of the effect of climate on landforms is to be established. This is difficult to achieve on a world-wide basis, owing to lack of adequate data. On a smaller scale, however, it is possible to illustrate some effects of climate by studying a small area.

There have been, for example, a number of studies of slope asymmetry, which give an indication of the effect of climate on a microscale. Even in a small area there are complex problems, and slope asymmetry has been shown to be due to different reasons, not all of them climatic. Nevertheless climatic effects are often important in accounting for slope asymmetry. In areas that have been or are influenced by periglacial processes the southwest facing slopes are often steeper than the northeast facing ones because they undergo greater thaw-freeze activity, than the colder, northeast facing, slopes. This argument has been put forward by Ollier and Thomasson (1957) to account for the asymmetry of the chalk valleys of the Chiltern Hills in southeast England. They consider that creep is more effective on the southwest facing slopes, which steepens them and causes more material on these slopes to move to the valley floor. The stream is forced to flow close under the gentler northeast facing slope, as long as periglacial impermeability allows it to flow. It migrates down the gentle slope as long as it can evacuate all the material and deepen the valley.

The work that has been done on a world-wide basis is more theoretical and general in nature, but it is at least an attempt to relate the landforms to the climate on a large scale. An early attempt to classify climate on a physiographic basis was that of A. Penck in 1910.

Penck defined two important boundaries, separating three climatic zones, the humid climates, the nival climates and the arid climates. These were separated by the snowline and the dry boundary respectively. In the humid climates N—V=F >0, N is the precipitation and V is the evaporation and F is the run-off. In the nival climate S—A=G>0, where S is snowfall, A is ablation and G is glacial ice. In the arid climate A—V<0. The snowline is where S=A, and the arid boundary TG is where N=V. Penck was more concerned with the delimitation of natural regions than with the detailed morphology of the features characteristic of these regions. His work is nevertheless a beginning of the appreciation of the effect of climate on landforms, and an attempt to quantify this effect.

Büdel (1948) considered that Penck's views were too simple, and he elaborated the concept of the relationships between climate and morphology. Table 3.5 illustrates his concept of climatic zones in relation to morphological processes and forms. The zones that he distinguished are related to processes and morphology. He considered that soil cover is important, and is

TABLE 3.5 : **Climatic-morphological zones of the earth after Büdel (1948)**

Submarine morphological province=sea floor zone

Subglacial morphological province=glacier zone

Subaerial morphological province—comprising :

1. Frost debris zone
2. Tundra zone
3. Temperate zone of mature soil (ortsboden)
    - (a) Oceanic zone
    - (b) Subpolor zone without tjaele (permafrost)
    - (c) Tjaele or permafrost zone
    - (d) Continental zone
    - (e) Steppe zone
4. Mediterranean transition zone
5. Dry debris zone
    - (a) Tropical hot deserts and marginal deserts
    - (b) Extratropical deserts
    - (c) High altitude (cool) deserts
6. Sheetwash zone
    - (a) Tropical
    - (b) Subtropical
7. Inner tropical zone of mature soils

related both to climate and to rock type. Areas of similar form, he argued, must be recognised empirically and then related to climate. He recognised that large areas are not related to present day processes.

The frost debris zone is one in which solifluction is active and streams are powerful during the short snow melt season. It is, therefore, a zone of effective processes that are not hindered by the vegetation, which is sparse. The capacity of erosion is great in this zone. Channel beds are debris-covered, and gentle slopes are striped, while steeper ones are gullied.

In the temperate zones soils are thick, vegetation cover dense, and as a result it is the least dynamic zone. Erosion capacity is small, and thus its morphology changes only slowly with time. The Mediterranean zone has similarities to the neighbouring zones. It is more active than the temperate zone and contains features common to the dry debris zone.

The dry debris zones consist of the deserts, in which mechanical weathering predominates over chemical. Features found in these areas include gently-sloping, debris-covered slopes, the pediments, and inselbergs. Water courses have low gradients with wide, flat, debris-covered floors.

The sheetwash zone is extensive and covers the seasonally moist tropics. The Monsoon lands are also included in this category. The two unique and effective processes characteristic of this zone are sheetwash and rill-wash, which occur during heavy precipitation. Surface denudation processes are predominant over linear ones, and only large rivers are found. The landforms characteristic of the area include stepped erosional plains, extensive rock pediments and inselbergs. It provides a clear morphological zone.

In the inner tropical zone of mature soils chemical weathering is strong and rock type differences are emphasised. Soils are deep and leached, and it has a thick forest cover. As slopes steepen subsurface slips and slumps occur, resulting in sharp ridges, especially in mountainous areas. These steep slopes can remain covered by thick soil and forest even at gradients of 55 degrees. The rivers are free of coarse debris and hence linear erosion is effective and vertical incision helps to give steep valley slopes. In quantitative terms erosion is not rapid in this zone as little debris is produced.

In his later contribution, Büdel (1963) lists five climatomorphological zones. The first is the glaciated zone. The second is the zone of pronounced valley formation, which is in non-glaciated polar regions, which are underlain by permafrost. The third zone is that of extratropical valley formation in mid-latitude regions. In this zone valley-forming processes are not at present important but fossil features of the glacial period are much in evidence. The fourth zone is that of subtropical areas with pediment and valley formation. It is transitional between the third and fifth zones. The fifth is the tropical zone of planation surface formation.

The second and fifth zones represent extreme types. They do, however, have one important feature in common ; they are both zones of deep weathering. In the second zone it is mechanical in nature, while in the fifth it is chemical.

Zone two of pronounced valley formation is associated with the Tundra and frost rubble zones. It was much more extensive in the glacial periods. Now it covers 8 million km^2, about half its former extent of 15 million km^2. The valley deepening of this zone is a rapid process, partly aided by the lowering of sea level in the glacial periods. The existence of permafrost in these areas is important in making the erosion more effective. Slope processes are very active in the upper layer that melts in summer. Ground ice also helps the rivers to erode effectively.

At the other extreme is the tropical zone of planation in which seasonal precipitation is very significant. It is also a zone of deep chemical weathering, extending to a depth of 300 m in parts of west Africa and India on gentle slopes. The extensive planation surfaces that extend right to the rivers are the major characteristic of this zone. These features are dominant in areas of low epeirogenic activity, and occur in large parts of tropical Africa, Guyana and the Deccan. In Africa the surfaces extend over 500 km by 5000 km, and occur at several levels. Inselbergs are the main interruption of the surface. They rise abruptly above the planation surfaces. Thus in this zone gentle and steep surfaces meet abruptly. The plains grow laterally and leave inselbergs as the steep slopes retreat.

Büdel's analysis emphasises the importance of process in determining the major landforms of the various zones. He does not stress the climatic controls quantitatively, although they are clearly stressed in qualitative terms.

Peltier (1950), on the other hand, stresses the quantitative approach. His delimitation of the intensity of processes relative to climatic values has already been mentioned. He then goes on to suggest morphogenetic regions on the basis of climatic controls. His morphogenetic regions are given in Table 3.6. Peltier does not elaborate the forms associated with these morphogenetic regions, apart from the periglacial one, for which he gives a possible erosion cycle. The later stages of this cycle must, however, be theoretical as sufficient time has not elapsed for a full cycle to run its course uninterruptedly. He suggests that wide surfaces of cryoplanation will eventually be formed by congelifraction and congeliturbation.

TABLE 3.6 : **Morphogenetic regions according to Peltier (1950)**

Morphogenetic region	Annual Temperature°C	Annual Precipitation cm	Morphological characteristics
Glacial	−18/−7	0/114	glacial erosion, nivation
Periglacial	−15/−1	13/140	wind action, strong mass movement, weak running water
Boreal	−9/+3	25/152	moderate frost action, moderate to slight wind, moderate running water
Maritime	+2/+21	127/190	strong mass movement, moderate to strong running water
Selva	+16/+29	140/229	strong mass movement, no wind, slight slope wash
Moderate	+3/+29	89/152	maximum running water, moderate mass movement, frost slight, no wind
Savanna	−12/+29	64/127	strong to weak running water, moderate wind
Semi-arid	+2/+29	25/64	strong wind, moderate to strong running water
Arid	+13/+29	0/38	strong wind, slight running water and mass movement

CONCLUSION

Climatogenetic geomorphology is a complex field of study that covers a wide range of possible approaches. Its essential aim is to elucidate the processes that have shaped the landscape over time, by studying the present landforms. The landforms should also provide evidence of present and past climates, through a knowledge of the relationships between processes and climate. Thus it should be possible to study the processes, and to establish under what climatic conditions they are most effective. One way in which this has been done is to measure the sediment yield of streams, which gives an overall value of the effectiveness of erosion, but no details of the processes involved. Empirical studies of morphometry provide measures that can be used to compare landforms quantitatively in different environments. The results of such studies, however, even when they are highly significant, must be interpreted with caution, as so many variables are involved.

The concept of climatic geomorphology has been discussed by Peltier (1975). He considers that climatic geomorphology is a component of regional geomorphology, together with tectonic geomorphology. In his view landforms are a function of geological material, rate of change of material with distance, rate of erosion, rate of uplift and total duration of the process. Lithology structure, erosion and tectonic activity are all involved. He argues that landform varies according to (1) the relief, (2) the geographical variation in lithology with respect to its vulnerability to weathering, (3) the rate of weathering and erosion, and (4) the duration of the assemblage of geomorphic processes.

He comes to the conclusion that most of the land surfaces have been formed or strongly modified in the last 50,000 years. Some show strong influence of processes inherited from the past, but truly old surfaces are rare. Maps that attempt to show polygenetic sequences of processes are more likely to be realistic than those showing only currently-acting processes.

Much of the work on climatogenetic geomorphology has been done in continental Europe. Tricart (1961, 1965), Tricart and Cailleux (1965), and Birot (1968) deserve more discussion than space has allowed, as they have written extensively on the topic. Tricart and Cailleux (1965) consider the dominant and subsidiary processes in the different morphogenetic zones. They recognise the importance of the zonal concept in discussing the relationship of climate to landforms zones, but also recognise azonal elements, again stressing the complexity of the problems.

The conclusion must be reached that climatogenetic geomorphology is a complex study. It aims to discover the effects of climatically controlled processes in creating distinctive landforms. In view of the recent large-scale fluctuations of climate, both present and past processes must be considered. The study is, therefore, a historical one, involving consideration of changes over time as well as over space.

SUGGESTED READINGS

Birot, P., 1968 : *The cycle of erosion in different climates.* Translated by C.I. Jackson and K.M. Clayton, Batsford, London, pp. 144 (original in French, 1960).

Büdel, J., 1948 : Das system der klimatischen Geomorphologie (Beiträge zur Geomorphologie der Klimazonen und Vorzeitklimate V.) *Verhandlungen Deutscher Geographentag.* Munchen, 27, pp. 65-100.

Büdel, J., 1957 : Die 'Doppelten Einebungsflächen' in den feuchten Tropen. *Zeitschrift für Geomorphologie* N.F. 1, pp. 201-28.

Büdel, J., 1959 : Climatic zones of the Pleistocene. *International Geology Review* 1(9), pp. 72-9.

Büdel, J., 1963 : Klima-genetische Geomorphologie. *Geographische Rundschau* 15, pp. 269-85.

Büdel, J., 1965 : Die Relieftypen der Flachenspulzone Sud-Indiens am Ostabfall Dekans gegen Madras. *Colloquium Geographicum* 8, pp. 1-100.

Chorley, R.J., 1957 : Climate and morphometry. *Journal of Geology* 65, pp. 628-38.

Corbel, J., 1959 : Vitesse de l'érosion. *Zeitschrift fur Geomorphologie.* N.F. 3, pp. 1-28.

Corbel, J., 1964 : L,Érosion terrestre, étude quantitative (méthodes—techniques—résultats). *Annales de Geographie* 73, pp. 385-412.

Cotton, C.A., 1942 : *Climatic accidents in landscape-making.* Whitcombe and Tombs, Christchurch, pp. 354.

Cotton, C.A., 1958 : Alternating Pleistocene morphogenetic systems. *Geological Magazine* 95, pp. 125-36.

Cotton, C.A., 1961 : The theory of savanna planation. *Geography* 46, pp. 89-101.

Cotton, C.A., 1962 : Plains and inselbergs of the humid tropics. *Transactions of the Royal Society of New Zealand* (*Geology*) 1, (18), pp. 269-77.

Davis, W.M., 1899 : The geographical cycle. *Geographical Journal* 14, pp. 481-504.

Davis, W.M., 1900 : Glacial erosion in France, Switzerland and Norway. *Proceedings of the Boston Society for Natural History* 29, pp. 273-322.

Davis, W.M., 1904 : Complications of the geographical cycle. *Reports of the VIII International Geographical Congress*, pp. 150-63.

Davis, W.M., 1905 : The geographical cycle in an arid climate. *Journal of Geology* 13, pp. 381-407.

Davis, W.M., 1906 : The sculpture of mountains by glaciers. *Scottish Geographical Magazine* 22, pp. 76-89.

Douglas, I., 1967 : Man, vegetation and sediment yield of rivers. *Nature* 215, pp. 925-28.

Douglas, I., 1968 : Natural and man-made erosion in the humid tropics of Australia, Malaysia and Singapore. *Publications of the International Association for Scientific Hydrology*, 75, pp. 17-30.

Emiliani, C., 1966 : Isotopic paleotemperatures. *Science* 154, pp. 851-7.

Emiliani, C., 1966 : Palaeotemperature analysis of Caribbean cores pp. 6304-8 and pp. 6340-9 and a generalised temperature curve for the past 425,000 years. *Journal of Geology* 74, pp. 109-26.

Flohn, H., 1964 : Grundfragen der Palaeoklimatologie in Lichte einer theoretischen Klimatologie. *Geologische Rundschau Stuttgart*, No. 54, pp. 504-515.

Fournier, F., 1960 : *Climat et érosion : la relation entre l'érosion du sol par l'eau et les précipitations atmosphériques*. Paris : Presse Universitaires de France, pp. 201.

Gellert, J.F., 1970 : Climatomorphology and paleoclimates of the central European Tertiary. pp. 107-111 in Pecsi, M. (Ed.) *Problems of relief planation*. Budapest Studies in Geography in Hungary, 8, Geographical Research Institute, Hungarian Academy of Science, Budapest.

Guilcher, A., 1965 : Questions de morphologie climatique en Mélanésie équatoriale (Nouvelle-Guinée, Nouvelle-Bretagne, Guadalcanal.) *Bulletin de l'Association de Géographes francais*, pp. 338-9, 28-40.

Judson, S. and Ritter, D.F., 1964 : Rates of regional denudation in the United States. *Journal of Geophysical Research* 69, pp. 3395-401.

King, L.C., 1962 : *Morphology of the earth*. London and Edinburgh, Oliver and Boyd, pp. 699.

Langbein, L.B. and Schumm, S.A., 1958 : Yield of sediment in relation to mean annual precipitation *Transactions of the American Geophysical Union* 39, pp. 1076-84.

Martonne, E. de, 1913 : Le climat-facteur du relief. *Scientia*, pp. 339-355.

Melton, M.A., 1957 : An analysis of the relations among elements of climate, surface properties and geomorphology. *Columbia University, Department of Geology, Technical Report* 11, pp. 1-102, New York.

Melton, M.A., 1960 : Intravalley variation in slope angles related to microclimate and erosional environment. *Bulletin of the Geological Society of America*, 71, pp. 133-44.

Milankovich, M., 1930 : Mathematische Kilmalehre and astronomische

Theorie der Klimaschwankungen, In *Handbuch der klimatologie*. (Eds. W. Koppen and R. Geiger) 1 Teil, A. Berlin.

Ollier, C.D. and Thomasson, A.J., 1957 : Asymmetrical valleys of the Chiltern Hills. *Geographical Journal*, 123, pp. 71-80.

Passarge, S., 1926 : Geomorphologie der Klimazonen oder Geomorphologie der Landschaftsgürtel? *Petermanns Mitteilungen* 72, pp. 173-5.

Peltier, L.C., 1950 : The geographic cycle in periglacial regions as it is related to climatic geomorphology. *Annals of the Association of American Geographers*, 40, pp. 214-36.

Peltier, L.C., 1962 : Area sampling for terrain analysis. *Professional Geographer* 14, pp. 24-8.

Peltier, L.C., 1975 : The concept of climatic geomorphology. Chapter 8 in *Theories of Landform Development*. (Eds. W.N. Melhorn and R.C. Flemal) Proceedings volume of the 6th Annual Geomorphology Symposia Series, Binghamton, S.U.N.Y. 1975, pp. 129-43.

Penck, A., 1910 : Versuch einer Klimaklassicification auf physiographischer Grundlage. *Preussen Akademie der Wissenschaft Sitze der physik-math*, Klasse 12, pp. 236-46.

Penck, A., 1905 : Climatic features in the land surface. *American Journal of Science*, Series 4, 19, pp. 165-74.

Penck, A., 1914 : The shifting of the climatic belts. *Scottish Geographical Magazine* 30, pp. 281-93.

Penck, W., 1924 : *Die morphologische Analyse*. Stuttgart, J. Engelhorn's Nachfolger, pp. 283.

Rapp, A., 1960 : Recent development of mountain slopes in Karkevagge and surroundings, Northern Scandinavia. *Geografiska Annaler* 42, pp. 65-200.

Schou, A., 1965 : Klimatisk geomorfologi. *Geografisk Tidsskrift* 64, pp. 129-61.

Schumm, S.A. and Lichty, R.W,, 1965 : Time, space and causality in geomorphology, *American Journal of Science* 263, pp. 110-19.

Stoddart, D.R., 1969 : Climatic geomorphology : review and re-assessment, *Progress in Geography* Volume 1. Eds. C. Board *et al.* Edward Arnold, London, pp. 159-222.

Strahler, A.N., 1952 : Dynamic basis of geomorphology. *Bulletin of the Geological Society of America* 63, 293-98.

Strahler, A.N., 1969 : *Physical Geography* 3rd Edition Wiley, New York.

Strakov, N.M., 1967 : *Principles af lithogenesis*. Volume 1. Tomkeieff, S.I. and Hemignway, J.E. Eds. Edinburgh and London, Oliver and Boyd, pp. 245.

Tanner, W.F., 1961 : An alternate approach to morphogenetic climates. *Southeast Geologist* 2, pp. 251-57.

Tanner, W.F., 1962 : Geomorphology and the sediment transport system. *Southeast Geologist* 4, pp. 113-26.

Thomas, M.F., 1965 : An approach to some problems of landform analysis in tropical environments. In J.B. Whittow and J.P.D. Wood Eds.

Essays in Geography for Austin Miller, University of Reading, Reading pp. 118-44.

Thomas, M.F., : The origin of borhardts. *Zeitschrift fur Geomorphologie* N.F. 10, pp. 478-80.

Thornthwaite, C.W., 1943 : Problems in the classification of climates. *Geographical Review* 33, pp. 233-55.

Thornthwaite, C.W., 1948 : An approach towards a rational classification of climate. *Geographical Review* 38, pp. 55-94.

Tricart, J., 1961 : Les charactéristiques fondamentales du systèms morphogénetique des pays tropicaux humides. *Information geographiquc* 25, pp. 155-69.

Tricart, J., 1965 : *Principes et méthodes de la géomorpholagie.* Paris, Masson, pp. 496.

Tricart, J., 1965 : *Le modelé des régions chauds, forêts et savanes.* Paris, S.E.D.E.S. pp. 322,

Tricart, J. and Cailleux, A., 1965 : *Introduction à la géomorphologie climatique.* Paris, S.E.D.E.S. pp. 306.

Visher, S.S., 1941 : Climate and geomorphology : some comparisons between regions. *Journal of Geomorphology* 4, 54-64.

Wilson, L., 1969 : Les relations entre les processus geomorphologiques et le climat moderne comme méthode de paleoclimatologie. *Revue de geographie physique et de geologie dynamique* 11, 303-314.

FOUR

C. R. Twidale

PLANATION SURFACES—THEORY AND REALITY

INTRODUCTION

PLAINS of low relief of one sort or another dominate continental topography, albeit in negative fashion. Not only are there extensive plains related essentially to modern baselevels, but remnants of palaeoplains are a common feature of most upland regions. King (1950) has claimed that these old surfaces "record in a relatively simple manner the geomorphological history of the continents" and has personally used them to good effect in order to unravel the development of the large parts of the earth's surface. Yet the evolution of planation surfaces, or surface of low relief due principally to erosional processes, remains confused, with glaring discrepancies between theory and reality.

Some early workers, notably Passarge (1895, 1904a, 1904b), and Keyes (1912) attributed regional planation to aeolian action, but, being inconsistent both with general theory and field evidence, the idea has never enjoyed wide acceptance. Others have interpreted ancient planate remnants as due to marine planation, a possibility that has exercised the minds of geologists and geomorphologists for many years. There have always been those, including such distinguished workers as Ramsay (1846, 1878), Barrell (1920), Bascom (1921), and Olmstead and Little (1946) who opined that certain surfaces of low relief are due to marine planation. The widespread development of shore platforms leaves no doubt as to the efficacy of marine processes in shaping planate forms, but unless very special conditions obtain, unless indeed there is a gradual secular relative rise of sealevel, the amount of planation that can be achieved is limited, for there is an inbuilt negative feedback in the process of marine planation. Though there are arguments as to the relative importance of wave action in shaping platforms, all agree that it is of major significance. But the effectiveness of wave action is limited by shallow water, and the broader a platform becomes the greater the distance waves have to travel through shallow water before they reach the real margin of the land and hence the less effective they are. Save in very special circumstances it is not possible for marine platforms more than 1½-2 km wide to be developed (King, 1963).

Most workers now agree that planation of regional extent is due to running water, with respect to which two major models of landscape evolution have developed over the years. One is Davis' (1909, pp. 249-278, 350-380) peneplain concept, based on earlier work by Powell (1975), Dutton (1880-81) and Gilbert (1877). The essence of the concept is that slope decline produces a rolling or undulating surface of low relief dominated by convexities and called a *peneplain* (almost a plain). The other is that involving scarp retreat and pedimentation so vigorously expounded by King (1942, 1949, 1953, 1957, 1962), and based on the work of Fisher (1866, 1872), Penck (1924, 1953), Lehmann (1933) and Wood (1940). These processes result in a plain made up of a coalescence of pediments, dominated by concave slopes, and called a *pediplain.*

PENEPLANATION

Though stimulated by and based on field work by various workers in various parts of the United States, Davis' model is essentially deductive. For the sake of simplicity of exposition and comprehension Davis made various assumptions : the uplift of the land mass took place virtually instantaneously with no time for landscape development during the earth movements; the mass was structurally homogeneous; there was climatic and tectonic stability. These conditions are of course unreal, but in effect Davis attempted to isolate one factor—

time—and then deduce the sequence of events and forms that would follow the uplift of a land mass.

According to Davis, in the *youthful* stage (Fig. 4.1) streams cut down rapidly and erode deep valleys that are V-shaped in cross-section. Areas of low relief—uplifted remnants of the earlier-formed plain or of the emerged seafloor, as the case may be—persist high in the landscape. As the rivers and streams approach baselevel they incise more slowly so that lateral corrosion and various process at work on the valley side slopes, always relatively more effective than vertical erosion (Johnson, 1932a), become even more dominant. Thus valleys are widened and flat valley floors or flood plains begin to develop. The river systems extend until the whole land mass is dissected. At this stage the valley sides are worn back so that the last remnants of the high plains are consumed. The valley floor flats are still a really insignificant and the landscape consists of all slopes. This is the stage of *maturity* (B in Fig. 4.1) when relief amplitude is at its greatest.

The rivers continue to erode their beds but now weathering and wash, as well as rills and streams, together cause the divides to be lowered more rapidly than the rivers which are nearing base level. Relief amplitude continues to diminish. The lateral migrations of the rivers cause flood plains of significant extent gradually to develop, and these together with the broadly rolling or gently undulating divides came to form a surface of low relief typical of *old age* or senility (See in Fig. 4.1). This is the peneplain which however continues to be worn down : there is less and less relief as weathering and wash continue to take their toll

Peneplains are not featureless, for apart from residual hills, remnants of circumdenudation called monadnocks after the mountain of that name in New Hampshire, there may be 50-75 m of relief per km^2. Slopes and rivers are however graded. Davis conceived the drainage system as extending over the entire land surface by way of rivers, streams, rivulets, rills and wash : the entire land surface was lowered. And because the end-product of this long-continued, predictable, evolutionary change caused by weathering, mass movement of debris and the work of rivers is similar to the postulated original surface of low relief, the whole sequence of events is referred

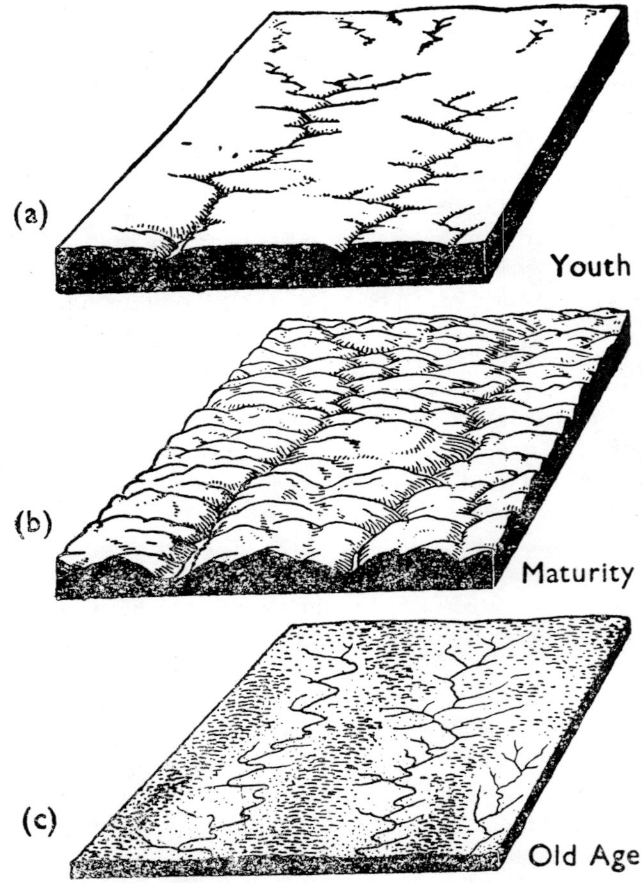

FIG. 4.1 : *Diagram showing Stages of Youth, Maturity and Old Age in peneplanation* (*after Holmes Principle of Physical Geology*)

to as a *cycle of erosion* or, and preferably, because more than weathering and erosion are involved, *a geomorphic cycle.*

Davis presented his evolutionary concept in an intellectual climate rendered propitious by Darwinism. Readers found attractive his use of anthropomorphic terms (youth, maturity, old age). Davis wrote persuasively, elegantly add at length over a long period. He was a dominating figure. The concept of peneplanation was widely accepted in North America, Europe and Australasia through the writings of such distinguished men as D.W. Johnson, H. Baulig, S.W. Wooldridge, C.A. Cotton and E.S. Hills. Of course the concept was modified, not least by Davis himself. Towards the end of his long life Davis anticipated the steady state ideas of such workers as Hack (1960) in his "old-from-birth" peneplain (Davis, 1922) and he also came to recognise differences in landscape development as between humid and arid regions (Davis, 1930). Also, Crickmay (1933) visualised adjacent flood plains merging as the low divides between were consumed to form a panplain (elements of which are indeed an integral part of the peneplain).

Nevertheless, the idea of the peneplain was widely and readily accepted; indeed, too readily, for peneplains became fashionable and remnants of them were identified on inadequate grounds. Consequently there was in the middle of the present century a reaction against Davis and the peneplain, though the concept never entirely lost favour and it is now appreciated that surfaces of low relief that accord descriptively with the Davisian peneplain are present in the landscape in many parts of the world (see Twidale, 1981a, also below), It is one of the mysteries of geomorphology that in his exposition Davis could point only to Montana and Siberia for modern examples of his penultimate landform, when there are so many modern peneplains in the Great Plains of Texas and adjacent parts of the southwest of the U.S.A., in the Ukraine, in various regions of Australia—Eyre Peninsula, the Wheat Belt of southwestern Western Australia, the Cobar-Wilcannia region of the Darling Basin, the Barkly Tableland and Carpentaria plains of northern Australia—and in many parts of southern Africa, for instance central western and northern Transvaal, and the western Karroo.

In addition many examples are known of small or local peneplains—areas of low rolling or undulating relief but only of limited areal extent.

Davis and his followers relied almost wholly for exemplification of the type on palaeoforms, on remnants of peneplains preserved in upland regions. Yet one of the major problems with the concept of peneplanation is that if, as Davis postulated, the entire land surface is lowered, no relics of palaeosurfaces ought to persist in the landscape for more than the duration of the cycle, estimated to involve between 11 and 33 ma depending on the extent and geology of the area concerned (Gilluly, 1955; Schumm, 1963, Twidale, 1976). Yet very old landforms and land surfaces survive in the landscape in many parts of the world (Twidale, 1976).

In relation to this problem, Crickmay (1932, 1959, 1968, 1969, 1971) has adduced evidence to suggest that far from being active over the whole land surface, stream erosion is restricted to stream channels and their immediate environs; and there is much to be observed in the field that appears to support this suggestion (Plate 1). Again, there is the possibility that a cycle may not run its course, but be interrupted by relative lowering of baselevel, resulting in the accentuation of structural contrasts during successive phases or epicycles, and the augmentation of relief (see e.g. Twidale, 1966a, 1969, 1976, 1980).

The Flinders Ranges may be cited as an example. Following early Cretaceous planation the area of folded Proterozoic and Lower Palaeozoic sediments was uplifted so that by the beginning of the Cainozoic a ridge and valley topography with a considerable relief amplitude had developed. Remnants of the Mesozoic land surface were (and are) preserved as summit bevels and high plains throughout the Ranges but particularly in the north. A further phase of stream rejuvenation has caused the early-middle Cainozoic valley floors to be dissected and there has been considerable basin filling, but the relics of the Cretaceous landscape remain a prominent feature of the contemporary scene (Twidale, 1976, 1980).

Furthermore, it is difficult to relate morphology to slope behaviour. There is no evidence that a rolling surface of relief is necessarily a consequence of slope decline. A peneplain may develop by backwearing, or the rolling relief may be due to a recent retouching of the surface and be unrepresentative of slope behaviour through time.

Many residuals projecting above the general peneplain level are so steepsided that they warrant being called inselbergs; and in any case most of them are structural forms. There is no demonstrable and necessary connection between slope behaviour and the resultant planate form; and peneplains are not restricted to temperate (those regions cited from northern Australia being examples of tropical forms that are developing at the

1. Deep narrow gutters cut in grus (weathered granite) near Paarl, cape province, South Africa. The dry cemented crust acts as a caprock and retards widening of the gutters

2. (a) Pediments cut in schist, Lucerne Valley, Southern California

2. (b) Pediments cut in siltstone and with coarse debris mantle fringing the Flinders Ranges near Brachina, South Australia

3. The granite platform bodering Corrobinnie Hill, on northern Eyre Peninsula, South Australia, is dumped and grooved and extends 700-800 m from the base of the residual. Patches of grus stand on the platform and there are also small blocks and boulders—earstwhile corestones

present time). Thus though the thought—linkage "slope decline peneplain—monadnock—temperate areas" may have wide currency, it has no basis in fact.

PEDIMENTS AND PEDIMENTATION

A major and early reaction to the seeming inevitability of Davisian dogma is based on two observed facts. The first is that in many parts of the world, but particularly in the arid and semi-arid tropics, plains are neither rolling nor undulating but instead take the form of little dissected remarkably smooth surfaces called pediments, so named after the architectural features of similar morphology. The second is that pediments extend headwards and laterally and so coalesce to form remarkably flat plains called pediplains.

Pediments are cut bedrock surfaces sloping gently (at angles in the range ½°—7° and most commonly 2°—4°) down from the adjacent uplands (Plate 2, a, b, c). The junction between hill and plain is abrupts and is known as the piedmont angle or pediment nick (see Twidale, 1967; Rahn, 1966; Mackin, 1970). Several type of pediment have been recognised (see e. g. Mackin, 1970; Whitaker, 1973, 1979; Twidale, 1981b). Some consist of bare exposed bedrock, and are known as *rock* pediments or platforms (see Twidale, 1978a). Most pediments however carry a veneer of debris which masks minor irregularities in the underlying bedrock surface and which is in effect responsible for the quite remarkable smoothness of the forms. In some the debris veneer is weathered rock *in situ* and these are called *mantled* pediments. Mantled pediments are shaped by rills and rivulets. They are typical of crystalline terrains and are most characteristically found fringing inselbergs (Fig. 4.2). The pediment is underlain by grus, and the base of this regolith is called the weathering front. Rock pediments are developed where this front is exposed; in other words they are etch forms. Most rock pediments are of limited area, though in wet sites they may extend several hundred of metres from the base of the inselberg (Plate 3). Some forms are resistant and tend to persist in the landscape, unlike mantled forms which do not long survive stream dissection.

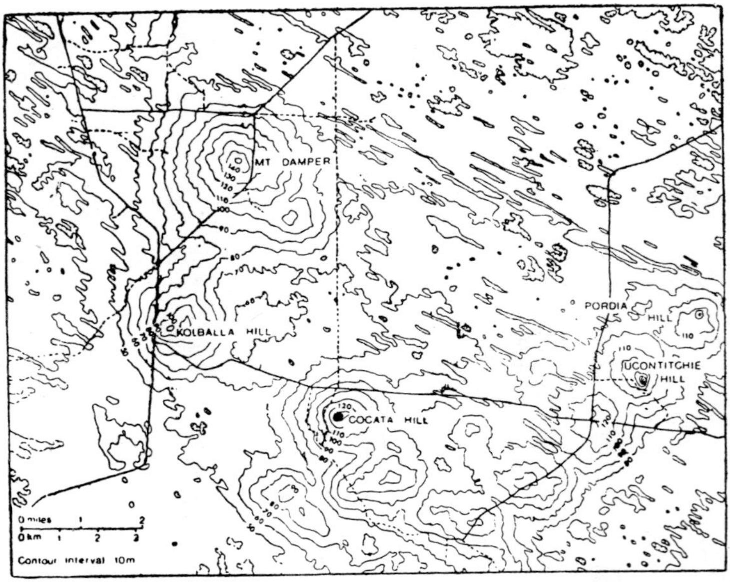

FIG. 4.2 : *Extract from South Australian Lands Department. Topographic Map Series 1:500000 Sheet Pordia 5921-V, showing granite inselbergs surrounded by lo angle Pediment cores that merge downslope with a rolling peneplain*

Some pediments carry a veneer of detritus that is introduced; it has been transported and deposited to form the debris cover. Such *covered* pediments are widely distributed and characteristically fringe uplands (Plate 4 a, b, c,). Seen in profile normal to the uplands they are frequently concave upward and rectilinear, though convex upward forms have been observed in several locations. Viewed from the plains the pediments form an apron fronting the uplands and are seen to comprise a series of low-angle rock fans that merge one with another along the mountain front. The apieces of the cones are coincident with the points of debouchment of rivers and streams from the upland; and in this they stand in marked contrast with tributary streams in "normal" situations which have carved secondary valleys (Fig. 4.3). These fringing covered forms typically occur in flights, with several levels of remnants preserved by their mantles above contemporary and still extending pediments.

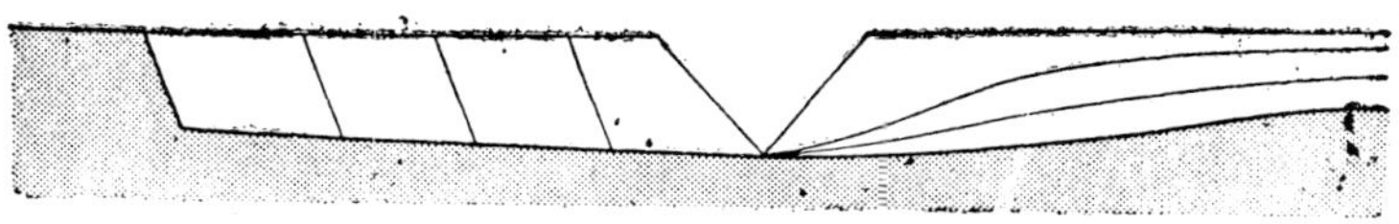

FIG. 4.3 : *Diagrammatic Section illustrating Contrast between backwearing (left) and downwearing (right)*

Pediments vary in thickness and completeness of cover, in the presence or absence of small residuals, in the character of the bedrock in which they are cut (consolidated rock or alluvium), and so on, but covered fringing pediments are by far the most common type and can be regarded as the basic form.

Explanations of pediments are offered at two levels. On the one hand they are seen in the context of general theories of landscape evolution and are regarded as concomitant of scarp retreat. Thus Howard (1942 p. 134) writes: "... development of the pediment depends on the recession of the base of the slope ..." King (1948, p. 85) makes the same point when he states that: "Retreat of the faces of a Bornhardt leaves a surrounding pediment at the foot", as does Pugh (1956, p. 28): "... a mountain mass with a well developed upper surface will shrink slowly by scarp retreat, with the development of bounding pediments". Oberlander (1972, p. 17) bases his assumption that parallel slope recession has produced pediments on the weight of American experience, and citing reviews by Tator (1952-3), Tuan (1959) and Hadley (1967) states: "most investigators of "pediments" in south-western United States have regarded them as the concomitant of the wearing back of a mountain front or escarpment".

The second sort of explanation involves the processes responsible for shaping pediments. Various suggestions have been made, each applicable in some instances. Weathering or mantle-controlled planation (Mabbutt, 1966; Twidale, 1978a) produces rock platforms or pediments. Workers like Johnson (1932b) and Howard (1942) favour lateral planation. Others are more specific and point to the evidence that many pediments at any rate are due to simultaneous planation and deposition of debris by distributing or divaricating streams (McGee, 1897; Blackwelder, 1931; Twidale,1978b, 1979). Other workers have emphasised rills and wash. King (1949) attributes pediments to sheetwash, and many investigators consider that combinations of these various processes, changing in relative significance from place to place and from time to time, are responsible for the shaping of pediments (Bryan, 1922, 1936; Gilluly, 1937; Sharp, 1940; Rahn, 1967).

There is, however, no doubt that lateral planation by divaricating streams is the major cause of covered pediment formation in many areas. As rivers in flood emerge from the uplands they break out from the confines of their channels and adopt a distributary habit. These several channels are less efficient than the single channel, so that more load is deposited and the river spreads laterally still more. Thus simultaneous planation and deposition is achieved.

Just as clearly such a mechanism cannot account for those many well-developed pediments that take the form of low-angle cones surrounding isolated granitic residual (Plate 4a, b, c and Fig. 4.2). The inselbergs are not of sufficient areal extent to generate major streams and here, as mentioned previously, rills and wash assume a considerable significance in lowering the plains eroded in weathered rock. Pediments are not necessarily associated with scarp retreat for they are well developed in areas where there has demonstrably been no recession of escarpments (Twidale & Bourne, 1975a; Twidale, 1978b). Pediments have formed in association with receding scarps in some areas but elsewhere are apparently due to downwearing.

There are as many misconceptions surrounding pediments as there are peneplains. Pediments, though well and most widely developed in the arid and semi-arid tropics, are not confined to those regions. Pediments have been described from such areas as Colorado (Scott, 1963; Ackermann, 1974), Alaska (see Wahrhaftig, 1970a, 1970b), the Pyrenees (Barrére, 1968, 1975) and the Cape Fold Belt of South Africa (Twidale & Van Zyl, 1980).

Pediments and peneplains are not distributionally distinct,

for they occur side by side and are contiguous in several areas. The covered pediments of the Cape Fold Belt merge laterally in several areas with rolling plains that descriptively resemble peneplains. Likewise in the Karroo, and on Eyre Peninsula (Fig. 4.2). Obviously the morphological contrasts cannot be attributed to climate, nor is lithology responsible for the pediment and peneplain forms are developed on similar if not identical bedrock. Pediments have been described from flat-lying sedimentary sequences, from crystalline and metamorphic terrains, from widely varied climatic settings. No pattern is obvious; yet common features are discernible.

Covered pediments are cut in weak rocks but there are always resistant outcrops close by : quartzite ridges, or inselbergs of fresh granite for example. These are source areas for the veneers that smooth, protect and preserve the covered pediments. They also form catchments and upland areas which spawn the streams and wash to carry the coarse debris to the piedmont zone.

Covered pediments occur in areas where for one of a variety of reasons stream discharges vary within wide limits. In the Pyrenees, in the Rockies and Alaska for instance it is the spring melt that produces flood conditions on the rivers. In north Queensland (Twidale, 1955), it is the monsoon rains. In Japan and Korea (Akagi, 1972, 1974) it is a combination of spring melt and summer monsoon that produces very high discharges. In desert regions it is episodic rains. But whatever the causation, the floods carry coarse debris from the uplands to the plains, where it is deposited in part in pediment mantles.

PEDIPLAINS

Pediments are alleged to extend headwards by scarp retreat, and coalesce to form pediplains. An evolutionary sequence or cycle of events is envisaged. King, for example, has stated that "residual uplands disappear, relief increases markedly and with coalescing of the ever-increasing pediments a bevelled landscape of low relief and of multi-concave form, a pediplain. . . . . . is produced" (1942, p. 53).

As has been implied there are many objections to the assertion that pediments are genetically related to scarp retreat.

Morphologically pediplains are said to differ from peneplains in that they consist wholly of concave upward slopes whereas peneplains are dominated by convexities. Unfortunately for the theory no such multiconcave surfaces exist in nature. Where, as on northwestern Eyre Peninsula and central Namibia, granite residuals fringed by pediments have been limited they are replaced by convexities that are indistinguishable from peneplain crests.

The planate surfaces labelled pediplains by some workers are not dominated by concave slopes, but are remarkably flat. Such features as the Bushmanland Surface of northwestern Cape Province, the Kuiseb Plain of central Namibia (Plate 5a), the Saharan plains featured by Peel (1966) and the Meekatharra surface of Western Australia (Plate 5b) are all of great antiquity and are due to long continued degradation of peneplains under conditions of tectonic stability. If peneplains are almost plains, these surfaces are the ultimate surfaces of low relief. They are the forms for which the term 'old land' was proposed by Hills (1955) and Opik (1961) though this term is unfortunately pre-empted, (see e.g. Waters, 1957) and it may be preferable to coin the term *ultiplain*. If peneplains are almost plains or plains of low relief, ultiplains are the extreme manifestation of weathering and erosion. They are plains of no relief, though stream rejuvenation, such as has occurred in parts of the Transvaal (Birot, Godard, Petit & Ters, 1974) induces a return to peneplain form.

However, in the present context, it is important to note that these pediplains despite their names have no necessary genetic connection with pediments, which are of only limited areal extent. They have a wide distribution and form on specific geological, topographical and climatic settings, though they are not restricted to any climatic region as conventionally defined.

OTHER PLANATION SURFACES

Over very wide areas planation surfaces, particularly peneplains have been deeply weathered and duricrusts of various types have developed. Such duricrusts presumably imply relatively slow degradation of the surface, and tectonic and climatic stability. Such *weathered land surfaces* are enduring features of the landscape in many tropical and subtropical regions, for the encrustations, whatever their minerological composition, are tough or hard and are virtually immune to weathering and the direct effects of erosion. They form caprocks in many areas of plateau, mesa and butte landforms.

These duricrusts are however susceptible to dissection and then to undermining, for the bedrock beneath the land horizon is frequently intensely altered and weakened. In this way the duricrust is striped away, exposing the former weathering front (Mabbutt, 1961a) or former limit of significant weathering. These exposed fronts are called *etch surfaces* and of planate form and etch plains (Wayland, 1934; but see Logan 1849, 1851). They are widely developed in many parts of the world having been described for example from Uganda (Wayland, 1934), Western Australia (where the New plateau of Jutson, 1934) is really of the etch plain genetically related to the lateritised Old plateau (see Mabbutt, 1961b) and in the Mount Lofty Ranges of South Australia (Twidale & Bourne, 1975b). More and more surfaces are being recognised as being of etch character.

Another type of land surface that increasingly is being appreciated at a wide variety of scales is the exhumed or resurrected land surface. An old surface whether planate or not, may be buried by younger sediments and then re-exposed. The old unconformity is exhumed.

Such *exhumed plains* of low relief have been recognised by tracing them into unconformity in many parts of the world. In northwest Scotland the pre-Torridonian and Lewisian landscape has been partly resurrected (Williams, 1969) and an exhumed surface of Proterozoic age has been described from Greenland (Cowie, 1961) and Canada (Ambrose, 1964). A surface of pre-Middle Cambrian age is locally exhumed in western Queensland (Opik, 1961). A mature late Cretaceous topography around Baji, in central India, was inundated by lava flows of the Deccan Traps, but has subsequently been re-exposed to form part of the contemporary scene (Choubey, 1972).

A land surface ranging in age between late Jurassic and early Cretaceous which was inundated by the transgressive seas and buried by marine sediments is widely exhumed on both eastern and western margins of the Great Artesian Basin in Queensland (Twidale, 1956, 1966b) and in South Australia (Wopfner, 1964; Wopfner & Twidale, 1967). It is upfaulted in the northern Flinders Ranges (Woodard, 1955; Twidale, 1969, 1980). Exhumed surfaces may of course be of either pediment or peneplain or etch type, or they may be of quite different origins. For instance many glaciated pavements of Precambrian or Permian ages have been exhumed in different parts of Australia (Perry & Roberts, 1968; Tate, Howchin & David, 1896). Other surfaces are of marine origin (see Twidale, Bourne & Smith, 1976) or retouched by waves during marine transgressions.

ALTERNATIVE MODELS

Some workers have suggested that open systems operate and that dynamic equilibrium or steady state obtained (Hack, 1960). Though theoretically conceivable this concept is difficult to prove and the whole of the field evidence argues against it (Bretz, 1963; Twidale, 1976).

The preservation in the landscape of the very old palaeosurfaces is inconsistent with both cyclic models described and indeed with steady state development. It has been suggested that, particularly in areas of tectonic instability, recurrent uplift has interrupted cycles before completion, resulting in increased relief amplitude and the survival of palaeoforms high in the landscape (Twidale, 1976). In the Flinders Ranges of South Australia for instance the region has not been baselevelled since the late Mesozoic. Throughout the Cainozoic rivers have etched out valleys and plains, leaving the ridges and ranges in higher and higher relief, and leaving palaeosurface remnants essentially beyond the reach of significant stream attack.

CONCLUSION

Planation surfaces, both ancient and modern, are due to running water. Given time, rivers can reduce large land masses to baselevel. The most common form of regional plain is the peneplain. Both valley side slopes and fringing pediments (which are merely a particular type of valley side slope) develop

4. (a) Isolated inselbergs surrounded low angle pediment cones near Broken Hill, western New South Wales. The inselbers are formed of discrete masses of quartz and the plains are cut in schist

4. (b) Pediment aprons cut in granite and surrounded by inselbergs of the same material in Lebowa, northern Transvaal, South Africa

4. (c) Pediment and inselberg landscape developed in crystalline rocks in Arizona (W.C. Bradley)

5. (a) The remarkably flat plains bordering the Kuiseb Canyon in central Namibia are cut in schista. The residual hills are granitic or gneissic

5. (b) The plains around Meekatharra in central Western Australia are cut in granitic rocks

and as valleys extend and widen divides are lowered and in due course an extensive rolling or undulating surface of low relief—a peneplain—is formed. Further degradation produces a flat erosional planation surface or ultiplain. Conversely, slight rejuvenation can cause a return to the peneplain form.

However a cycle may not run its course and instead there is repeated rejuvenation and revival, and in such regions as the Appalachians and the Flinders Ranges, relief amplitude has increased at least through the Cainozoic and remnants of very old land surfaces are preserved in the landscape. But given tectonic stability planation surfaces of regional extent develop.

SUGGESTED READINGS

Ackermann, H.D., 1974 : Shallow seismic compressional and shear wave refraction and electrical resistivity investigations at Rocky Flats, Jefferson County, Colorado. *U.S. Geol. Surv. J. Res.* 2, pp. 421-430.

Akagi, Y., 1972: Pediment morphology in Japan. *Fukuoka Univ. Educ. Bull.* 21, pp. 1-63.

Akagi, Y., 1974 : Pediment on the Taean Peninsula and the Yeogsan River Basins, Korea. *Sci. Repts. Tohoku Aniv 7th Sev. (Geogr)* 24, pp. 183-197.

Ambrose, J. W., 1964 : Exhumed palaeoplains of the precambrian shield of North America. *Am. J. Sci.* 262, pp. 817-857.

Barrell, J., 1920 : The piedmont terraces of the northern Appalachians. *Am. J. Sci.* 49, pp. 227-258, 327-362, 407-428.

Barrére, P., 1968 : Le relief des Pyrenées Centrales Occidentales. *J. d'Etudes Pau-Biarritz.* 194, 31-52.

Barrére, P., 1975 : Terrasses et glacis d'érosion en roches tendres dans les montagnes du Haut-Aragon. *Etudes Géogr.* (Melanges à Georges Viers) Univ. Toulouse, 29-43.

Bascom, F., 1921 : Cycles of erosion in the piedmont province of Pennsylvania. *J. Geol.* 29, pp. 540-559.

Birot, P. A. Godard, M. Petit & M. Ters., 1974 : Contribution a l'étude des surfaces d'aplanissement et de l'érosion differentielle dans le Transvaal septentrional et oriental (Afrique du Sud). *Rev. Géogr. Phys. Géol. Dynam.* 16, pp. 421-554.

Blackwelder, E., 1931 : Desert plains, *J. Geol.* 39, pp. 133-140.

Bretz, J.H., 1963 : Dynamic equilibrium and the Ozark landforms. *Am. J. Sci.* 260, pp. 427-438.

Bryan, K., 1922 : Erosion and sedimentation in the Papago country, Arizona. *U. S. Geol. Surv. Bull.* 730, pp. 19-90.

Bryan, K., 1936 : Processes of formation of pediments at Granite Gap, New Mexico. *Z. Geomorph.* 9, 125-135.

Choubey, V. D., 1972 : Pre-Deccan trap topography in central India and coastal warping in relation to Narmada Rift structure and volcanic activity. *Bull. Volcanol.* 25, pp. 660-685.

Cowie, J.W., 1961 : Contributions to the geology of north Greenland. *Medd. Gronland*, p. 164.

Crickmay, C.H., 1932 : The significance of the physiography of the Cypress Hills. *Can. Field Nat.* 46, pp. 185-186.

Crickmay, C.H., 1933 : The later stages of the cycle of erosion. *Geol. Mag.* 20, pp. 337-347.

Crickmay, C.H., 1959 : *A Preliminary Inquiry into the Formulation and Applicability of the General Principle of Uniformity.* Calgary.

Crickmay, C.H., 1968 : *Some Central Aspects of the Scientific Study of Scenery*, Calgary.

Crickmay, C. H., 1969 : *The Art of Looking at Broad Valleys.* Calgary.

Crickmay, C. H., 1971 : *The Role of the River.* Calgary.

Davis, W. M., 1909 : *Geographical Essays*, Boston.

Davis, W. M., 1922 : Peneplains and the geographical cycle. *Geol. Soc. Am. Bull.* 33, pp. 587-598.

Davis, W. M., 1930 : Rock floors in arid and humid climates. *J. Geol.* 38, 1-27, pp. 136-158.

Dutton, C. E., 1880-81 : The physical geology of the Grand Canyon District. *U. S. Geol. Surv. 2nd Ann. Rept.* pp. 47-166.

Fisher, O., 1866 : On the disintegration of a chalk cliff. *Geol. Mag.* 3, pp. 354-356.

Fisher, O., 1872 : On cirques and taluses. *Geol. Mag.* 8, pp. 10-12.

Gilbert, C. K., 1877 : Report on the geology of the Henry Mountains Utah. *U. S. Geogr. & Geol. Surv. Rocky Mt. Region.* 160.

Gilluly, J., 1937 : Physiography of the Ajo region, Arizona. *Geol. Soc. Am. Bull.* 48, 323-345.

Gilluly. J., 1955: Geologic contrasts between continents and ocean basins. *Geol. Soc. Am. Spec. Paper.* 62, 7-18.

Hack, J. T., 1960 : Interpretation of erosional topography in humid temperate regions. *Am. J. Sci.* 238A, 80-97.

Hadley, R. F., 1967 : Pediments and pediment-forming processes. *J. Geol. Educ.* 15, pp. 83-89.

Hills, E. S., 1955 : Die landoberfläche Australiens. *Die Erde.* pp. 7, 195-205.

Howard, A.D., 1942 : Pediment passes and the pediment problem. *J. Geomorph.* 5, pp. 3-31, 95-136.

Johnson, D.W., 1932a : Streams and their significance. *J. Geol.* 40, pp. 481-497.

Johnson, D.W., 1932b : Rock plains of arid regions. *Geogr. Rev.* 22, pp. 656-665.

Jutson, J.T., 1934 : The physiography (geomorphology) of Western Australia. *Geol. Surv. W. Aust. Bull.* 95 (2nd Ed.)

Keyes, C.R., 1912 : Deflative systems of the geographic cycle in an arid climate. *Geol. Soc. Am. Bull.* 23, pp. 537-562.

King, C.A.M., 1963 : Some problems concerning marine planation and the formation of erosion. *Trans. & Papers Inst. Brit. Geogr.* 33, pp. 29-43.

King L.C., 1942 : *South African Scenery*. Edinburgh.

King, L.C., 1948 : The pediment landform: Some current landforms. *Geol. Mag.* 86, 245-250.

King, L C., 1949 : A Theory of Bornhardts. *Geogrl. J.* 112, pp. 83-87.

King, L.C., 1950 : The Study of the world's plainlands. *Quart. J. Geol. Soc. London.* 106, pp. 101-131.

King, L.C., 1953 : Canons of landscape evolution. *Geol. Soc. Amer. Bull.* 64, pp. 721-752.

King, L.C., 1957 : The uniformitarian nature of hillslopes. *Trans. geol. Soc. Edin.* 17, pp. 81-102.

King, L.C., 1962 : *Morphology of the Earth*. Edinburgh.

Lehmann, O., 1933 : Mophologische Theorie der Verwitterung von Steinschlagwänden. *Vierteljahrschr. Naturforsch. Gesellsch. Zürich.* 73, pp. 83-126.

Logan, J.R., 1849 : The rocks of Palo Ubin. *Genoofs. Kunsten Wetenschoppen.* (Batavia) 22, pp. 3-43.

Logan, J.R., 1851 : Notices of the geology of the Straits of Singapore. *Quart. J. Geol. Soc. Lond.* 7, pp. 310-344.

Mabbutt, J.A., 1961a : 'Basal surface' or 'weathering front'. *Proc. Geol. Assoc. London.* 72.

Mabbutt, J.A., 1961b : A striped land surface in Western Australia. *Trans. & Papers Inst. Br. Geogr.* 29, pp. 101-114.

Mabbutt, J.A., 1966 : Mantle-controlled planation of pediments. *Am. J. Sci.* 264, pp. 78-91.

Mackin, J.H., 1978 : Origin of pediments in the western United States, pp. 85-105 *in Problems of Relief Planation.* (Ed. M. Pecsi). Budapest.

McGee, W.J., 1897 : Sheetflood erosion. *Geol. Soc. Am. Bull.* 8, 87-112.

Oberlander, T.M., 1972 : Morphogenesis of granite boulder slopes in the Mojave Desert, California. *J. Geol.* 80, pp. 1-20.

Olmsted, E.W. & L. S. Little., 1946 : Marine planation in Southern New England. (Abs) *Geol. Soc. Am. Bull.* 57, 1271.

Opik, A.A., 1961 : The geology and palaeontology of the headwaters of the Burke River, Queensland. *Austr. Bur. Min. Res. Geol. Geophys. Bull.* 53.

Passarge, S., 1895 : *Adamana*. Berlin.

Passarge, S., 1904a : Rumpfflache and Inselberge. *Z. Deutsch. Geol. Gesell.* 46, pp. 195-213.

Passarge, S., 1904b : Die Inselberglandschaft in tropischen Afrika. *Naturwiss. Wochenschr*, 42.

Peel, R.F., 1966 : The landscape in aridity. *Trans. Inst. Brit. Geogr.* 38, pp. 1-23.

Penck, W., 1924 : *Dié Morphologische Analyse*. Stuttgart.

Penck, W., 1953 : *Morphological Analysis of Landforms*. (Transl. H. Czeck & K.C. Boswell). London.

Perry, W.J. & H.G. Roberts., 1968 : Late Precambrian glaciated pavements in the Kimberley region, Western Australia· *J. Geol. Soc. Aust.* 15, pp. 51-56.

Powell, J.W., 1875 : *Exploration of the Colorado River of the West (1869-1872)*. Washington.

Pugh, J.C., 1956 : Fringing pediments and marginal depressions in the inselberg landscape of Nigeria. *Trans. & Papers Inst. Brit. Geogr.* 22, pp. 15-31.

Rahn, P. H., 1966 : Inselbergs and nickpoints in southwestern Arizona. *Z. Geomorph.* 10, 217-225.

Rahn, P. H., 1967 : Sheetfloods, streamfloods and the formation of pediments. *Assoc. Am. Geogr. Ann.* 57, pp. 593-604.

Ramsay, A. C., 1846 : On the denudation of South Wales. *Geol. Surv. Great Britain Man.* 1, 297-335.

Ramsay, A. C., 1878 : *Physical Geology and Geography of Great Britain* (5th Ed.). London.

Schumm, S. A., 1963 : Disparity between present rates of denudation and orogeny. *U. S. Geol. Surv. Prof. Paper* 454.

Scott, G. R., 1963 : Quaternary geology and geomorphic history of the Kassler Quadrangle, Colorado. *U.S. Geol. Surv. Prof. Paper* 421-A.

Sharp, R. P., 1940 : Geomorphology of the Ruby-East Humboldt Range, Nevada. *Geol. Soc. Am. Bull.* 51, pp. 337-372.

Tate, R. W. Howchin & T. W. E. David., 1896 : Report of the Research Committee appointed to collect evidence as to glacial action in Australia. *Rept. Austr. Assoc. Adv. Sci.* 6, pp. 315-320.

Tator, B. A., 1952-53 : Pediment characteristics and terminology. *Assoc. Am. Geogr. Ann.* 42, pp. 295-317; 43, pp. 47-53.

Tuan, Y. F., 1959 : Pediments in southeastern Arizona. *Univ. Calif. Publ. Geogr.* 13, pp. 1-164.

Twidale, C. R., 1955 : Pediments at Naraku, north-west Queensland. *Austr. Geogr.* 6, pp. 40-42.

Twidale, C. R., 1956 : Chronology of denudation in northwest Queensland. *Geol. Soc. Am. Bull.* 67, pp. 867-882.

Twidale, C. R., 1966a : Chronology of denudation in the southern Flinders Ranges, South Australia. *Trans. Royal Soc. S. Aust.* 90, pp. 3-28.

Twidale, C. R., 1966b : Geomorphology of the Leichhardt-Gilbert Area of northwest Queensland. *Land Res. Ser.* 16.

Twidale. C. R., 1967 : Origin of the piedmont angle, as evidenced in South Australia. *J. Geol.* 75, 373-411.

Twidale, C. R., 1969 : Geomorphology of the Flinders Ranges, pp. 57-137 *in Natural History of the Flinders Ranges*. (Ed. D.W.P. Corbett). Adelaide.

Twidale, C. R., 1976 : On the survival of palaeoforms. *Am. J. Sci.* 276, pp. 77-94.

Twidale, C. R., 1978a : Granite platforms and the pediment problem, pp. 288-304 *in Landform Evolution in Australasia.* (Ed. J. L. Davis & M. A. J. Williams). Canberra.

Twidale, C. R., 1978b : On the origin of pediments in different structural settings. *Am. J. Sci.* 278, pp. 1138-1176

Twidale, C. R., 1979 : The character and interpretation of some pediment mantles. *Sediment Geol.* 22, pp. 1-20.

Twidale, C.R., 1980: Landforms, pp. 13-41 *in A Field Guide to the Flinders Ranges.* (Ed. D. W. P. Corbett) Adelaide.

Twidale, C. R., 1981a : Pediments, peneplains and Ultiplains. In press.

Twidale. C. R., 1981b : Origin and environments of covered, mantled and rock pediments. *Trans. Inst. Brit. Geogr.*

Twidale, C. R. & J. A. Bourne, 1975a : Episodic exposure of inselbergs. *Geol. Soc. Am. Bull.* 86, 1473-1481.

Twidale, C. R. & J. A. Bourne, 1975b : Geomorphological evolution of part of the eastern Mount Lofty Ranges, South Australia. *Trans. R. Soc. S. Aust.* 99. pp. 197-209.

Twidale, C. R., J. A. Bourne & D. M. Smith., 1976 : Age and origin of palaeosurfaces on Eyre Peninsula and the southern Gawler Ranges, South Australia. *Z. Geomorph.* 20, pp. 28-55.

Twidale, C. R. & J. A. Van Zyl., 1980 : Poorls and pediments in the western Little Karroo, South Africa. *S. Afr. Geogr.*

Wahrhaftig, C., 1970a : Geologic map of the Fairbanks A-2-3-4-5 Quadrangles, Alaska. *U. S. Geol. Surv. Maps* CQ-808-811.

Wahrhaftig, C., 1970b: Geologic maps of the Healy D-2-3-4-5 Quadrangles, Alaska. *U. S.Geol. Surv. Maps* CQ-804-807.

Waters, R. S., 1957 : Differential weathering and erosion on oldlands. *Geogrl. J.* 123, pp. 503-513.

Wayland, E. J., 1934 : Peneplains and some erosional landforms. *Geol. Surv. Uganda Ann. Rept. Bull.* 1, pp. 77-79.

Whitaker, C. R., 1973 : *Pediments: a Bibliography.* Norwich.

Whitaker. C. R., 1979 : The use of the term 'pediment' and related terminology. *Z. Geomorph.* 23, pp. 427-439.

Williams, G. E., 1969 : Characteristics and origin of a Precambrian pediment. *J. Geol.* 77, pp. 183-207.

Woodard, G. D., 1955 : The stratigraphic succession in the vicinity of Mt. Babbage Station, South Australia. *Trans. Royal. Soc. S. Aust.* 78, pp. 8-17.

Wood, A., 1940 : The development of hillside slopes. *Proc. Geol. Assoc. London.* 53, pp. 28-138.

Wopfner, H., 1964: Permian-Jurassic history of the western Great Artesian Basin. *Trans. Royal Soc. S. Aust.* 87, pp. 117-128.

Wopfner, H. & C. R. Twidale., 1967 : Geomorphological history of the Lake Eyre Basin, pp. 118-143 *in Landform Studies from Australia and New Guinea* (Ed. J. N. Jennings & J. A. Mabbutt) Canberra.

FIVE

L. I. Maruashvili

PENEPLANATION PLANES IN AN ALPINE COUNTRY
Georgia as an Example

THE TERRITORY of the Georgian SSR, located at the Caucasian Isthmus, is included into the Alpine orogenic belt. Georgia occupies the middle part of the Kavkasioni mountainous system, that raised from an independent geosyncline of the northern lateral line of Thetis, as well as a part of the northern branch of the above orogenic belt's main body.

GEOLOGICAL BACKGROUND OF THE RELIEF REMOVAL IN THE LATE KAINOZOIC

The most ancient geological history of Georgia is as follows. In the Late Oligocene and Early Miocene a low island, circumfluous by so called the Maikop Sea, was located at the place of present day Kavkasioni mountainous system. The island was bordered by the Precaucasian and Transcaucasian straits in the North and South correspondingly. The Anticaucasus Upland (present day Caucasus Minor and Transcaucasian or Armenian Upland) still was not present on the South of the Caucasian Isthmus. The Neogenic orogenesis changed drastically the relief of Georgia : The Kavkasioni became a mountainous country; during the Attic orogenic phase (the end of Miocene) the Transcaucasian or Rioni-Kura strait disappeared. It was cut by the Dzirula horst into Colchis and Alban* Bays. The Caucasus

* Albania is the ancient (mentioned by Strabo in the 1st century A.D.) name of the Eastern part of Transcaucasus where present day Azerbaijan SSR is located. It should not be confused with todays Albania at the Balcan Peninsula.

Minor was bulging up, and inside of its semi-circle, being convex to the north, an affusive volcanism have started (lava streams and volcanic ash outbursts have started to build up the Transcaucasian Upland). The existence of Precaucasian strait came to the end much later. It was periodically dried and restored from the Upper Pliocene (firstly as a normal sea strait and later, from the Middle Pleistocene, as a unilateral discharge bed of excess water from the Caspian to Black Sea basin). During Pliocene the Kavkasioni was transferred into mountainous country, while the Caucasus Minor into mid-mountain one.

In fact, Quaternary was the period of continued rise of mountain structures and volcanism. In total, the Anthropogen can be characterized (during the last 3-4 millions of years) by terrace formation and glaciation (the latter affected only the Alpine zone of Kavkasioni and the highest tectogenic and volcanic mountain massives of the Anticaucasus). In the Quaternary period the Transcaucasian Upland was characterized by conjugation of the Alpine plains (plateaus) with deep lava canyons and impressive volcanic mountain chains, while at the Transcaucasian intermontane region the Colchis and Albanian Bays gave way to alluvial plains, hill country and rolling uplands.

It is interesting to note the difference in anthropogenic history of the Western (located in the Black Sea basin) and the Eastern (sloping to the Caspian Sea) parts of Georgia and the whole Caucasian Isthmus as well. Even in the Middle Pliocene the levels of Black and Caspian Seas have started to differ from each other. At the beginning of the Upper Pliocene temporary equilibrium of the both sea levels could take place (The Kujalnizkoe and Akchagilskoe Seas could equilibrate their levels at the certain epoch), but later starting from the Gurijsko-apsheronian time, their fluctuations became sharply different. The level fluctuations and horizontal displacement of the Caspian Sea shores in the Quaternary period had larger amplitude than that of the Black Sea. In particular, the total of fluctuations of the Caspian Sea level exceeded 150 m, while that of the Black Sea was hardly up to 80 m. Moreover, distribution of transgressions and regressions of the both sea basins in time, was different and even reverse. The Caspian Sea transgressions have invaded the Volga and Kura river valleys at a distance of several hundreds of kilometres, while that of the Black Sea—just for some tens of kilometres at Kolkhida. Naturally, during the Anthropogene, erosion and accumulative processes in the West and East Georgia have proceeded in different ways, that had affected the character of terrace series and the youngest peneplanation planes.

GENERAL CHARACTERISTICS OF PENEPLANATION PLANES OF GEORGIA AND THEIR STUDY

Comparative newness and dynamism of Georgian relief was not favourable, in general, for planation process and preservation of denudation surfaces in the country. Despite of this, numerous fragments of erosion and denudation surfaces of three genetic types, such as exposed, fossilized and excavated ones were spared. Most of the surfaces are results of partial planation.

Peneplanation planes are present in all four basic geomorphological zones or provinces of Georgia: (1) The Kavkasioni, (2) The Intermontane, (3) The Caucasus Minor and (4) The South Georgian Upland* (Fig. 5.1.). Most of the surfaces are dated as the Neogenic. More ancient erosion and denudation surfaces are distributed much less, and are more difficult to study.

There is a difference of opinions among scientists on the genesis and geological age of peneplanation planes of the Caucasus. Most of them believe that erosion and denudation surfaces were formed at low hypsometric levels and later, due to uprising tectonic movements, were shifted to different heights. A part of the authors is adherent to "planation from the top" and believe, that the main factors of relief removal were glacial and periglacial processes (Kosov, Scherbakova, Rastvorova). Others, consider the fragments of surfaces as the parts of single vast peneplanes being raised at different levels (S. Kuznetsov, L. Vardanjanz). The rest, stick to the idea on consecutive origin of fragments (the top are older than bottom ones) at low altitudes and at a different times (Dumitrashko, Lilienberg, Astakhov, Safronov and others).

* A part of the Transcaucasian Upland within the boundaries of Georgia.

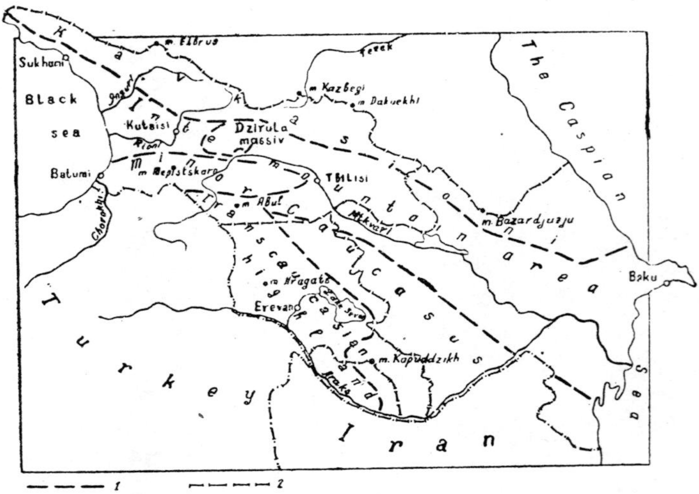

FIG. 5.1 : *Scheme of the main orographical parts of Transcaucasian (Scale 1:500000)*

1. **Boundaries of Orographical Parts.**
2. **Boundaries of Soviet Republics.**

The majority of scientists date the greatest part of peneplanation planes from various epochs of the Neogene (Miocene and Pliocene). L.A. Vardanjanz believed, that the Caucasian mountains were subjected to total peneplanation in the Upper Pliocene and that the mountainous character of the Caucasus was finally formed during the Quaternary period. Some scientists mark out surfaces of the pre-Neogene and Miocene age being excavated from under the cover of later deposits (Maruashvili).

EXCAVATED PENEPLANE OF THE DZIRULA MASSIF

The above mentioned Dzirula horst, elevated at the end of Sarmat, due to Attic mountain building movements, is a kind of "a bridge" between mountain structures of the Kavkasioni and the Caucasus Minor. The geologists are used to call it "Dzirula massif" and consider it as an uplifted, middle part of the Transcaucasian tectonic block. Here, the ancient crystalline rocks forming the basement of Transcaucasian block being subsided to the East and West are outcropped.

Considerable fragments of ancient peneplane are preserved at the Dzirula massif. The peneplane was worked out in the Mesozoic (Upper Jurassic-Lower Cretaceous), then was covered with sedimentary rocks of the Upper Cretaceous-Middle Miocene, and from the Upper Sarmat, starting to remove this mantle, it has become gradually destroyed. Presently, the remnants of peneplane are spread at elevation from 600-700 to 1200-1500 m (Fig. 5.2.). The largest part of peneplaned relief is at the crest of Likhi Ridge (the watershed of Black and Caspian Seas) at elevation 1200-1300 m. The part of peneplane 4-6 km. wide is stretched along the watershed at a distance of 20 km., and is rugged by heads of gorges and ravines into narrow (200-300 m) branched stretches. At the higher Getsamanian Ridge peneplanation plane is obliterated, but the ridge itself preserves straight, subhorizontal longitudinal profile that outlines the highest part of the peneplaned country. Westwards, in the vicinity of village Gezruli, the height of peneplaned crests is 600-700 m above sea level.

The above surfaces are developed on the Paleozoic crystalline substratum and occupy the South-East part of Dzirula massif. At the north-west part of the region, the Mesozoic peneplane is still covered by the Upper Cretaceous and even younger strata. Its existence in the depth is proved experimentally by presence of the single subterranean water plane under the karst plateau (G. Gigineishvili, 1979).

Since Dzirula peneplane is dated from the Mesozoic, it belongs to the early alpine geomorphological cycle and is of the same age as the Gondwana Land and Lawrasia peneplanes. It is strange, that such an ancient surface is exposed just in the very heart of an alpine geosyncline, in the encirclement of mountain structures of the latest geomorphological cycle. Probably, the peculiar tectonic regime of the region has favoured preservation and later exposure of the above surface within the present day relief. The Dzirula massif is a bending site (anteclase) of the Transcaucasian tectonic block. However, after the Middle Jurassic, it was neither sunk too deeply nor uplifted too high, up to the Upper Miocene. The massif was subjected just too weak, oscillating movements that periodically

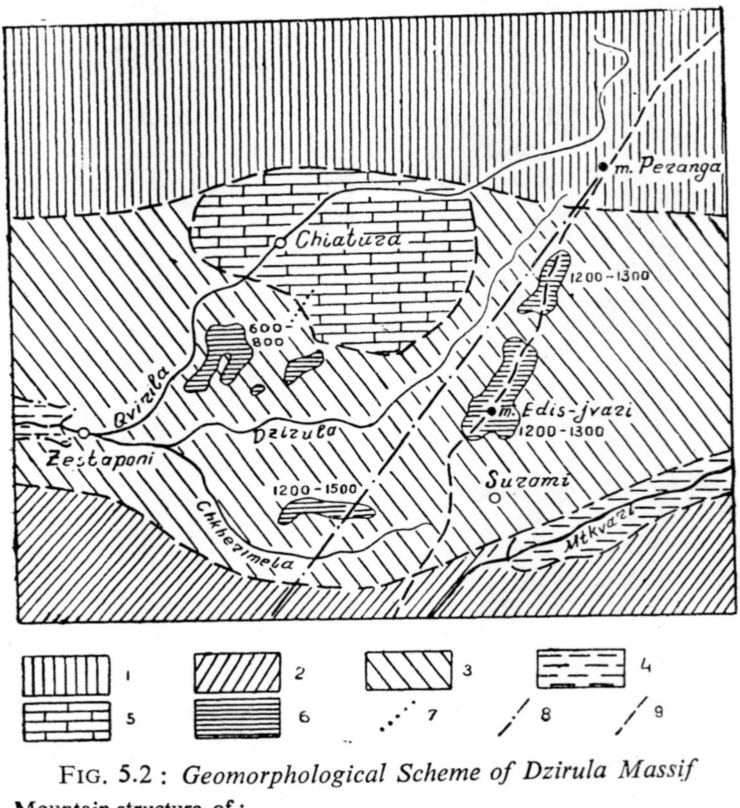

FIG. 5.2 : *Geomorphological Scheme of Dzirula Massif*

Mountain structure of :

1. Great Caucasus ; 2. Minor Caucasus ; 3. dissected by erosion parts of Dzirula Massif and piedmont ; 4. alluvial plain ; 5. Chiatura Structural Plateau ; 6. preserved fragments of global peneplain ; 7. Former course of a river ; 8. axis of arch morphostructure ; 9. Black Sea-Caspian Watershed; Figures on the scheme show the absolute height of peneplain.

covered it with thin layer of deposits, or removed it, thus exposing relief to the action of denudation processes.

I.P. Gerassimov (1964, 1970) has called—"the geomorphological stage" that part of geological past, during which the features of the earth crust, being preserved up today, were formed. The stage embraces Later Mesozoic and the whole Cenozoic periods—the last 150-160 millions of years, according to geological time-scale, and is subdivided into three epochs (cycles):

1. The epoch of global peneplanation (Late Mesozoic and Paleogene), 150-25 millions of years.
2. The epoch of interrupted, partial peneplanation (End of Paleogene and Neogene), 25-4 millions of years.
3. The epoch of terrace building (Anthropogene), last 4 millions of years.

Thus, peneplane of Dzirula massif is included into the first epoch of the geomorphological stages, and may represent a fragment of the supercontinent (Gondwana Land or Lawrasia) that was blocked in the Late Cenozoic orogene, as tectonic fate has willed it.

The minor sections of the Late Mesozoic peneplane occur within the belt of jurassic deposits, and similar to Dzirula peneplane relate to the excavated type, for example—a plateau to the West from Lankori Pass at the Lechkhumi Ridge (2400 m above sea level), an area of Tsutskhvati village at the Okriba region (500-600 m above sea level), etc.

NEOGENIC PLANES OF PARTIAL PLANATION

Preservation of peneplanation planes depends on the following basic factors:

1. Age of a plane.
2. Climate (amount of precipitations).
3. Composition of the substratum (geological structure).

All in all, peneplanation planes of the Kavkasioni mountainous system are poorly preserved, and the nearer to the tectonic and orographic axis of the system, the rarer they occur. However, limestone regions are the exception, where karst has favoured the preservation of ancient relief.

The intermontane zone of Georgia has favoured preservation of the above relief in its eastern part, due to relatively very dry (semi-humid and semi-arid) climate. In the West Georgia, where excessive humid conditions prevail (precipitations fall equals 1000-4000 mm/year), the situation for existence of ancient relief is less suitable.

In the Caucasus Minor, peneplanation planes are preserved much better than in the Kavkasioni. This phenomenon can be explained by youthfulness of the former's relief in comparison with that one of the main part of the Kavkasioni, as well as by widely spread Eocene volcanogenic series, in the latter, that possess denudation stability and permeability characteristics at the same time.

The vast area of South-Georgian Upland is covered with anthropogenic lavas, lacustrine and fluvial deposits (due to lava partitioning of negative forms of relief) and was subjected to the latest folding. Hence, the ancient denudation and erosion surfaces almost do not occur there.

The author's idea on erosion and denudation planation process of the relief is based on the following consideration. In mountainous countries, the so called "planation at the top" or "suspended planation" do not exist. All planation surfaces are formed at low altitudes and then are moved by tectonic movements, either upwards—developing exposed surfaces of the plain relief, or downwards—under accumulation level, and are with covered deposits. Planation of mountain ridges and massives is hindered by intensity of denudation and slopes erosion. Neither glacial cover, nor periglacial processes of destruction are able to develop a plane relief above the steep slopes, since it will be immediately subjected to weathering, denudation and erosion, and thus will be added to the slopes.

Intensive destruction processes also do hinder preservation of the planed relief, which was developed down and then was uplifted tectonically. It is favoured by permanent position (on plane) of erosion valleys and gorges in an intermontane and alpine country. The auspicious conditions of the planed relief preservation, take place under deviation of mountain relief development from "the normal" course, i.e., under rivers displacement in certain direction on plane, under interception of one river by another, and so forth.

Peneplanation planes of different age in the mountains of Georgia, usually occur as large as or small fragments scattered sometimes far away from each other, that hampers study of their genesis, number and age. However, there are some regions where instead of fragments the whole series (steps) of

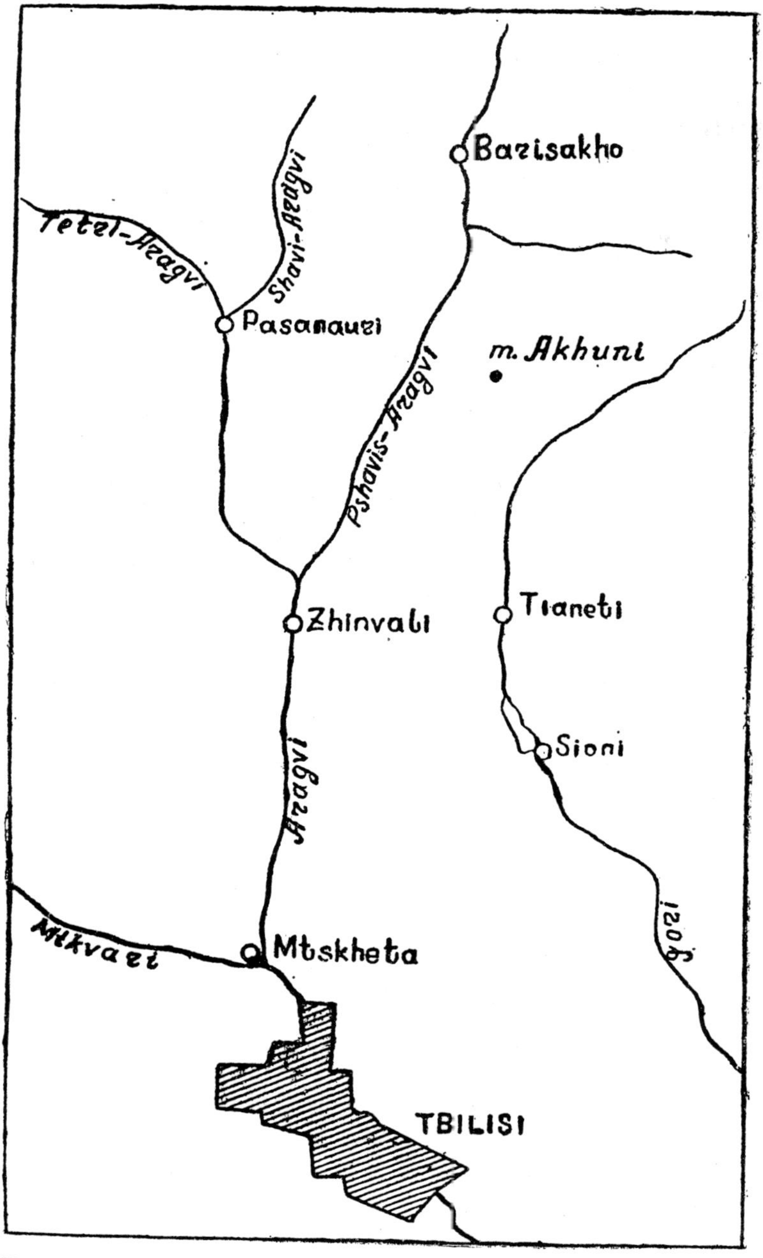

FIG. 5.3 : *Location of Akhuni Mountain and Mountain Range (Stairs of Erosional Surfaces)*

the planes are represented. By analogy with key sections in stratigraphic investigations, these regions can be called—key profiles. To study the genesis, number and age of peneplanation planes at these regions is much easier.

A key profile in the Kavkasioni is the Akhuni Ridge, located within 65-70 km. to the North from Tbilisi—the capital of Georgia (Fig. 5.3). The Akhuni Ridge is western spur of the Pshavi Ridge, that is a southern spur of the Main Watershed Ridge of Kavkasioni*. The highest summit of the ridge is mount Akhuni (2268 m above sea level). The ridge stretches for 8 km from the east to west, where it drops to the left bank of river Pshavis, Aragvi, at an elevation 950 m above sea level.

The longitudinal profile of the Akhuni Ridge is—"step-like" (Fig. 5.4.) Its five "steps" are descending from the east to west (from the saddle on the east of Mount Akhuni summit to river Pshavis Aragvi), the total length is 6 km., relative height 1150 m (above river Pshavis Aragvi).

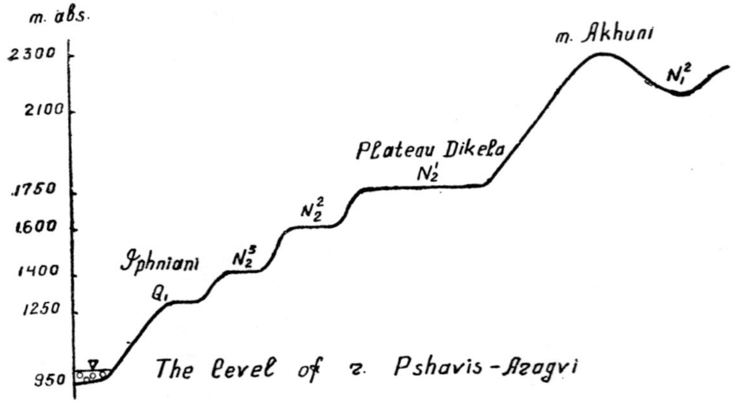

FIG. 5.4 : *Longitudinal profile of Akhuni Range*

The suppositional geological age of the erosional surfaces is littered.

* The ethnographic province of Georgia, where the Akhuni Ridge is located is called Pshavi. V. Pshavela the famous poet who glorified the mountain nature and life of mountaineers was born and lived there.

The inter-location, altitudinal distribution and horizontal extension of the Akhuni Ridge steps, their names as well as geological age and connection with orogenic phases are given in the following Table 5.1.

TABLE 5.1 : **Erosion Steps of the Akhuni Ridge**

No. from bottom	Name of a place	Relative H above r. Pshavis Aragvi level m	Age of a step	Orogenic phase that has stipulated a downcut into a step	Distance from the Pshavi Ridge
I	Ifniani	350	Early Pleistocene	Pasadena	7
II	—	450	Late Pliocene	Valakhian	6, 5
III	—	650	Middle Pliocene	New Rodanian	6
IV	Dikhela	850	Early Pliocene	Ancient Rodanian	4, 5
V	Akhuni	1150	Late Miocene	Attic	2

The morphometric indices of steps are as follows : the height of scarps between adjacent steps varies from 150 to 350 m. The highest scarp is between Akhuni (V) and Dikhela (IV) steps, while the largest in plane is Dikhela Plateau (1 × 0.4 km.). The other steps are much smaller.

The steps development is visualized in the following way. They are not connected with disjunctive dislocations, i.e., with faults and differential displacements of blocks. It is obvious from the fact, that the fault lines separating them are almost perpendicular to the direction of tectonic structures (folds, thrusts, lineaments) prevailing in the area. Meanwhile, the relation of steps and of the whole "step-like" structure to the valley of river Pshavis Aragvi, indicates that they had been worked out by the river, during the process of its displacement from the ESE to WNW, and the downcut of its valley. In other words, here are present bulging up terraces—the fluvial terraces being formed under conditions of intensive, but interrupted uplifts (L. Maruashvili, 1959, 1975). However, if the steps of Akhuni Ridge bear slight resemblance with fluvial terraces, that is a consequence of considerable age of the steps (see below) multiplied by intensity of geodynamic processes (tectonic, denudation and erosional), that are peculiar for the Kavkasioni.

The following is to be said about geological age of the Akhuni Ridge steps. Above all, "the erosion steps" cannot be squeezed into the narrow chronological frames (Quaternary Period) that were ascribed to the Kavkasioni relief by L.A. Vardanjantz and his supporters. Not a step bears any trace of alluvial deposits. The morphology of steps and terraces became so different, that it is extremely difficult to determine their primary, terrace origin. Large intervals between the levels of steps are undoubtedly indicative of five considerable tectonic uplifts, that have taken place during the period of steps development. The above said proves that Neogene was the period, when mountainous relief of the Kavkasioni was formed.

The age problem of the Kavkasioni mountainous relief can be solved by the entity of methods. The geological and paleobotanic data prove that this relief exists at least since the Upper Miocene, and that during this time no peneplanation has taken place there. The foot of Kavkasioni is engirdled with rudaceous molasses of the Sarmat and Quaternary periods as well as with the Upper Pliocene ones. The fossil fauna complexes found in intermontane deposits are of the same age. They are composed of ecologically different vegetation species, being originated at various elevation zones, that prove existence of dissected mountainous topography with vegetation belts of relatively thermophilic and cold-enduring species during the mentioned period (Dolidze, 1968, Isaeva-Petrova, 1972).

The succession and relative dimensions of steps and their scarps make it possible to date the separate steps, from the point of view of historical geology and paleogeography. It is known, that the most intensive orogenic phase of the Neogene was the Attic one in the Upper Miocene (at the boundary of the Middle and Late Sarmat, according to stratigraphic scheme of the Crimea-Caucasian province). The explorers of the Northern Caucasus and Mediterranean believe, that the most prolonged and morphologically effective planation phase of the Neogene is dated from the Lower Pliocene (Birot, Dresch 1958, Safronov, 1969). Starting from relations between the lengths of Akhuni steps and the heights of their scarp, the author attributes formation of the large scarps, between steps V and IV to the Attic orogenic phase, while that of IV step (or Plateau Dikhela) to the Lower Pliocene. The above is agreed with the appearance of rudaceous molasses in the Transcaucasian intermontane depression, starting from the Upper Sarmat.

The rest of stepsa re to be dated correspondingly : V step (Akhuni) Middle-Early Miocene, III step—Middle Pliocene, II step—Late Pliocene and I step—Early Pleistocene. The orogenic phases, that have stipulated erosional downcut into steps IV, III, II and I are identified with the Ancient Rodanian, New Rodanian, Valakhian and Pasadena ones.

Thus, the planation planes at the crest of Akhuni Ridge are none other than the forms of intraplain pedimentation—the result of alternating phases of river beds expansion during tectonically quite epochs, and of erosional downcut of a river bed during mountain building movements of the Earth's crust.

The "step-like" ridges of Akhuni type represent the only form, probably the most expressive one among the forms of relief, which is developed as a result of long-term erosional and denudation transformation of intraplain pedimentation surfaces (bulging up terraces) in mountainous regions. There is a great variety of other forms, being developed during the evolution process of mountainous relief, out from the surfaces of the above mentioned type of planation.

Considering morphology of the Kavkasioni and Caucasus Minor, one may notice some other types of forms being developed due to denudation and erosional transformation of intraplain pedimentation forms. They are as follows :

1. Sections of wide flood plains, deprived of their streams and suspended over the living, actively developing gorges. (Type of Gostibe*).
2. Broad saddles (passes) at the crests of mountain ridges, that are relics of ancient plains. (Type of Atkveri).
3. Ridges of "step-like" profile, that unlike the Akhuni Ridge have no watershed planes or plateaus. (Type of Churta).
4. A plateau set into erosion basin and cut by new gorges. (Type of Tusheti).

* The names of the types are given according to the toponims of the regions where corresponding forms of relief were found.

5. The spared upper erosion steps and the fragmented lower ones. (Type of Ksani).
6. Crestal plane ridges. (Type of Khidotani).
7. Flat summits. (Type of Borocha).

It goes without saying that the above types do not exhaust a great variety of forms developed on mountainous relief as a result of morphological transformation of plain pediments. Definitely, the future investigations will reveal a lot of new types.

N.E. Astakhov (1955), while studying peneplanation planes of the Trialeti Ridge at Caucasus Minor, came to the following conclusion :

1. The planes are not remnants of a single peneplane, but morphological traces of various stages of river drainage development. The planes represent bottom fragments of ancient valleys that came off unhurt by erosion, and in some cases the whole sections of that valleys with both slopes expressed morphologically.
2. At the Trialeti Ridge, six levels of such planes can be differentiated. The most ancient one is dated from the Miocene, while the newest one (the lowest) from the end of Pliocene-Early Pleistocene.

Having correlated the both schemes (Astakhov's for the Trialeti Ridge and his own for the Akhuni Ridge and the Kavkasioni province), the author has found marked similarity. The basic difference is just a number of levels being differentiated (Astakhovs-6, the author's-5). It can be explained by location of the Akhuni Ridge at the outlying province of Kavkasioni (at a distance of 40 km. from the tectonic axis of the mountainous system), where the most ancient surfaces are not available. The latter existed at the inner parts of Kavkasioni, but were demolished by intensive processes of destruction.

Thus, peneplanation planes of the Kavkasioni and the Caucasus Minor were mainly formed by intraplain pedimentation, and then transformed by denudation and erosion (partially by neotectonics, gravitation and volcanism) into various forms.

Since the above mentioned Neogene peneplanation planes of the Kavkasioni and the Caucasus Minor have not been covered with deposits after their formation, they should be referred to the type of exposed surfaces. However, in Georgia with her intensive neotectonic regime, there were some places and conditions for burial of planes of that age. The places are located at intermontane zone, where tectonic settlings have taken place in the post-Oligocene. The certain intermontane zones, at different times during the Neogene and Quaternary periods were subjected to inversion of tectonic regime (submersion was followed by elevation), followed by denudation which exposed the buried peneplanation planes.

The excavated Neogene planes of this type were indicated at the Gombori Ridge (Maruashvili, 1955). An area of this range sank intensively from the Sarmat to Pont and accumulated conglomerate-psammatic strata of the Tsivi suite, under which the Middle-Miocene planes were buried. In the Early Anthropogene an inversion of tectonic regime took place and the Gombori Ridge was uplifted over 2000 m. The Tsivi suite was affected by erosion processes that have exposed its substratum here and there and displayed denudation and erosion planes on its surface.

CONCLUSION

Bearing in mind general theoretical theses of geology and geomorphology, the study of peneplanation planes of Georgia and the Caucasus leads to the following conclusions :

1. The Neogenic planes of partial planation were connected with interrupted course of development of mountainous regions in the late Alpine orogenic cycle. The planes were formed during intervals between the global orogenic phases.
2. Peneplanation planes developed in various parts of the Earth are synchronous in general, since the phases of planation were stipulated by the same tectonically quite epochs.
3. The number of levels of planation depends on the age of a mountainous area, i.e., on the period of its transition into the stage of continental development. However, the

ancient planes are destroyed in the course of time, and usually only 4-6 levels are spared (depending on the structure and composition of substratum and the intensity of geo-dynamic processes).

4. The altitudinal position of planation levels is determined by regional tectonic regime, and is characterized by different hypsometric indices in various regions. Sometimes the common character of altitudinal distribution of levels is counted in the relation between extension of altitudinal intervals of different planation phases (the maximum interval of an area may correspond in age to the similar one of another area, and so forth).
5. Development of area of peneplanation planes is determined, first of all by duration of the tectonic phase, as well as by the local factors (capacity of a river, composition and structure of substratum, intensity of exogenetic destruction processes). However, in the course of the following destruction of planes these relations can be changed and came to naught.
6. Almost all planes of denudation and erosion planation, being formed during the Neogene and Anthropogene in the Alpine type mountainous countries, are the effects of intra-plain pedimentation and the following, secondary morphological alterations. The conception of W. Penck on "piedmont steps" is totally abstract scheme, worked out on the examples of Mid-European Hercynian and other mountains of non-Alpine type. Hence, it cannot be assumed as a basis for understanding the genesis of peneplanation surfaces at such regions as the Caucasus. As for the idea on "planation of crests" by glacial and cryonival processes, mentioned by some authors, it is fully groundless.
7. The main Alpine type mountainous systems that have 5-6 planation levels are permanently present during the last 25-30 millions of years (since the end of Paleogene). At ridges that were formed during the Anthropogene (as Gombori Ridge) the number of denudation surfaces do not exceed 1-2.

SUGGESTED READINGS

Astakhov, N.E., 1955 : On peneplanation planes in Trialeti mountains (The East Georgia) and their relation with River Kura Valley Development. *Trudy Instituta Geografii im. Vakhushti AN GSSR*, Vol. 6, pp. 75-85.

Birot P. Dresch J., 1953 : *La méditerranee et le Moyen-Orient*. I-II. Paris, 1956.

Dolidze, J. Sh., 1968 : For the study of akchagilean flora. Kvabebi *Soobschenija AN GSSR*, 50, No 2, pp. 469-474.

Dumitrashko N.V., Lilienberg D.A., Budagov B.A., 1961 : The relief and latest tectonics of the south-east Caucasus. Moscow. Published by AN SSSR, pp. 3-117.

Geomorphology of Georgia; 1971 : Publishing House "Metsniereba", Tbilisi, pp. 5-609.

Gerasimov, I.P., Mescherjakov, Ju. A. The geomorphological Stage in the Earth Development, 1964 *Izvestija AN SSSR. Serija geografiches-kaja*, No. 6, pp. 3-12.

Gerasimov, I.P., 1970 : Three main cycles in the history of geomorphological stage of the earth development. *Geomorfologija*, No 1.

Gigineishvili, G.N., 1979 : Karst waters of the Great Caucasus and the basis problems of Karst Hydrology. Publishing House *Metsniereba*, Tbilisi, pp. 5-224.

Isaeva-Petrova, L.S., 1972 : Reconstruction of vertical vegetation belt in the eastern part of the Great Caucasus in the apsheronian time. Collection *Palinologija pleistocena*, Moscow, "Nauka", pp. 193-211.

King Lester, C., 1967 : *The morphology of the Earth. A Study and Synthesis of World Scenery.* Edinburgh-London.

Kosov, B.F., 1948 : Denudation levels of the Central Caucasus. *Vestnic Moskovskogo Universiteta*, No 2.

Kuznetsov, S.S., 1935 : On development of the Trialeti Relief (The Transcaucasus). *Problemy fizicheskoi geogrofii* II.

Kuznetsov, S.S., 1937 : *The Achara-Trialeti Folded System*. Published by Academy of Sciences of the USSR. Moscow-Leningrad.

Lebedeva, N.A., 1954 : The step-like relief of the northwest Caucasus. Collection *Voprosy Geografii*, 36, Geomorphologija, Moscow, Geografgiz, pp. 207-214.

Machatschek. Fr., 1955 : Das relief der Erde. Versuch einer regionalen morphologie der Erdoberfläche. I-II. Berlin.

Maruashvili, L.I., 1955 : Geomorphological essay of the Gombori or Kakheti range in the eastern Georgia. *Trudy Instituta Geografii im. Vakhushti AN GSSR* Vol. VI, p. 5-24.

Maruashvili, L.I., 1959 : Regularities of the terraces development in orogenes. Geomorphologicheskaja komisija AN SSSR. *Materialy geomorphologicheskogo soveschenija*, Moscow.

Maruashvili, L.I., 1966 : The Likhi peneplane. *Uchenye zapiski Azerbai-*

janskogo Gosuniversiteta, Serija geologicheskikh i geograficheskikh nauk, Baku, No 5.

Maruashvili, L.I., 1972 : The Dzirula massif—A relic of ancient peneplane encircled with young folded structures. *Geomorfologija*, No. 4, pp. 69-74.

Maruashvili, L.I., 1975 : The Valley planation of uplifted mountain systems. The Caucasus as an example. *Geomorphologija*, No 4, pp. 12-21.

Penck, W., 1924 : *Die morphologische Analysé. Ein Kapitel der phisikalischen Geologie*. Stuttgart.

Peneplanation planes and weathering crusts at the Territory of the USSR. *Nedra*, Moscow, 1974.

Rastvorova, V.A., 1963 : On the ancient denudation surfaces of the central Caucasus. *Buleteń' Moskovskogo Obschestva ispytatelei prirody, otdel geologii*, Vol. 38, Vyp. 6. pp. 65-83.

Safronov, I.N., 1969 : On the surfaces of weathering at the north Caucasus. *Trudy po geologii i poleznym iskopaemym Severnogo Kavkaza*, Vyp. 9.

Scherbakova, E.M., 1960 : The influence of periglacial processes in the development of relief at the Northern slope of the Great Caucasus. Collection *Perigljtsialńye javlenija na territorii SSSR*, Moscow, Published by Moscow University, pp. 231-248.

Vardanjantz, L.A., 1948 : *The post-pliocene history of the Caucasian-Black sea-Caspian sea region*. Erevan.

SIX

Pramila Kumar and *R. K. Rai*

HAS DECCAN FORELAND REJUVENATED

GEOMORPHOLOGICAL studies of the different parts of the Deccan Foreland and references in geological publications should make it possible to synthesise certain facts which seem to be universally present in this region. The recent geomorphic study of the Sonar-Bearma basin (23°8′ N to 24°20′ N and 78°30′ to 80°0′ E) has further brought to light the geomorphic characteristics of yet another part of the Central Highland which falls almost in the middle of the Deccan Foreland[1]. Unlike an old peneplain, the Deccan Foreland has the characteristics of a rejuvenated land with interrupted cycles of erosion. These are reflected in the landforms, the slopes and the processes and features associated with rivers and may lead one to certain logical conclusions.

The terms Deccan Foreland here refers to that northern zone of the Deccan plateau (22°0′- 28°20′N and 72°15′-88°0′E) which faces the Himalayas and also the adjoining ranges of the northwest and the northeast. This zone seems to have acted as one of the forelands in the Himalayan orogeny and has in turn been affected by this colossal activity[2]. Geomorphologically this area is drained by the right-hand tributaries of the Ganga. Besides, the highest elevation in the Deccan Foreland is attained in its extreme south and has a gentle slope northwards. Field studies by geologists have proved that the Vindhyans and the Bundelkhand gneiss extend below the Gangeiss alluvium and that the northern margin of the old Gondwanaland was involved in the Himalayan mountain building. The Deccan Foreland comprises nine physiographic regions (Fig. 6.1), which for the convenience of discussion have been grouped into three major units (a) the Chota Nagpur plateau and the Baghelkhand plateau, (b) the Aravalli region, and (c) the Central Highland. The Chota Nagpur and the Baghelkhand plateaus mainly comprise the pre-Cambrian granites and gneisses, belonging to the Dharwar and the Gondwana, the Central Highland is essentially composed of the Vindhyan and the Deccan Trap; and the Aravalli range consists of pre-Cambrian sedimentaries. Except for the Gondwanas and the Deccan Trap, the Deccan Foreland is pre-Cambrian landmass which has not experienced any long period of marine transgression, and has a history of continuous erosion and degradation throughout the Paleozoic and Mesozoic era. Logically, this region should have been a completely worn-out peneplain, but the highest elevation in the Chota Nagpur (1160 m), the Baghelkhand (1130 m), the Central Highland (850 m) and the Aravalli range (above 1160 m) is a definite evidence of recent rejuvenation. Besides, the following regional characteristics further support the above view.

Chota Nagpur and Baghelkhand Plateau

Dunn[3] has identified four peneplains in the Chota Nagpur plateau which have been further studied intensively by Chatterjee,[4] Bagchi and Sengupta,[5] Ahmad,[6] Singh[7] and Varma.[8] These peneplains are noticed at the heights of 1000 m, 700 m, 550 m, and 450 m. The following evidences may further be added in support of the view of rejuvenation which have been collected and analysed by the above-mentioned authors :

1. Well-defined scarps in between the erosional surfaces which have been identified as either the erosional scarps or composite scarps.
2. Zones of intense dissection in Kodarma and east and south of Ranchi plateau.
3. Mis-fit streams in the Netarhat plateau and the 'Pat'country of the Baghelkhand plateau.
4. The undulating surface of the Ranchi and Hazaribagh plateaus identified as an uplifted peneplain having old remnants as monadnocks. Besides, the hills of the Gondwanas are almost in equal height to those of the gneissic plateau.

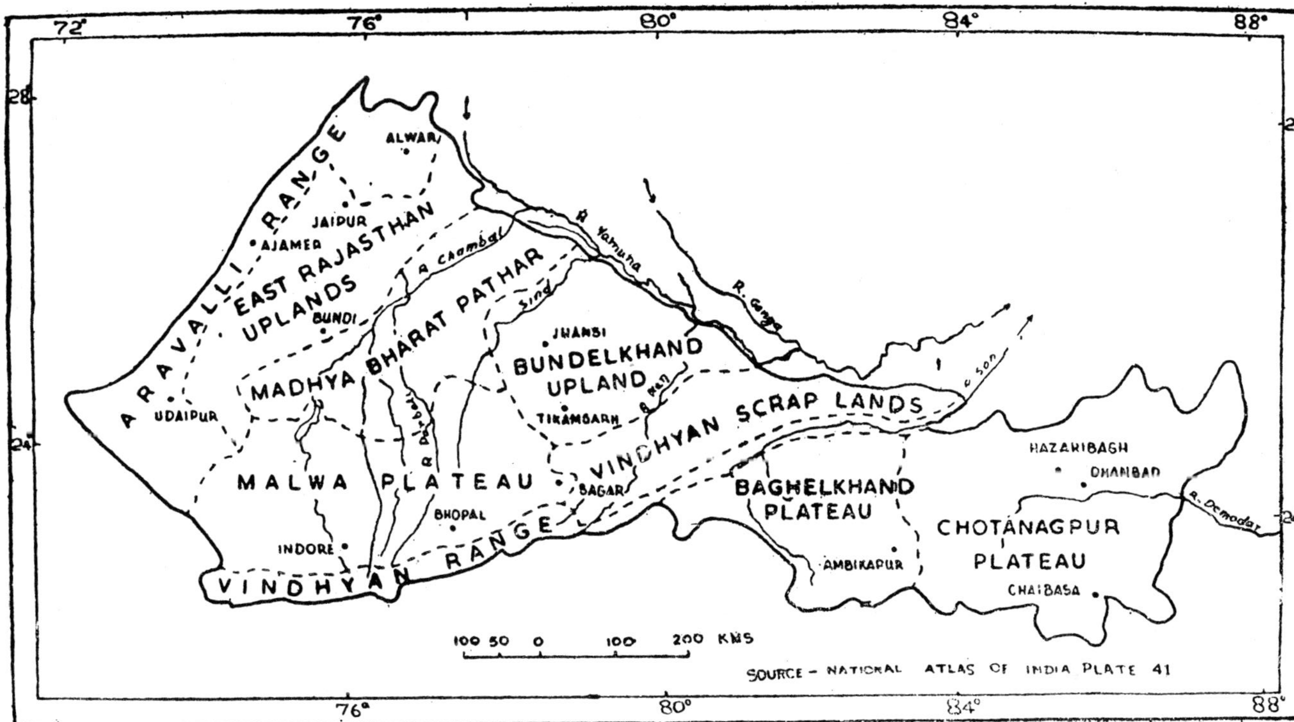

FIG. 6.1 : *The Deccan Foreland*

5. Active erosion by streams with steep gradients, entrenched valleys meanders, river terraces, boulder deposits, along the streams and ungraded long profiles dotted with waterfalls[8-9].
6. Superimposed drainage, particularly of Subernarekha river over the Dalma hills.

The Aravalli Region

The geomorphic studies of the Aravalli region are rather meagre.[10-11] However, the geological publications[12] refer to certain characteristics of the topography and the landforms which are revealing. The Aravalli range had reached the old age sometime in the late Cretaceous period. The present ranges are the stumps or roots of the synclines and are a good example of 'reversal of the topography'. Here three erosional surfaces have been identified at elevations of (i) 1200 m which is marked by the highest peaks, (ii) 600-680 m around Udaipur in the valleys and the plain towards the east of the range and (iii) 300-430 m. An intensive study of the whole Aravalli range is awaited to throw further light on the erosional surfaces and the related geomorphic features.

The following evidences further support the idea of uplifts and the rejuvenation of the region:

1. The uniform skyline of the ranges.
2. Hard formations like quartzites, conglomerates and limestones stand as ranges while the phyllites and schists form the lowlands.
3. Numerous examples of superimposed drainage in the Udaipur region and of the Chambal in the Bundi hills and the Triple scarps.
4. Erosional scarps facing the northwest bank of the Chambal river.
5. The ravine formation of the Chambal and its tributaries : It is generally stated that the ravines develop due to the defective land uses and the nature of the soil. But the studies by Sharma[13] in the lower Chambal Valley reveal that due to the regional rejuvenation the Chambal has entrenched into its own alluvial deposits. Once the river cuts a deeper valley, the water from the adjoining banks rushes down to the level of the main river; thereby the high

banks are dissected into a wilderness of ravines. The field observations of the present authors further support this view.

6. The incised meanders identifies in the Chambal and its tributaries.
7. The morphometric analysis of the Lower Chambal valley has revealed knickpoints, particularly between the Malwa plateau and the Triple scarps at the elevation of 375 m, 275 m and 225 m.
8. The erosional surfaces have also been identified in the Lower Chambal Valley at the elevations of 480-600 m, 360-450 m, 180-305 m. (Sharma).[14]

   The recent studies by Sharma[15] in the Morel Basin (East Rajasthan) has further indicated the entrenching of the river course, thereby, suggesting the tendency of rejuvenation of the Morel Padmaja[16] working in the Mej Basin (tributary of the Chambal) has also shown the evidence of rejuvenation in this Basin.

The Central Highland

The Central Highland is the triangular plateau between the Aravalli range, the Chota Nagpur-Baghelkhand plateau and the Narmada-Son axis. It has been frequently referred to as the highland comprising the undisturbed horizontally reposing Vindhyan formations which remained a passive landmass throughout the Paleozoic and Mesozoic eras. Nonetheless the study of the Sonar-Bearma basin by the authors reveals almost the same geomorphic characteristics as have been enumerated in the other adjoining parts of the Deccan Foreland.

The Sonar-Bearma Basin

After the emergence of the Central Highland as a landmass, the landmarks in the history of this region were the formation of the Lametas and the volcanic activity of the Cretaceous period flooding of western Deccan tableland giving rise to the Deccan Trap. The Sonar-Bearma basin falls on the contact of the Deccan Traps and the Vindhyans which have enabled the study of not only the landforms developed on these two formations but the exhumed topography an the superimposed drainage as well. The Sonar-Bearma basin in the south extends

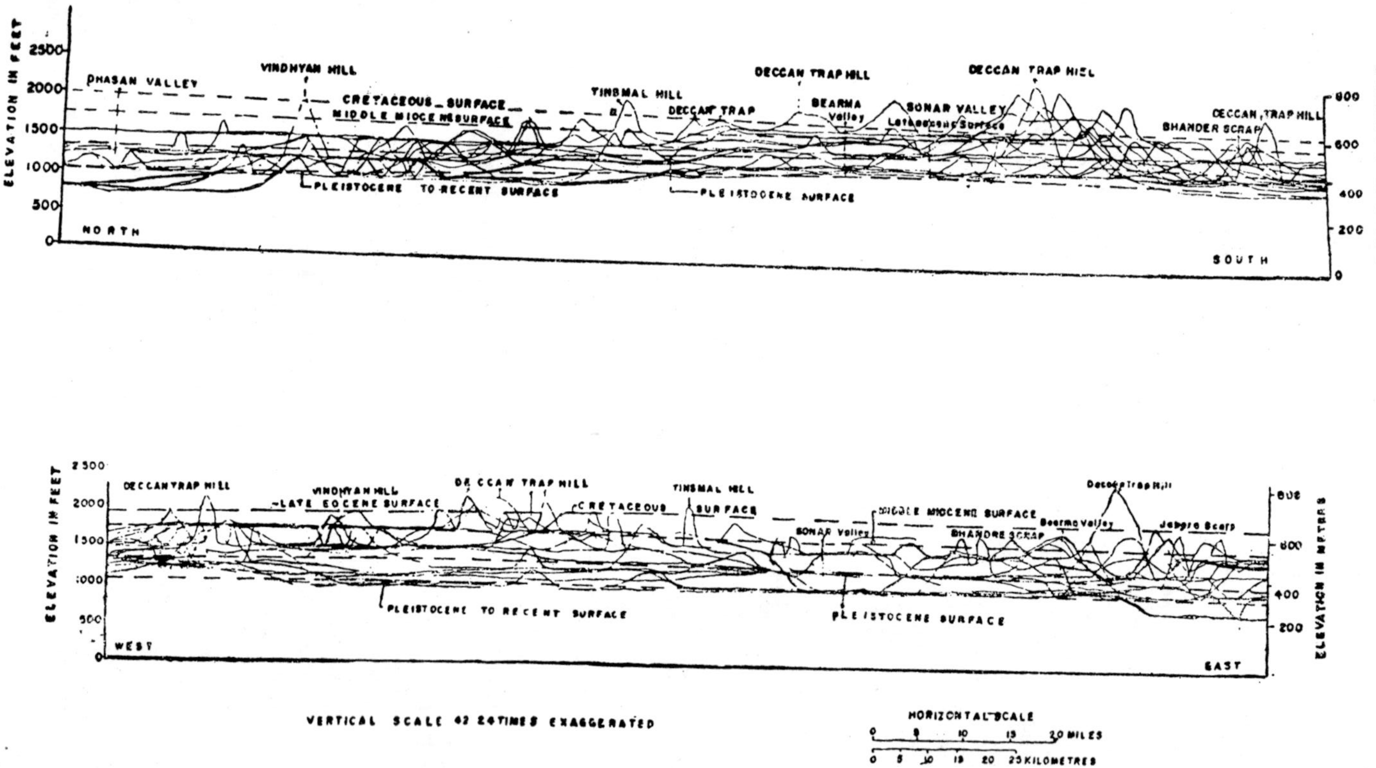

FIG. 6.21: *The Sonar-Bearma Basin and Adjoining Area Superimposed Profiles*

upto the watershed between these and the Narmada. It runs along the Vindhyan range which strictly speaking is a scarp. On the north-west another scarp bounds the basin. The boundary fault between the Archaeans and the Vindhyans runs along the Vindhyan range. Since this is the downthrow side, the Vindhyan range is an erosional scrap.[14] Similarly the northwestern scarp has also been carved out due to the differential erosion between the Bhanders and Rewa formations. Besides this, the rocks in the marginal zone of the Vindhyans along the boundary fault are dipping, folded and shattered. Such features have also been observed north of Hoshangabad along the Vindhyan range.

The analysis of the superimposed profiles and the altimetric frequency curve (Fig. 6.2 & 6.3) and the field evidences such as the presence of deep gorges along the small streams, the waterfalls, breaks in the slope and the thick mantle of the weathered material indicate that there are at least five erosion

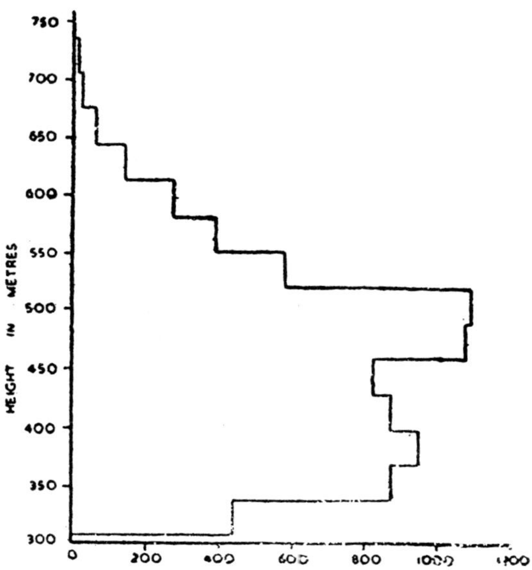

FIG. 6.3 : *The Sonar-Bearma Basin and Adjoining Area—altimetric frequency curve*

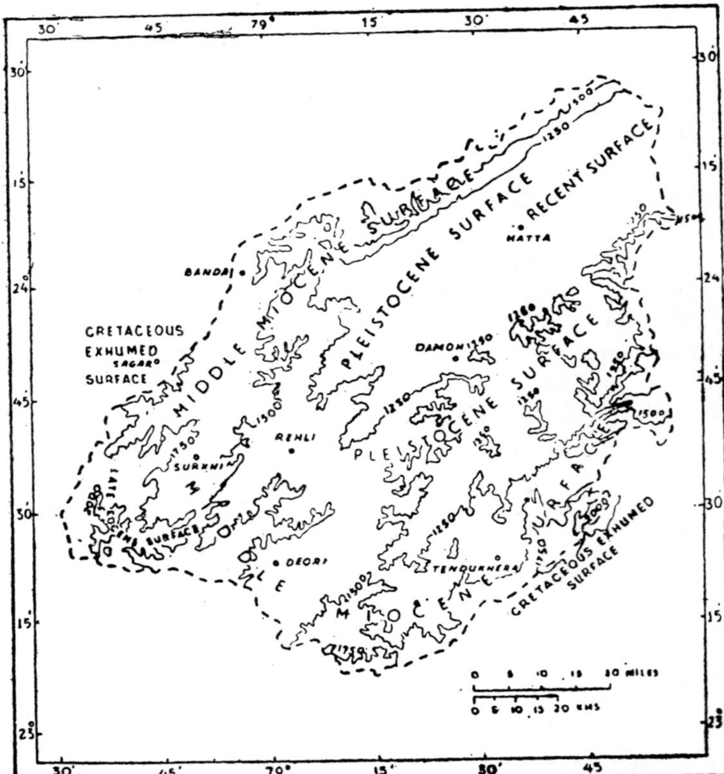

FIG. 6.4 : *The Sonar-Bearma Basin—Erosion Surfaces*

surfaces in the region. These have been identified at the heights of 580 m and above, 442 to 533 m, 365-396 m and 305 m. Two erosional surfaces, one on the Vindhyans and the other on the Deccan Trap, are at the elevation of 580 m and above (Fig. 6.4).

A careful study of the longitudinal profiles of the Sonar, Bearma, Bewas and Kopra show that in the upper courses their profiles are concave upwards. The gradient is steep and the profile is not smooth, showing thereby that these rivers have sufficient erosive power and they have not yet reached a graded condition. The cross profiles are narrow, deep and V-shaped. On the contrary, in the lower reaches the valleys are open and shallow, with thick deposits along their banks.

The erosion surfaces, almost corresponding to the Sonar-Bearma basin, have also been identified in the Rewa plateau by Dube.[15] He has also emphasised that the Bhander scarp, the Panna scarp and Binjh range scarp north of the Rewa plateau, facing the Gangetic alluvium, are all erosional scarps. These scarps have developed due to the differential erosion either along the fault or due to the varying rock resistance amongst the various members of the Vindhyans. This active erosion seems to have occurred due to rejuvenation giving rise to a typical topography of the scarpland. The presence of high waterfalls all along the Binjh range is also an indication of rejuvenation of the region in the recent past.

The erosional surface/peneplains found in various parts of the Deccan Foreland and their respective heights could now be tabulated as follows :

Area	Erosion surface	Author	Height of erosion surface
(1)	(2)	(3)	(4)
Western plateaux	Early Tertiary Peneplain	J. A. Dunn	915 m
Netarhat surface	— —	S.P. Chatterjee	100 m
Rewa plateau	Bhander surface	R.S. Dube	600-750 m
Sonar-Bearma basin	Cretaceous exhumed Vindhyan surface	R.K. Rai	580 m and above
Lower Chambal Valley	Pre-Cretaceous erosion surface	H.S. Sharma	480-600 m
Western Chota Nagpur plateau	Middle to late Tertiary peneplain	J.A Dunn	700 m
Ranchi	— —	S.P. Chatterjee	700 m
Rewa plateau	Panna surface	R.S. Dube	400-500 m
Sonar-Bearma basin	Middle Miocene surface	R.K. Rai	442-533 m
Lower Chambal Valley	Middle Miocene surface	H.S. Sharma	360-450 m

(1)	(2)	(3)	(4)
Northeastern part of Chota Nagpur plateau	Late Tertiary peneplain	J.A. Dunn	305 m
Basia plateau	— —	S.P. Chatter-jee	550 m
Rewa plateau	Rewa surface	R.S. Dube	300-350 m
Sonar-Bearma basin	Pleistocene surface	R.K. Rai	365-396 m
Simdega plateau		S.P. Chatter-jee	450 m
Rewa plateau	Trans-Jamuna surface	R.S. Dube	100-150 m
Sonar-Bearma basin	Pleistocene to Recent	R.K. Rai	305 m
Lower Chambal Valley	Pleistocene surface	H.S. Sharma	180-305 m

The preceding discussion including the observations of various authors in the different parts of the Deccan Foreland is convincing enough to accept the suggestion that the Deccan Foreland has been rejuvenated in the recent geological past and the present topography has been carved out by accelerated stream action. Besides, there are ample signs in the landforms and streams to believe that the Deccan Foreland has multi-cycle landscape. The regions are in various stages from late youth to old stage, and the present streams still have the signs of active erosion.

Age and Causes of Intermittent Uplifts

Accepting these arguments one is faced with the questions of (i) age of rejuvenation, and (ii) cause of intermittent uplift.

The geological evidences in the Aravalli region and the Chota Nagpur plateau place the Great Boundary Fault and the faults astride the Damodar, in the Cretaceous period. For the rejuvenation of the Aravalli range, Heron has remarked: "The range appears to have been peneplaned during the Mesozoic at sometime not very much earlier than the Cretaceous. This may well have been the first peneplanation since the range arose in pre-Vindhyan times. The peneplain has since been warped in such a way that its central portion near Udaipur city is now some 4000 ft. above sea level, . . .". This 4000 ft. surface may be taken as the Cretaceous surface. The second peneplain (2000-2250 ft.) has been correlated by Heron with the Tertiary peneplain of the Chota Nagpur plateau.

About the peneplains of the Chota Nagpur, Dunn is of the opinion that the rejuvenation of the Western Highland (including adjoining Baghelkhand plateau) occurred during the Cretaceous period, and that of the Ranchi plateau in the later part of the Tertiary period.

Regarding the cause of this rejuvenation, Dunn[16] while analysing the causes of earthquakes in north Bihar, has explained the mechanism of subsidence of the Indo-Gangetic trough and rising of the Himalayas and the Deccan Foreland. Wadia,[17] on the basis of the studies of the Himalayas, has postulated three intermittent uplifts of the Himalayas. "The first of these was post-Nummulitic, i.e., towards the end of the Eocene, culminating in the Oligocene, . . . greater intensity about the middle of the Miocene . . . The last stage was mainly of post-Pliocene age, later than the deposition of the greater part of the Siwaliks and did not cease till after the middle of the Pleistocene". According to Mithal[18] the uplift of the Himalayan chain was initiated during the Cretaceous and continued up to the Pleistocene. Dunn is of the view that the intermittent uplift of the Himalayas has corresponding periods of rejuvenation in the Chota Nagpur plateau, which may be correlated to the interruptions in cycles of erosion. Consequently, various other characteristics of rivers and landscape have developed. Ahmad seems to agree with this view and writes "Probably as the side effects of mighty Himalayan mountain movements, the southern plateau of Bihar appears to have undergone three successive intermittent uplifts". It seems all through the Deccan Foreland the zones of weakness, i.e., the faulted zone like the Boundary Faults east of the Aravalli range, along the Narmada Valley and the Damodar Valley have been affected in this rejuvenation. This has also found expression in a number of scarps in this region.

Once we take this premise that alongwith the Himalayan orogeny the Chota Nagpur plateau and the Aravalli range have been involved in the diastrophic action, there is no reason to ignore such a possibility for the Central Highland. For, the whole Foreland in front of the Himalayas should have acted as one unit. The erosional surfaces of the Sonar-Bearma basin have therefore been correlated with the intermittent uplift of the Himalayas and have been placed in the following periods:

1.	The Cretaceous (Pre-Deccan Trap) exhumed Vindhyan surface	580 metres and above
2.	The late Eocene surface (on the Deccan Trap)	580 metres and above
3.	The middle Miocene surface	442 to 533 metres
4.	The Pleistocene surface	365 to 396 metres
5.	The Pleistocene to Recent surface (in developing stage)	305 metres

The regions of the various erosional surfaces have been examined from the point of view of other related geomorphic features in the various stages in the cycle of erosion and these have been found in harmony.

CONCLUSION

The preceding synthesis of the geomorphic evidences from widely scattered areas of the Deccan Foreland leads one to the possible conclusion that the Deccan Foreland must have been involved in diastrophic activity in the recent geological past. The study of the landforms further suggested that the rejuvenation was intermittent. In the geological history of the Deccan Foreland since the rejuvenation seems to have started from almost the same period as the Himalayan orogeny and continued up to the recent period, their is a point in correlating the two. The Himalayan orogeny seems to have occurred intermittently and such a colossal activity may not have left it's foreland untouched.

REFERENCES

1. R.K. Rai,, 1980 : Geomorphology of the Sonar-Bearma Basin (M.P.), Concept Publishing Company, New Delhi.
2. S.W. Wooldridge & R.S. Morgan., 1959 : *An outline of geomorphology*: *The physical basis of geography*, Longmans, p. 73.
3. J.A. Dunn., 1942 : Economic geology and mineral resources of Bihar province, *Mem. Geol. Surv. Ind.* Vol. 78, pp. 6-14.
4. S.P. Chatterjee., 1940 : Gneissic topography of the Ranchi plateau. *Cal. Geog. Rev.* 3(2) ; 1952 : Geomorphology of the Ranchi plateau. *Proc.*

I.G.U.; (1963) : *Fifty Years of Science in India* : *Process of Geography*, p. 46.

5. K. Bagchi, & A.K. Sengupta., 1958 : Physiographic regions of the Ranchi plateau, *Geog. Rev. Indi.* 20.
6. E. Ahmad., 1965 : *Bihar—Physical, economic, regional geography*, Ranchi University, Ranchi.
7. R.P. Singh., 1957 : Landscape of southeast Chota Nagpur and its evolution, *Nat. Jour. Ind.* 3 (3 & 4).
8. P. Varma., 1958 : Ranchi plateau, its geomorphology and human settlements, Ph.D. Thesis (unpublished) Allahabad University.
9. F.K. Khan., 1954 : Waterfalls of the Chota Nagpur plateau, *Ind. Geog. Jour.* 29, (2 & 3).
10. S.P. Chatterjee, & K. Bagchi 1957 : Geomorphic regions of east Rajasthan Pathar, *Geog. Rev. Ind.* 19(4).
11. P. Varma, 1960 : Udaipur region—A study in geomorphology, *Ind. Geog.* 5(1 & 2).
12. A.M. Heron., 1953 : The geology of the central Rajputana, *Mem. Geol. Surv. Ind.* Vol. 79, pp. 37-42.
13, H.S. Sharma, 1979 : The physiography of the lower Chambal valley and its agricultural development, New Delhi, Concept Publishing Company.
14. H.S. Sharma., 1972 : Erosion surfaces in the lower Chambal valley, India, *International Geography*, Vol. 1, pp. 61-62 Canada.
15. H.S. Sharma, 1976 : Morphology of the Ravines of the Morel Basin, (Rajasthan), India, *International Geography*, Vol. 1, pp. 219-223 Moscow.
16. G. Padmaja., 1976 : Geomorphology of the Mej Basin (Rajasthan), unpublished Ph.D. Thesis. University of Rajasthan, Jaipur.
17. W.D. West, & V.D. Choubey., 1964 : The geomorphology of the Country around Sagar and Katangi, M.P.—An example of superimposed drainage, *Jour. Geol. Soc. Ind.*, Vol. 5.
18. R.S. Dube., 1968 : The Study of Erosion surfaces in the Rewa plateau. *Presented at the International Geographical Congress*, New Delhi, 1968.
19. J.A. Dunn, 1939 : Post-mesozoic movements in the northern part of the peninsula in the Bihar-Nepal earthquake, *Mem. Geol. Surv. Ind.* Vol. 73, p. 142,
20. D.N. Wadia,, 1953 : *Geology of India*, Macmillan and Co. Ltd., London, p. 306.
21. R.S. Mithal, 1968 : The Physiographical and structural evolution of the Himalayas., presented in the *Mountains and Rivers of India*, National Committee for Geography, Calcutta, p. 67.

SEVEN

Nageshwar Prasad and *Md. Naseemuddin*

POLYGENETIC EROSION SURFACES OF THE BARAKAR BASIN

INTRODUCTION

THE MAIN objective of this study is to determine erosion surfaces in the Barakar Basin, and also to analyse them genetically. The study leads to the conclusions that the Barakar Basin displays roughly four erosion surfaces—each surface characterized by minor landforms. These erosion surfaces are of polygenetic nature. They still preserve form of ancient surface in modern relief, although the surfaces appear to be uplifted epeirogenetically.

The minor landform, such as waterfalls, repids, entrenched valleys, incised meanders, gullies, tors, hummocky residual hillocks or tohgari, etc., are no less significant as they bear testimony to the past events. Almost all the erosion surfaces the scarp zone, planation levels, degradational and dissected plains, and Damodar Trough-bear scars of the epeirogenetic uplift of the area.

EROSION SURFACES

Almost synonymous with the term, planation surface, erosion surfaces of the Barakar Basin are of polygenetic nature. The Basin, at the present time, displays roughly four erosion surfaces (Figs. 7.1 & 7.2).

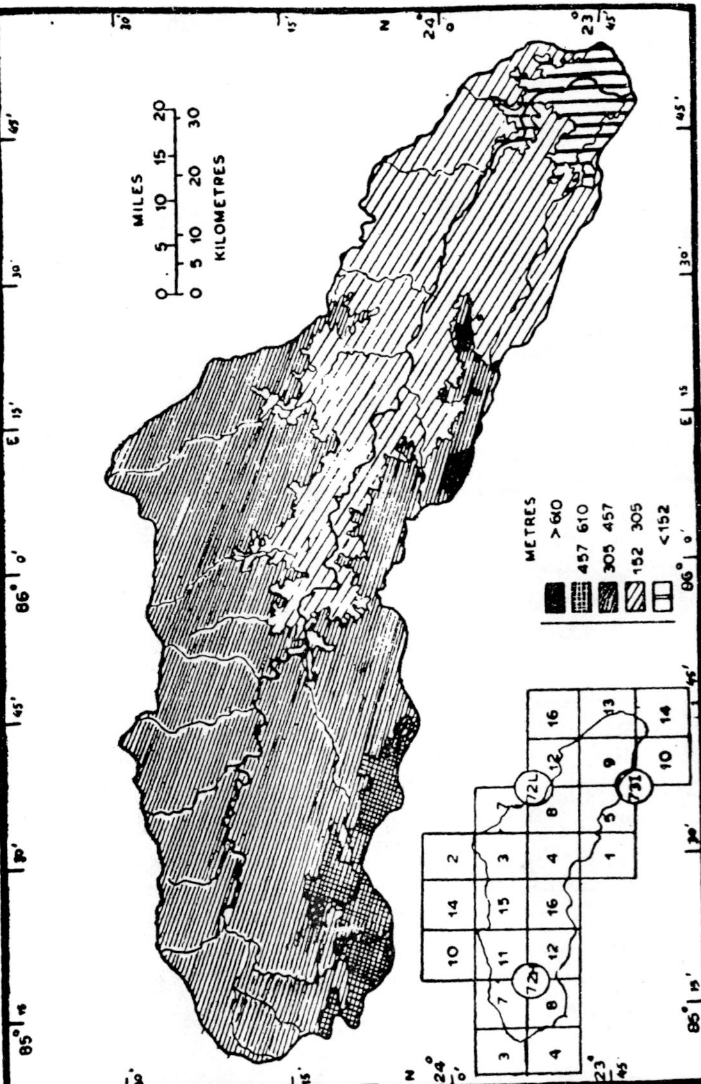

FIG. 7.1 : *The Barakar Basin—Physiognomy*

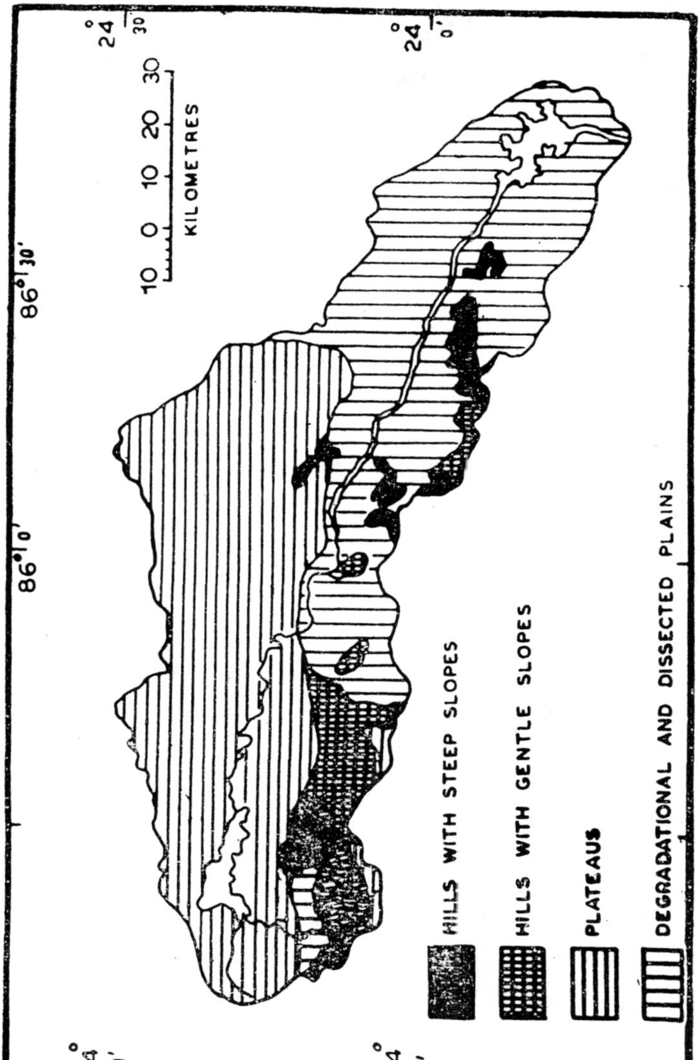

FIG. 7.2: *The Barakar Basin—Erosion Surfaces*

1. The Scarp Zone of Upper Hazaribagh Planation surface and lofty Parasnath Hill rising above 457 metres with scars of hills characterized by steep and gentle slopes;
2. Lower Hazaribagh Planation level rising roughly between 305 m and 457 m and sculptured as a continuous upland tract with interlocking spurs high ridges, and intricate valleys;
3. Degradational and dissected plain with an average elevation of 265 m excavated in the gneissic surface of the lower Hazaribagh Plateau; and
4. Damodar Trough of the extreme southeast lying below 152 m and presenting the appearance of a series of rolling downs, dotted here and there with isolated residual hills.

SCARP ZONE

The Scarp Zone rising above 457 m, is the highest erosion surface and is a part of the Upper Hazaribagh planation surface. This surface has scars of hills with steep slopes and gentle slopes. These hills also form the higher rims. Many of these hills appear to be the monadnocks of the ancient erosion surface which has been elevated in the geological past. The Tertiary uplift during the Himalayan Orogeny is recent.

This region is an ancient geological land surface; and this is also known that each of the series of plateaus in Chota Nagpur descend to the next lower plateau not be gradual slopes but by rather narrow and steep slopes. These narrow and steep slopes are known as scarps and the region, i.e.,, the south-western corner of the Barakar Basin is a part of the scarps, as stated above. Its average length (east-west) and width within the Barak ar Basin is 50 and 56 kilometres respectively. Though the averag e height of the scarp within the basin is about 600 metres above sea level, it has various points above the average height. The dominant height of Mahabar Pahar (85°35′ : 24°11′) on the left side of the Grand Trunk Road between Barkatha and Sakrej is 686 metres above sea level. In the west Mahuda or Mahudi Pahar (85°12′ : 24°12′) rises 676 metres above sea level. In between these two hills, the spot height rises from about 460 m along the Hazaribagh-Barhi Road to the east with an increasing height up to 684 m (85°28′ :

24°12′) and 686 metres (Mahabar Pahar) above sea level. The area possesses a waterfall (85°23′ : 24°09′) 8.5 metres high along the Kewta Nala a tributary of the Barakar. The river Barakar rises in this part of the Basin near Ichak, a small town seven miles north of Hazaribagh. "But the gorge it has notched in the northern escarpment of the plateau has reached a point only six miles (9.6 km.) from the source near Padma. Here the river tumbles over a number of small falls and rapids in a wild setting, locally known as the Tiger pool. And affluents of the river form a broad tree—like pattern on the scarps." (Roy Choudhary, 1957). But some of these streams have developed valleys parallel to the edge of the escarpment such as that of the Kewta. "On the scarp zone the rivers flow through dark V-shaped gorges and nearly all the roads which ascend from below one or the other river-valley and go up by hair-pin bends near these points called 'ghats' " (Bose, 1948).

Thus the scarp-zone is a dark wall of rocks rising abruptly nearly 305 metres or more from the lower plateau of the basin below. The top of the escarpment as usual in this region, is very often "a vertical pallisade of bare rocks while the base is covered up by typical tallus slopes" (Bose, 1948). The most formidable part of the Mahabar range near Barkatha, which resembles a huge knife edge is ten kilometres long and 687 metres high. Near it is Surajkund, a hot spring from which sulphurous steam and hot water pours out continuously. The Barsoti Nadi and some of its branches at this place cut remarkable valleys into and parallel to the escarpment. It is one of these which is responsible for the knife like shape of Mahabar Pahar. The escarpment ends in the Sindralli and Mahuda peaks on the water shed, 734 metres high, beyond which is the valley of river Mohana. "A number of blocks have been carved out from the plateau edge and stand as detached jungle-clad conical peaks with summit levels about 2000 feet (610 metres). Six such blocks occur between Bagodar and Barhi" (Bose, 1948). The entire scarp-zone in the south-western corner of the Basin consists of an ancient geological land surface and is composed of gneisses and granites.

The lofty Parasnath Hill is the highest point in Bihar in general and the Barakar Basin in particular. In the south of the Basin on the border of Dhanbad, the great mass of

Parasnath (1365 metres) which overhangs the Grand Trunk Road for many Kilometres, lies between the Barakar and Jamunia and is the most imposing physiographic feature of the whole region. Parasnath Hill has a central narrow ridge about 6.5 kilometres long and 3.25 kilometres wide with many rocky peaks, irregular in shape, but taking the general configuration of a crescent with its ends pointing to the north east and north. In these directions, the principal spurs of the hill extend. On the south west, there are no spurs and the greatest continuous rise occurs; to the north and west the spurs are very extensive; to the south east there is only one spur of importance.

The valley-widening processes resulting from the Barakar in the north and Jamunia in the south, have produced the sharp concave crest of the hills running from the west to east. The concavity of the slopes is caused by humid tropical climate.

From the top of the mountain one has a grand panoramic view of nearly the whole of the central part of the Basin. Most remarkable of the features seen in it are numerous sharp conical peaks clustering on all sides, but specially to the south and west. Another feature which catches the eye "is the wriggling Jamunia river to the south west with its omega shaped meanders". A distant silhouette of the hills from the north shows a step-like appearance, revealing the presence of horizontal and vertical joint planes. A closer view of the hills shows two types of crags; a set of vertical joints more intensively developed and a set of slightly inclined horizontal slabs, where the horizontal joint-planes have a more prominent development. The latter constitute the highest peak and the Jain Shrine is perched upon these slabs (Bose, 1948).

"The Parasnath Hill represents a gneissic pillar left by post-Gondwana erosion and probably indicates the highest previously attained by the gneissic plateau."

(Roy Choudhry, 1957).

This hill rises "abruptly above the surrounding gneissic country. Such steep-sided hill in a moist tropical area may be due to a body of igneous intrusion whose form the present hill represents. The fact that Parasnath Hill standing abruptly like a huge pillar above the surrounding gneissic country is an igneous intrusive structure is proved by the fact that it consists of hard, compact granite or augite-enstatite-granite (Ahmad,

1965)." C.S. Fox holds the view that it is not an worn-down stump of an ancient mountain chain. According to Sinha and Dave (1963) "The entire group has been intruded by an epidioritic sill, which is also folded. The effects of two main igneous intrusions, in addition to these epidiorites, are visible in the metasediments; one is mafic, represented by metadolerite and metanorite in the south eastern portion of the present area, and the other is folsic, represented by a granitic pluton cutting a part of the older rocks in the northern portion."

Such deepseated intrusive bodies of hard igneous rocks like granite exposed by long continued erosion have resisted erosion with reference to the less hard gneiss of the surrounding country, and are known as residual hills (Ahmad, 1965).

LOWER HAZARIBAGH PLANATION LEVEL

This planation level is an elevated tract of comparatively flat or level ground. Lying roughly between 305 m and 457 m this erosion surface is a continuous upland tract with interlocking spurs, high ridges and intricate valleys. Thus the surface is a gently undulating country with occasional residual ridges, locally known as 'tongaries'. In the south west it is flanked by steep scarps.

The lower Hazaribagh Plateau is the 'third erosion surface'. After the early and later Tertiary block uplift which gave rise to the Netarhat Plateau and the resurrected peneplain or the Central Chota Nagpur Plateau respectively there was stillstand and recession of scarps. Then there was uplift practically of the whole of Chota Nagpur Plateau excepting the eastern peneplain surface in Manbhum and the country east of Rajmahal hills by about 700′—1000 ft. (214-305 m) in Pleistocene to Recent time (Ahmad, 1958). The Lower Hazaribagh plateau of the Barakar Basin in particular has been uplifted by this later movement and this final uplift of the whole mass is responsible for rapidly downcutting streams, headward erosion and undulating surfaces. The surface is still undulating. The jungle is rarely better than scrub and the grazing lands are barren, and scattered rocky ridges disclose the severity of the denudation to which this ancient land surface has been exposed. From one of the ridges at Simradhab on the Dhanwar road an extensive prospect of the Barakar Basin clearly reveals its physiographical character. The land is seen inclining gently eastward with the rivers course, but north and south west it rises slowly as though to a shallow saucer's rim. On the south east and south the horizon is shut in by the masses of the Parasnath and the buttresses of the higher plateau, but in other directions a few isolated peaks along interrupt the skyline. The terraces are seen in sharp relief.

Thus this erosion surface is an open rolling plateau, strewn with boulders and embossed by crumbling peaks with pleasing Mahuda glades here and there. This is due to the greater energy of the affluents. A number of peaks stand out as sentinels on the watershed, such as Banda Pahar near Kodarma, Bamani beyond Domchanch and Ghoranj at Kharagdiha near the source of river Usri.

"The valley of river Usri, however has the fine open aspect with crumbling peaks scattered over the area. Khandauli, the peculiar saddle shaped hill made up of granitoid gneiss dominates the landscape near Giridih, the well known mica-splitting centre. Its peak is like a volcanic cone and the great saddle attached to it consists of massive boulders." (Bose, 1948). These boulders are formed by intruded granitoid gneiss. Diaclase and exfoliations have so much reduced it that it gives a view of a mere collection of huge boulders. Christian Hill is another feature of the same nature, but here disintegration is of greater degree. Hence it is difficult to climb. From the top of this hill one can note many low patches af bare rock, which are in the last stages of being peeled off by exfoliation, to the surrounding level. To the south of Giridih coal basin lie the Chepo Hills which are "peaky hills" carved out of hornblende. The dip and strike of the rocks in these hills are the same as those of surrounding gneissic terrain.

The Barakar in this portion of the Basin passes through a number of small Gondwana basins, the biggest and the most important of them near Giridih. The area near Giridih is drained by Khakho Nadi, a tributary of the Barakar. "Doterite dykes, quartz reefs and an island of crystallines are the igneous rocks found in this sedimentary basin". The dykes and the reefs form wall-like features. The sandstones form characteristic hills (Bose, 1948). The most fantastic shapes, in courses of streams include miniature cliffs, and beds decorated by patterns of miniature pyramids.

Among other Gondwana patches, the Cheri coal field in the north west is important.

Usri river passes north of Giridih coal field. Nine miles (14.5 km.) down Giridih occur the Usri falls. This is another beauty spot for tourists. Here the character of the gneissic complex of this region is well exposed. "Shattering along the joints has split up the country rock into blocks of huge dimensions, whose polished and clean surfaces are chequered by pegmatitic veins of various hues." (Bose, 1945). The cross profile of the river-bed changes abruptly after passing down the Fall. The lower area is almost flat slightly undulating country which gives way to gorge like form. This clearly indicates rejuvenation of the area and the Fall is a knick point. Since the nature of the rock is the same in both the upper and lower area, there is no question of differential erosion (Chatterjee, 1949) producing this fall. The fall is only a temporary feature of the thalweg in the process of river grading, and must recede and disappear in time. A discussion with the regular observers has made clear that the fall has gradually been lowered down and receded.

DEGRADATIONAL AND DISSECTED PLAIN

In the eastern and south eastern section of the Barakar Basin lies the degradational and dissected plain with an average elevation of 265 m. This has been excavated in the gneissic surface of the Lower Hazaribagh plateau and is flanked by residual hills very often mounted on projecting spurs which abut on the river-valleys.

In this part, the gradient of the surface is relatively low so that both vertical dissection and headward erosion are limited and the rivers are rather statically placed in the channels and valleys. This smallness of gradient and relative flatness of the surface has restricted the powers of the lower Barakar as a land-eroding stream and even minor topographic irregularities have to contain the stream within a limited belt. This appears to magnify the role of the manadnocks and other hills that occur in the margins of the lower part of the Basin. In the south west ern portion a rugged topography is seen along the eastern limit of the Lower Hazaribagh Plateau (305 m) because here the gradient is relatively high.

The outer erosion fringe of the basin does form a part of the Chota Nagpur Plateau. In Dhanbad District, this part of the Basin is characterized by greater variation in relief. The Parasnath Hills (1365 m) send out spurs to this unit, one of which passes through the district of Dhanbad via Topchanchi and Tundi. Thus the more or less continuous ridges of the Parasnath range run for a long distance eastward from this place and from the northern hilly portion of the Dhanbad district. These ridges vary in altitude from 366 metres to 745 metres above sea level. Tundi (745 m), Dhangi (384 m) and Kenja (360 m) are the important hills in the lower Barakar Basin. Tundi Hill (86° 23′ : 23° 57′) is the highest point in this region which is located in the west of Tundi. On the west of the hill there is a waterfall. From this hill small streams radiate in all directions. At some distance from the hill towards the south and east, small hills are seen which may be called monadnocks. Dhangi hills run from Pradhankhanta to Gobindpur in Dhanbad district. They lie between the Grand Chord line of the Eastern Railway and the Grand Trunk Road. The highest peak (384 m) in these hills is at Dhangi. Kenja hill (360 m) is located near Durgapur, some 8 kilometres away west from Maithon. Besides these hills, Murgatona Pahar (232 m) near Mihijam and Bodra Pahar (340 m) both in Santal Parganas are monadnocks.

Now it is notable that most of the hills and ridges of the area owe their preservation to the greater hardness of the rocks. "The common types of rocks which give rise to outstanding hills are the metamorphic rocks like epidiorites, amphibolites, metadolerites and metanorites. Some of these hills may thus be regarded as relict type of ridges which have suffered less erosion than the surrounding area. The veins of quartz (often brecciated) which are so common in their metamorphic terrain of the area, also form low ridges due to their resistance to denudation. These ridges may, however, be regarded as tectonic in origin as they usually indicate faultzone (Sharma, 1964).

In the eastern portion of the lower basin, the Gondwana

patches and other recent deposits are found which are the continuation of the Raniganj coal field.

DAMODAR TROUGH

In the extreme south east, the Basin occupies a very small area of the Damodar trough, formed during Bermo Carboniferous earth movements. Here the relief is subdued and undulations are less pronounced. This is why the country is more open than any other region of the Basin. It presents the appearance of a series of rolling downs, dotted here and there with isolated residual hills.

In this part of the Barakar Basin, there are no scarps, no hills and no undulations. It is a peneplain which merges with the Purulia peneplain in the south. "This floor is often interrupted by Gondwana (sandstone) residuals" (Ahmad, 1965). The Lower Gondwana sedimentary rocks of the coal-basins as (i) Sampur and Chanch-Dumarkanda seams, (ii) Rangamati-Nirsa-Gobindpur seams, and (iii) Pusal-Merthadih seams etc. lie unconformably over the older crystalline rocks of Dharwar age. "The recent formations which form superficial deposits are the alluvial sand of rivers and streams and the residual soils derived from the weathering of bedrock. (Sharma, 1964). Thick deposits of recent alluvial sand occur adjacent to the banks of the Barakar, Khudia and Pusal streams. Small deposits of 'kankar' (calcium carbonate) occur on the bank of the Barakar river.

CONCLUSION

The steep scarps standing above planation level in the Basin, their dissection by erosion into mountainous ranges of peaks and valleys ; and the reduction of many parts of the Basin to undulating lowlands or plains characterised by long, deep and tarraced river valleys, waterfalls and rapids, residual hill, (monadnocks, fors, etc.) entrenched meanders, gullies prove that the surfaces bear testimony to the past events. So, the erosion surfaces of the Basin are of polygenetic nature.

SUGGESTED READINGS

Ahmad, E., 1958 : Geomorphic outline of Chota Nagpur, *Geographical Outlook*, Vol. II, p. 18.

Ahmad, E., 1965 : *Bihar : A Physical, Economic and Regional Geography*, Ranchi Univ., p. 26, 371.

Ahmad, E., & Devi, P., 1965 : Origin of Chota Nagpur scarps, *Geographical Outlook*, Vol. IV, pp. 84-85.

Bose, S.C., 1945 : Land form development in the Giridih-Parasnath region, *Calcutta Geog. Rev.* Vol. VII, No. 4.

Bose, S.C., 1948 : *The Damodar Valley Project*, Phoenix Press Ltd. Calcutta, Reprinted in D.G, Hazaribagh, p. 8-9, 13-14.

Chatterjee, S.C., 1949 : Physiographic evolution of Chota Nagpur plateau, *Cal. Geog. Rev.*, Vol. IX, pp. 41-42.

Roy Choudhary, P.C., 1957 : *District Gazetteer, Hazaribagh*, Secretariat Press, Bihar, Patna, p. 4, 13.

Sharma, N.L., 1964 : *District Gazetteer Dhanbad*, Secretariate Press Patna, p. 12, 17.

Sinha, R.C. & Dave, V.K.S., 1963 : Effects of granitic employment on epidiorites of Parasnath area (Bihar), *Bull. Geog. Soc. Am.* Sept. p. 1145.

EIGHT

C. R. Twidale

GRANITIC INSELBERGS

INTRODUCTION

GRANITIC inselbergs are of three basic types. Bornhardts (Plate 1) are steep-sided, bald, domical hills (Bornhardt 1900; Willis, 1934, 1936). Castle koppies, known in Britain and elsewhere as tors (but see Twidale 1971, pp. 14-17), are also steep-sided but they are angular and blocky, though the blocks are *in situ* and display an orderly arrangement determined by orthogonal joint sets (Plate 2). Nubbins are more or less conical hills that seemingly consist of a chaotic mass of large blocks and boulders (Plate 3).

There are of course many variations on these three major themes. Many bornhardts for example are stepped (Twidale and Bourne, 1975a) and stepped topography (Wahrhaftig, 1965) is well developed in blocky and gneissic out crops. Some residuals consist of part dome and part nubbin, some part koppie and part nubbin and so on.

Some inselbergs are single-unit forms standing in isolation above the plains. Others, however, are in reality complexes or multiple unit forms consisting of several or many of the single unit features standing in close and ordered juxtaposition (Fig. 8.1). But whether they stand in isolation or whether they occur in orderly complexes, the three forms, bornhardts, castle koppies and nubbins are basic. Each is widely distributed and indeed they are coexistent in many parts of the world, though bornhardts and nubbins appear to be more numerous than are castle koppies.

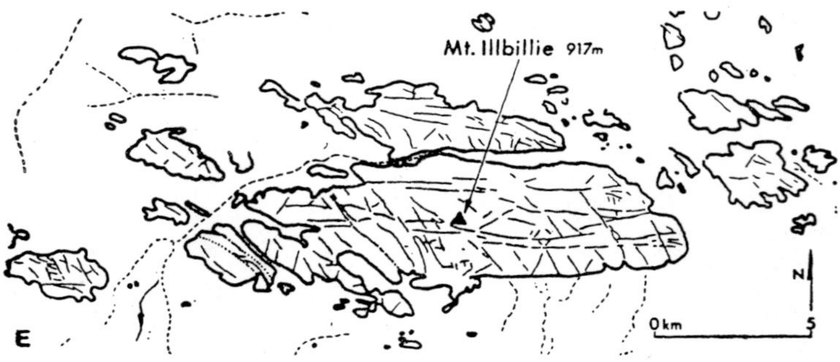

FIG. 8.1 : *The Everard Ranges, Northern South Australia, showing large fracture delineated blocks*

Are the three forms genetically related and if so, how?

The contrast between bornhardts and castle koppies seems at first sight merely to reflect joint patterns, with sheeting joints (Gilbert, 1904; Dale, 1923; Twidale, 1973) dominant in the former, and orthogonal sets in the latter. Quarry exposures show that sheet structure is well developed within Paarlberg, near Cape Town, and domical forms dominate the outcrop. Mount Kobe, in the Memba district of Mozambique, displays both domical and castellated slopes (Holmes, 1918) as does Blackingstone Rock, on eastern Dartmoor (Jones, 1859). The northern slope of the latter is a half-dome formed by arcuate, convex-upward massive sheeting slabs. By contrast large quadrangular blocks are exposed in the blocky southern face, showing that the two sets of joints most commonly developed in granitic rocks coexist in some areas at least, but again demonstrating the close relationship that there is at some sites between structure and surface.

However, there is no necessary connection between fracture pattern and surface form. Sheet structure is well developed in the Rocklin Quarry in the foothills of the Sierra Nevada in central California, and in the Sidobre of Southern France, but in both areas the granite hills are boulder-strewn. Nubbins, and not castle koppies, are typical of the Mount Bundey area, southeast

1. Granite bornhardts standing abruptly above pediplain near Uwerus, central Namibia

2. (a) Castle koppie a few kilometres northeast of Salisbury, Zimbabwe/ Rhodesia

2. (b) Haytor West, a tor or castellated residual on eastern Dartmoor, southwestern England

3. Block strewn granite nubbin near Naraku, northwest Queensland

4. Remarkable Rocks, Kirkpatrick Point, southwestern coast of Kangaroo Island, South Australia, showing angular blocks standing on bare dome protruding from aeolianite (wind-blown calcarenite)

5. Stock of gneissic granite emplaced in schists, Namib Game Park SSW of Karibib, central Namibia

6. Isolated domes of porphyritic foliated granite near Nanutarra, western Pilbara region, Western Australia

7. Domes exposed in escarpment below African land surface (X) near Witrivier, eastern Transvaal. The Bushman's Kop, the prominent residual at right clearly stood above the general level of the African plain

8. The Armchair and other granite domes in the northern Flinders, Ranges, South Australia, occur below the level of a pronounced summit surface which is of exhumed type and of Lower Cretaceous age

of N.T. Darwin, despite the perfect orthogonal sets revealed in quarry faces. Thus though fracture patterns play an important part in the determination of the morphology of granite inselbergs, other factors obviously come into play in some areas. It is suggested below that climate is one of these, but it cannot be argued that it alone controls the morphology of the residuals, for the various forms coexist: two forms or elements of both bornhardts and castle koppies are found side by side developed in the same rock type in Blackingstone Rock and Mount Kobe, and complete examples of contrasted forms, again developed in similar materials, occur in the Swakop Valley of Central Namibia (Fig. 8.2) and in eastern Zimbabwe. Domes and nubbins are found in close proximity in the western Pilbara of Western Australia and the Llano of Central Texas.

FIG. 8.2 : *Field sketch of the Swakop Valley, South of Karibile, Central Namibia, showing palaeosurface (X) and domes on nubbins exposed in the Valley Floor*

At the Devil's Marbles, N.T. (Twidale, 1980a) at Remarkable Rocks on Kangaroo Island, South Australia, (see Daily, Milnes, Twidale & Bourne, 1978a) and in the Mount Manypeaks region of the southwest of Western Australia, blocky forms rest on plinths or foundations that are large radius domes (Plate 4). Furthermore, castle koppies occur in very different climatic regions: the Rocky Mountains of Colorado and Dartmoor on the one hand, and Zimbabwe on the other; and bornhardts have an even wider distribution, so that neither the domes nor the castellated form is confined to particular climatic regions as usually defined.

It is suggested here that bornhardts are the basic inselberg form, and that the other two, castellated and block-strewn, are derived from the weathering of bornhardts. In this paper the origin of bornhardts is first discussed, and the disintegration of bornhardts into the other two types is then considered.

THEORIES OF DEVELOPMENT

General Statement

Many hypotheses have been advanced in explanation of bornhardts. Some have specific local application. For example some of the *morros* and sugarloafs of southeastern Brazil are simply upfaulted blocks (Barbier, 1957; Birot, 1958). Some domes in Mozambique (Holmes, 1918) and others in Central Namibia (Plate 5) are stocks of granite gneiss exposed through the erosion of the country rock into which the igneous masses were emplaced. Stone Mountain, Georgia (Lester, 1938) is similar in that it is upstanding because the granite of which it is composed is more resistant than the surrounding (schistose) country rock. Some domes in Mozambique are shaped from a grey granulitic or porphyritic granite in contrast with strongly foliated or banded rocks of the plains (Holmes, 1918). But most bornhardts are not bounded by large-displacement faults and are underlain by granite or gneiss which is mineralogically and texturally similar, if not identical, to that beneath the surrounding plains. In explanation of these, the vast majority of bornhardts, two major competing hypotheses have been put forward.

Scarp Retreat

For many workers, notably L.C. King (1949, 1953, 1962) but including also Howard (1942), Mabbutt (1955), Pugh (1956), Louis (1959), Ollier and Tuddenham (1962), Oberlander (1972) and Selby (1977), bornhardts are the last remnants of circumdenudation (*Ferlinge, monadnocks de position*) remaining when scarp retreat and pedimentation have all but baselevelled the land mass (Fig. 8.3a).

Scarp recession has undoubtedly occurred and is still taking place in certain structural settings (Tricart, 1957; Twidale, 1960,1978a), and there is no theoretical reason why scarp retreat should not develop on granitic bedrock, though the rationale has not been examined by the proponents of the concept. Slopes develop and maintain a maximum inclination

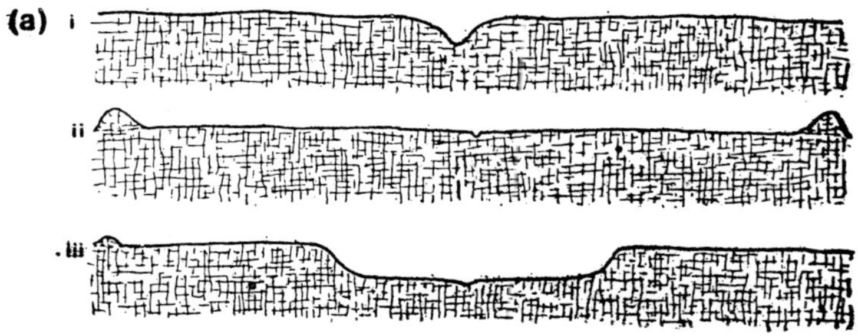

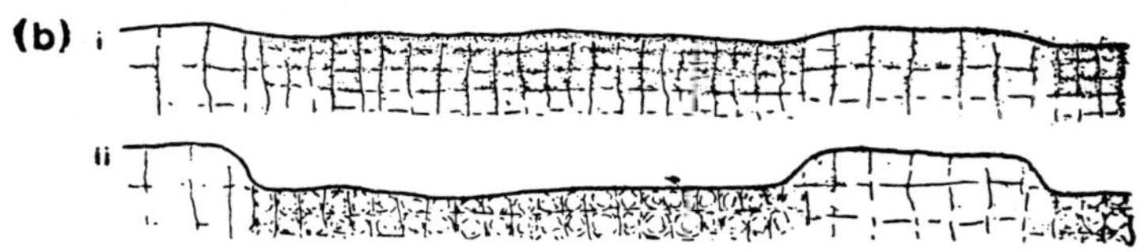

FIG. 8.3(a) : *Stages in the baselevelling of land mass by scarp retreat and pedimentation and (iii) continued erosion following Stream rejuvenation*

FIG. 8.3(b) : *Two Stages development of inselbergs showing (i) Deep Fracture controlled subsurface compartment weathering and (ii) Differential Erosion of the compartments*

commensurate with stability where basal slope erosion exceeds that on upper sectors. This situation is prevalent in caprock situations and in arids lands, but as was noted many years ago (Barton, 1916; Bain, 1923: see also de Saussure, 1796, I, p. 318) granite, more than most rocks, exhibits the property of being resistant when dry but weathering rapidly in the presence of moisture. As water gravitates towards lower slopes the latter ought to be weathered and hence eroded more rapidly than the former, so that steep slopes ought in theory to be developed and maintained. Scarp retreat in granitic terrain is theoretically feasible provided that the rock is essentially homogeneous.

Two-Stage Model

Field observations in many parts of the world show that fracture density varies from compartment to compartment within granitic masses (see, e.g., Mennell, 1904). Using this compartmental variation several workers have argued that the well-fractured compartments are readily penetrated by water and hence are weathered more rapidly and more deeply than are the massive compartments (Fig. 8.3b). Then, when stream incision takes place, differential erosion occurs, leaving the massive compartments in relief (Linton, 1952, 1955 ; Handley, 1952 ; Büdel, 1957 : but see also Hassenfratz, 1791 ; Twidale, 1978b). This concept is based on the structural inhomogeneity of the bedrock, and the bornhardts are regarded as *Hãrtlinge* or *monadnocks de résistance.*

Minor objections can be made against the simple statement of the hypothesis. For instance the two stages implied need not necessarily be temporally distinct (Lewis, 1955), but observations in many parts of the world show that the essential prerequisite of the two-stage concept, namely the occurrence of compartments of contrasted fracture density, is widely developed.

Thus both of the major hypotheses offered in explanation of bornhardts developed in granitic rocks is theoretically feasible. Which of the two is preferred has been largely a subjective matter and arguments and evidence brought forward to sustain one point of view have either been fairly readily dismissed or reversed to support the opposed view. The field evidence and arguments now adduced, however, clearly points to most bornhardts having evolved according to the two-stage mechanism, (Twidale, 1980b).

EVIDENCE AND ARGUMENT

Distribution of Bornhardts

Relationship to Drainage

If scarp retreat had taken place in a homogeneous medium, the last remnants surviving planation ought to be found on the crests of major divides. If on the other hand the landscape evolved in accordance with the two-stage model the distribution of bornhardts can be expected to be related to the stress field and

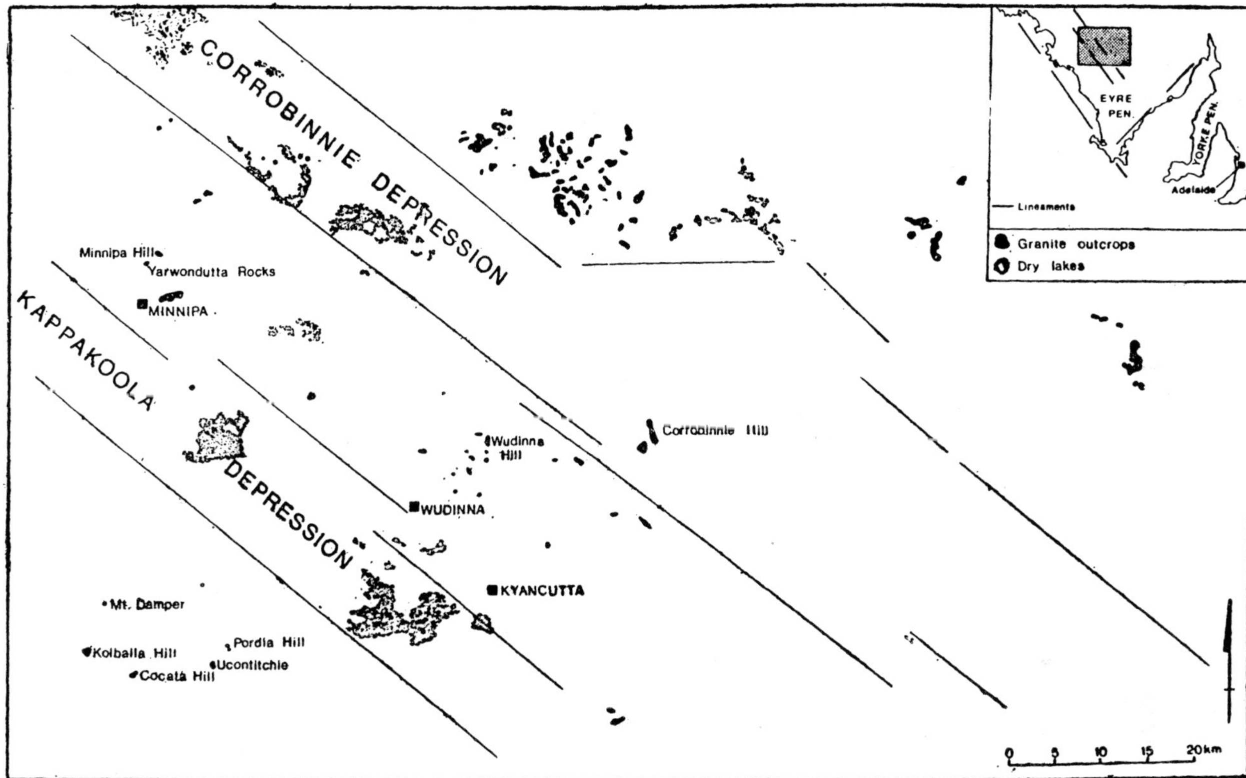

FIG. 8.4 : *North western Eyre Peninsula showing distribution of granite outcrops major structural lines and (inset) regional lineaments*

regional tectonic style rather than with drainage patterns, though of course structure may well have influenced the latter.

Hugo (1970) noted that the distribution of inselbergs in the Pietersburg area of northern Transvaal bears little or no relation to drainage (Brook, 1978) reached the same conclusion, evidently in ignorance of Hugo's work, and other examples can be cited from southern Africa and Australia. For instance if it is assumed that the Corrobinnie and Kappakoola lineaments of northwestern Eyre Peninsula are shears, then zones of compression ought to have developed along north-south or east-west axes, depending on the sense of shear. The theoretical picture can be expected to be complicated by variations in rock composition and strength, but in fact the pattern is consistent with the concept of shearing (Fig. 8.4). The size and number of the resistant compartments is small suggesting that fractures are well developed, and that zones of weathered granite are far more extensive than are compartments of fresh rock.

On the other hand though most of the granite outcrops occur on the broad topographic rises, they are not restricted to the crestal areas or the central axes of the highs. Many occur in the marginal areas and some, including the extensive outcrops in and around Corrobinnie Hill and Peella Rock, are in the fault-line depressions.

Though many of the inselbergs of the southwest of Western Australia stand above the level of the lateritised land surface (see below) many, including some of the largest bornhardts in the region are exposed in ancient river valleys (Fig. 8.5a). They include Mount Stirling, Kokerbin Hill and Hyden Rock. Likewise, in southern Zimbabwe, Shashe inselberg stands close to the Shashe River, and many other residuals of note occur in close proximity to river courses (Fig. 8.5b). The deepest weathering due to moisture attack occurs beneath valley floors and with the continued downcutting of the rivers the results of differential weathering are exposed and evident there.

Another aspect of this supposed relationship between bornhardts and drainage is that if bornhardts really are the last remnants of circumdenudation they ought to occur in plains settings. Some do (Plate 6) but many are exposed in the sides of narrow valleys, for example in the Valley of a Thousand Hills,

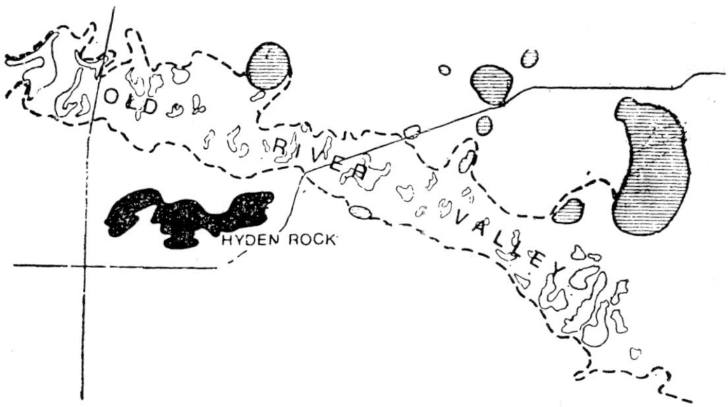

FIG. 8.5(a) : *Association of inselbergs and palaeodrainage in part of the wheat Belt of southwestern Australia*

Natal; the Krokodil River valley of eastern Transvaal; Domeland and the Yosemite Valley of California (Matthes, 1930); in the Colorado Rockies (Plate 8); in the coastal Greenland (Soen, 1965) ; in the Darling Ranges of Western Australia. They can in no way be interpreted as the last remnants surviving after long-distance scarp retreat; they are massive resistant compartments that have been exposed as a result of valley erosion.

Fracture-Controlled Margins

According to the scrap retreat hypothesis the shape and size of bornhardts ought to be independent of such structural features as fracture patterns. They are not. The shape and size of all bornhardts that have been subjected to detailed investigation are to a considerable, frequently large, extent determined by rock fractures (see Twidale and Bourne, 1978 and Fig. 8.6). This is consistent with the two-stage model of development.

Occurrence in Multicyclic Landscapes

As Obst (1923), King (1949), Twidale (1964), and others have noted, bornhardts invariably occur in multicyclic landscapes, that is in areas where there is evidence of palaeosurfaces high

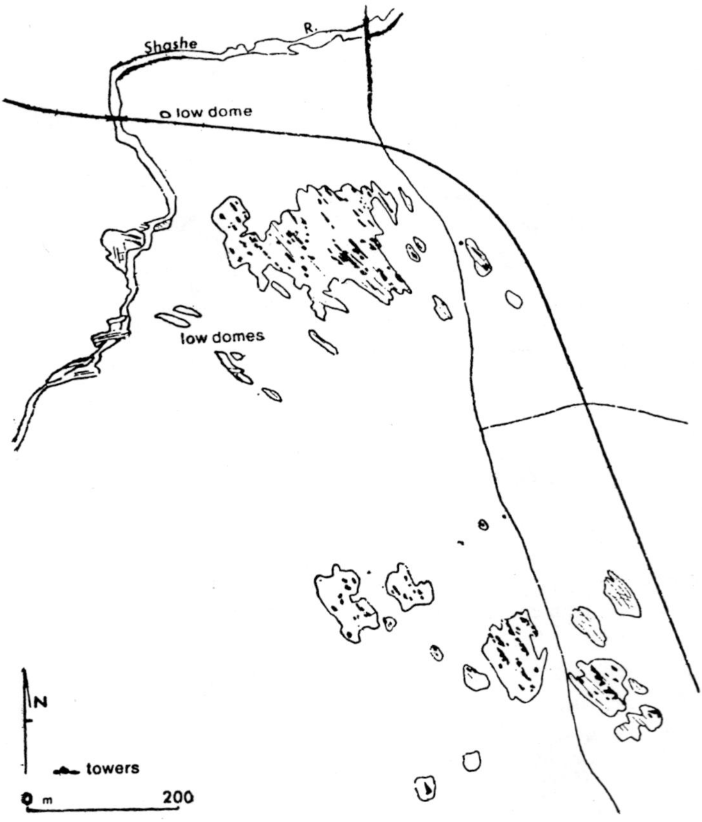

FIG. 8.5(b) : *Shashe River and Castle Koppie, Zimbabwe*

in the relief. Most bornhardts occur at levels lower than those of the old planation surfaces. Thus the bornhardts of the Pilbara region of Western Australia occur below remnants of the Hamersley surface preserved on Precambrian strata and of as yet undetermined age. In the Swakop River valley, and indeed throughout Central Namibia most of the bornhardts are exposed beneath the level of a wide-spread master planation surface developed in Precambrian schists and gneisses and again of unknown age (Fig.8.2). In Zimbabwe more than 90 per cent of the many bornhardts occur in the eastern and southeastern regions

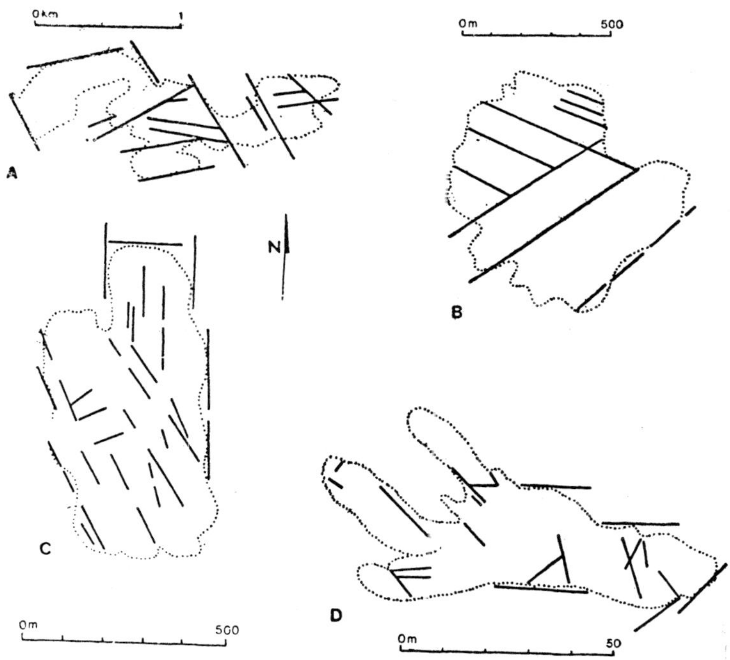

FIG. 8.6 : *Fracture Controlled Margins of some granite bornhardts*

where the African Surface (King, 1962 ; Lister, 1976) has been dissected by rejuvenated rivers (Whitlow, 1980). In the eastern Transvaal also, the African Surface is prominent and most of the bornhardts occur below it (Plate 7) though there are again exceptions. The many domes of the Valley of a Thousand Hills occur in a complex valley cut by the Umgeni River below the African Surface. In the northern Flinders Ranges such domes as the Armchair (Plate 8) developed in Ordovician granite occur beneath widespread remnants of an exhumed early cretaceous planation surface (Woodard, 1955 ; Twidale, 1969, 1980c). So one could go on giving example of bornhardts in multicyclic landscapes.

The genetic relationship is twofold. First a planation surface implies relative standstill, and time for deep, intense, and, some would say, differential subsurface weathering. Second, multiple planation surfaces implies deep erosion. The significance of this is that if as has been suggested bornhardts are in compression and sheet structure is an expression of the resultant antiformal structure, the upper zones of such structures are in tension, and readily weathered and eroded (Fig. 8.7).

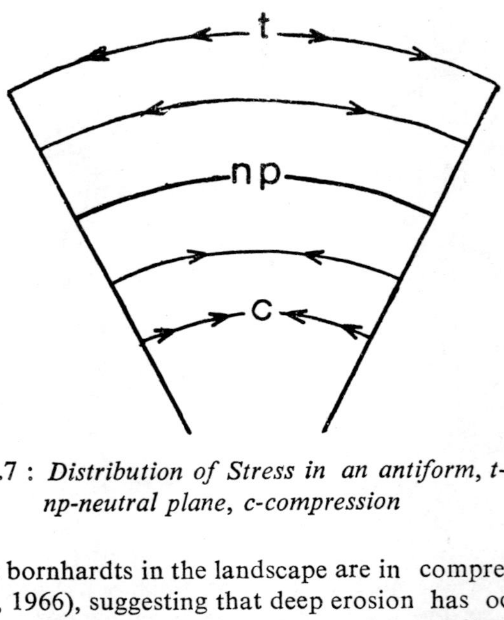

FIG. 8.7 : *Distribution of Stress in an antiform, t-tension, np-neutral plane, c-compression*

Most bornhardts in the landscape are in compression (but see Hack, 1966), suggesting that deep erosion has occurred to expose the deeper compressional zones of the antiforms (Price, 1966 ; Twidale, 1964,1973). Thus the apparent anomaly between structure, state of stress and field evidence is resolved.

The scarp retreat hypothesis offers no explanation of this association of bornhardts and palaeosurfaces, whereas it is a necessary corollary of the two-stage model. Indeed whereas survival through more than one geomorphic cycle is impossible in terms of scarp retreat and the completion of any form of cycle development, it is both feasible and to be anticipated in terms of multiple stage development (Twidale & Bourne, 1975a; see also below).

Weathering and Minor Landforms

Weathered Granite beneath Plains

Deeply weathered granite has been reported from many parts of the world (see Ollier, 1965; Thomas, 1966). Its presence is equivocal so far as the present discussion is concerned, for it can be argued with equal force that the weathered compartments represent the first of the two stages necessary for the development of inselbergs; or that the weathered rock is an incidental feature, being merely the material across which scarp retreat and pedimentation has migrated ; or that the deep weathering developed after pediplanation.

On the other hand some deeply weathered profiles offer further information, for example near Marandellos, eastern Zimbabwe; at Paarl, in the Cape Province; at Albany in the southwest of Western Australia; at Gemencheh, near Tampin, Malaysia; at Shatin, New Territories, Hong Kong ; in the Estes area of the Colorado Rockies. For obvious reasons exposures of the bedrock beneath the plains are scarce, but the evidence available shows that though there are some massive compartments that are apparently incipient domes (see below). most of the granite is well-jointed.

At Ucontitchie Hill on northwestern Eyre Peninsula, the granite exposed in a reservoir excavated at its northwestern margin consists of corestones set in a matrix of grus (Plate 9). That this is typical of the region is suggested by borelog data which shows that weathered granite up to 40 metres and commonly 30 metres thick underlies the plains, and that corestones are common in the near-surface horizons. In a general way the size of the corestones gives an indication of the original joint spacing (see Linton, 1955; Wilhelmy, 1958; Twidale, 1971; Godard, 1977) for corestones are developed by joint-controlled subsurface weathering within joint blocks. At Ucontitchie the diameter of the corestones is varied but the mean is less than one metre, which is of quite a different order from the adjacent bornhardts where the fractures are commonly 10-15 metres apart. It can of course be argued that jointing is in some respects a result of weathering and erosion (Chapman, 1956; Twidale, 1977) rather than a determinant of its location and extent, but the least that can be said is that this evidence is not inconsistent with one of the major precepts of the two-stage mechanism namely that there are marked difference in joint spacing between adjacent compartments.

Subsurface Initiation of Forms

If scarp retreat were responsible for the development of bornhardts not only the major forms but all of the minor features found on the host masses ought also to be of exogene origin. If the bornhardts were prepared in the subsurface major and minor forms ought also to be initiated there.

Incipient bornhardts with no surface expression or marked by only rock platforms, and bounded by weathered granitic indicative of differential subsurface weathering have been observed at many sites. One of the best exposures, at Ebaka in southern Cameroun (Plate 10), has been investigated by French workers (Boyé and Fritsch, 1973; Boyé and Seurin, 1973), but others have been observed at halfway, between Pretoria and Johannesburg; near the Leeukop and in the Vredefort Brickpit, both in the Vredefort Dome structure of western Transvaal; at Buccleuch, north of Johannesburg; a few kilometres southeast of Pietersburg, in the northern Transvaal; between Nuwerus and Vanrhynsburg, in Namaqualand; at the Pomona Quarry, near Salisbury, Zimabwe; at several sites around Kondinin, in the southwest of Western Australia; and so on—there are many examples in the landscape (Plate 11). They are not unreasonably interpreted as massive compartments of granite that either have been just exposed or about to be. It cannot be argued that they are the stumps of bornhardts that have finally been reduced to baselevel, for their surface are pitted, which is indicative of subsurface weathering (Twidale and Bourne, 1976) and are also grooved and dimpled due to the development of saucers and gutters initiated in the subsurface (Twidale and Bourne, 1975b; see also below). Moreover, these incipient domes are bounded on all sides by deeply weathered granite indicative of compartment weathering.

Notwithstanding Thomas' (1967, pp. 258-259) conclusion that most minor landforms found on bornhardts are developed after the exposure (not exhumation, *pace* Thomas, 1967, p. 253) of the rock mass, there is convincing evidence that several of them are initiated at the weathering front. In brief, saucer-shaped depressions and gutters are commonly present when the regolith is removed by either natural or artificial means (Logan, 1849, 1851; Twidale, 1971, p. 90; 1976a, pp. 203-204; Boyé and Fritsch, 1973; Twidale and Bourne, 1975b). It is suggested that these are the forerunners of weather pits or gnammas (Twidale and Corbin, 1963) and *Granitrillen, Silikatrillen* or *Pseudokarren* respectively. These forms certainly develop after exposure but they are initiated at the weathering front (Plate 12). Similar though less conclusive evidence of subsurface initiation has been described concerning tafoni, A-tents and even rock doughnuts (Twidale and Bourne, 1975b, 1977; Twidale, 1978c, 1980d).

Scarpfoot Depressions

Scarpfoot depressions (linear depressions, *Bergfussniederungen*) are well developed in West Africa (Cotton, 1942, pp. 92-96; Clayton, 1956) on Eyre Peninsula (Twidale, Bourne and Smith, 1974), in Egypt (Dumanowski, 1960), in the Mojave Desert, in Namibia, in Central Australia and elsewhere. As other workers (Pugh, 1956; Clayton, 1956) have appreciated also they pose problems for the scarp retreat hypothesis because they are basically developed as a result of intense scarpfoot weathering. Thereafter, the depression develops either as a result of volume decrease consequent on weathering, settling and compaction, and lowering of the surface; or through excavation of the scarpfoot weathered zone by streams; but they argue a period of standstill during which the scarpfoot weathering took place. This is in keeping with the two-stage model but difficult to explain in terms of scarp retreat, for the depressions are not restricted to small residuals which because of their limited area do not provide much runoff hence became stabilised (see Twidale, 1978a, pp. 1160-1162 and his Fig. 8.10).

Stepped Bornhardts

Many bornhardts are stepped : superimposed on the overall domical shape are steps and treads, many of them minor but at some sites quite pronounced (Twidale and Bourne, 1975a). The treads consist of gently sloping rock platforms and frequently carry residual blocks and remnants of soil or regolith. The steps comprise assemblages of concave slopes known as flares (or flared slopes), tafoni, scarpfoot depressions, narrow platforms and breaks of slope ,(Plate 13). Flared slopes develop as a particular form of weathering front (Twidale, 1962, 1967, 1968a,1971, pp. 90-96; 1972, 1976a, pp. 7-17). They are in no way related to flat-lying or sheeting joints, which they transect. They can be observed in incipient form before exposure in articial excavations (Plate 14) and there are many lines of evidence to support the contention that they are particular forms of weathering front, initiated in the subsurface (Fig. 8.8). The other forms associated with flares also originate in the subsurface at the weathering front and where horizontal or subhorizontal zones of such assemblages occur above the present piedmont they surely mark the locations of former hill-plain junctions.

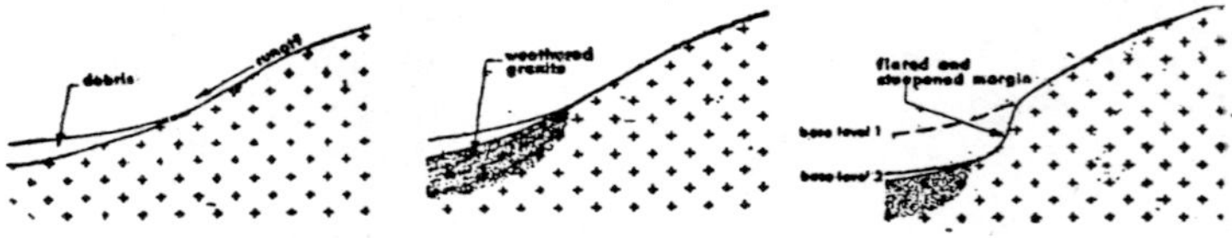

FIG. 8.8 : *Development of flared slopes by subsurface moisture attack in the piedmont, followed by erosion and exposure of the Concave Weathering front*

The general argument (Twidale, 1980e) is well illustrated by Yarwondutta Rock, a low granite residual on northwestern Eyre Peninsula (Fig. 8.9, Plate 15). The residual is markedly stepped. On the northern side for example a flared slope some 4 metres high leads up from the present plain to a narrow platform on which stands a large residual block, hollowed out by tafoni development. This platform is in turn backed by another flared slope at the base of which is a shallow scarp-foot depression. This concave slope, some two metres high, leads up to a rolling dimpled and grooved summit surface. Bearing in mind the origin of flared slopes the shoulders marked 1 and 2 at the upper limit of the two concavities each represent a former hill-plain junction, where scarpfoot weathering occurred.

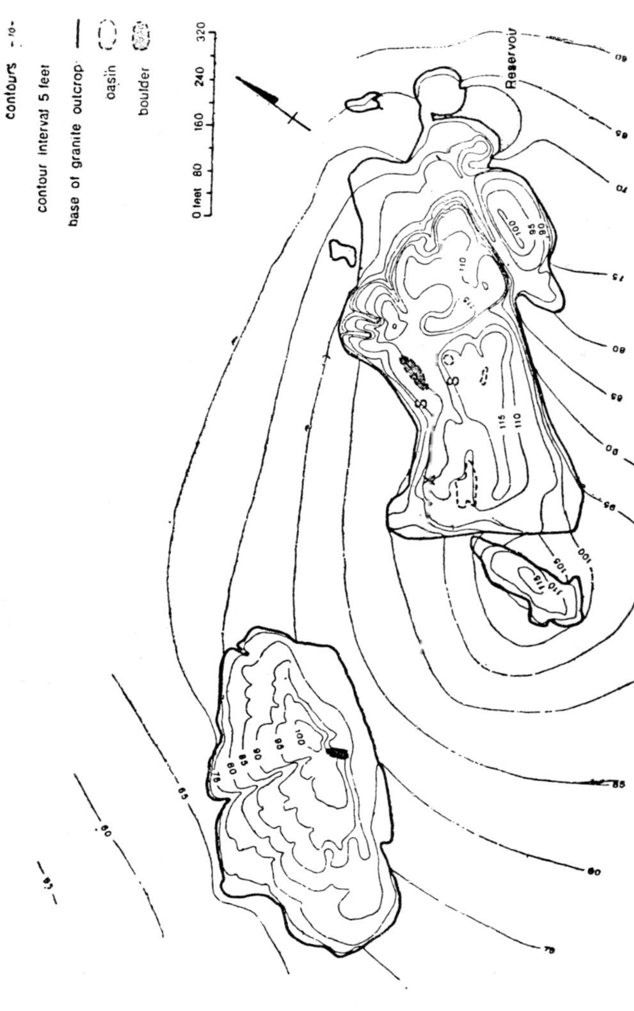

FIG. 8.9 : *Contour Plan of Yarwondutta Rock with X the Site from which Plate 15 was taken, looking northeast, and S the shoulders of the flared slopes (cf Plate 15).*

If this interpretation is correct then :

1. Yarwondutta Rock has grown as a relief feature as the surrounding plains have been lowered.
2. This growth has taken place not gradually but in distinct phases or episodes separated by periods of standstill during which subsurface weathering occurred.
3. Though there has been some backwearing of bounding slopes as a result of subsurface scarpfoot weathering it amounts to only a few metres, (not much more than the width of the platforms) whereas in terms of the scarp retreat model the inselberg landscapes of Eyre Peninsula, the Pilbara (Plate 6), Namaqualand, Namibia (Plate 1) atc the Mojave and elsewhere demand recession of kilometres and even scores of kilometres.

Yarwondutta Rock is not a special case. Many stepped inselbergs have been noted (Twidale and Bourne, 1975a) and some of them mapped (Table 8.1) in southern, western and central Australia, in southern Africa, and in California (Fig. 8.10). In gneissic areas resistant gneissic bands act as local baselevels on valley side slopes giving rise to pseudostepping (cf. Wahrhaftig, 1965) but many genuine stepped inselbergs have been observed in the regions mentioned.

Episodic exposure of bornhardts is entirely consistent with two-stage development, and as has been pointed out elsewhere (Twidale and Bourne, 1975a) fully answers King's (1966) argument against the model relating to disparities between the height of inselbergs and depth of weathering in the areas in which they occur.

Age of Bornhardts

A further corollary of the arguments arising from stepped bornhardts is that none of the residuals ought in terms of scarp retreat to be of very great antiquity. Given uninterrupted conditions, every last residual ought to be eliminated by the end of a geomorphic cycle. Even if baselevel were lowered and the streams responsible for planation were rejuvenated, the upper shoulders of the nick points marking the junction of streams graded to old and new baselevels would act as local baselevels for the streams and surfaces above them and consumption of the landmass, including any residual remnants,

9. A reservoir at the northwestern edge of Ucontitchie Hill, northwestern Eyre Peninsula, South Australia, showing the contrast in joint density implied by the small corestones exposed in the excavation and the massive rock of the inselberg

10. Quarry face at Ebaka, southern Cameroun, showing domical compartment of massive granite, overlain and bordered laterally by weathered

11. The Leeukop, in western Transvaal, with, in the foreground, the crest of an incipient dome (X) of which no sign existed in the natural landscape

12. *Rille* or gutters extent beneath the natural soil level, and indeed, here at Dumonte Rock, near Wudinna, northwestern Eyre Peninsula, coverged in the subsurface. X-X is the natural soil level, and the gutters are exposed in an excavation intended for water storage

13. Flared slope and platform, side of Ucontitchie Hill, northwestern Eyre Peninsula

14. Yarwondutta Rock and reservoir, northwestern Eyre Peninsula, showing flared marginal slope of the residual, and the concave weathering front exposed in the artificial excavation. The shoulders (S) mark the present natural soil level and hill-plain junction and the former piedmont

15. Part of northwestern slope of Yarwondutta Rock showing the stepped morphology and the shoulders (S) marking former hill-plain junctions

16. The foliation of the rock is expressed in well developed dipping cleavage which is exploited by weathering. Below the shoulder (right) which is a remnant of a palaeosurface above which stand the major domed forms, the weathering is more pronounced presumably because it took place in the subsurface

17. The Humps, in the southwest of Western Australia is a group of granite domes surrounded by a plain on which laterite is well developed

18. Subangular to subrounded corestones are being exposed on Tcharkulda Hill, northwest Eyre Peninsula, as the grus matrix is washed away. Some (X). *in situ* are seen sitting on the platform (foreground) and below the large residual boulders

19. Part of the Enchanted Rock complex in the Llano of central Texas, with, on the lower slope, a thick sheet of rock that has broken up into large polygonal slabs. A few remnants of the sheet stand in isolation on the upper slope face

20. Eastern side of Ucontitche Hill, Eyre Peninsula, showing multiple flares and exposed end of sheeting joint

21. Castle Rock, east of Albany in the southwest of Western Australia is an angular blockly residual standing on the crest of a board granite dome. The flares on its sidewalls indicate that it has been exposed in at least two phases, each comprising subsurface piedmont weathering followed by lowering of the plain and exposure of the erstwhile weathering front

TABLE 8.1 : **Many Stepped Inselbergs have been noted in Southern-western and Central Australia**

	Level of steps (in metres above hill base)	Criteria : (rp - break of slope fs - flared slope t - tafoni p - platform nt - truncated shed dp - scarp foot depression
Mount Wudinna, S.A.	45-50	rp
	30-35	rp, fs, t, p
	10-15	fs
	3- 5	p
Ucontitchie Hill, S.A.	35	rp, nt
	3- 5	fs, p
Yarwondutta Rock, S.A.	7- 8	fs, p, dp
	4- 5	fs, p, dp
	0- 3	fs
Corrobinnie Hill, S.A.	15	fs, p
	5	fs, p
	0- 2	fs, p
Kokerbin Hill, W.A	175-180	fs, t, p
	150	p
	80	rp, nt
	10	fs
	2	fs, p
Mount Stirling, W.A.	210	p
	80	p
	40	p
	25	fs, nt
	10	fs
The Humps, W.A.	65	fs, p
	40	rp, nt
	4-5	fs, p
Hyden Rock, W.A.	45-50	fs, p
	35-40	fs, p
	25-30	fs, p
	1-2	fs, p
Ayers Rock, N.T.	100	rp
	30-65	rp, t
	3-5	fs, p, dp
The Olgas, N.T.	100	t, rp
	30	p, rp
	3-5	fs, p

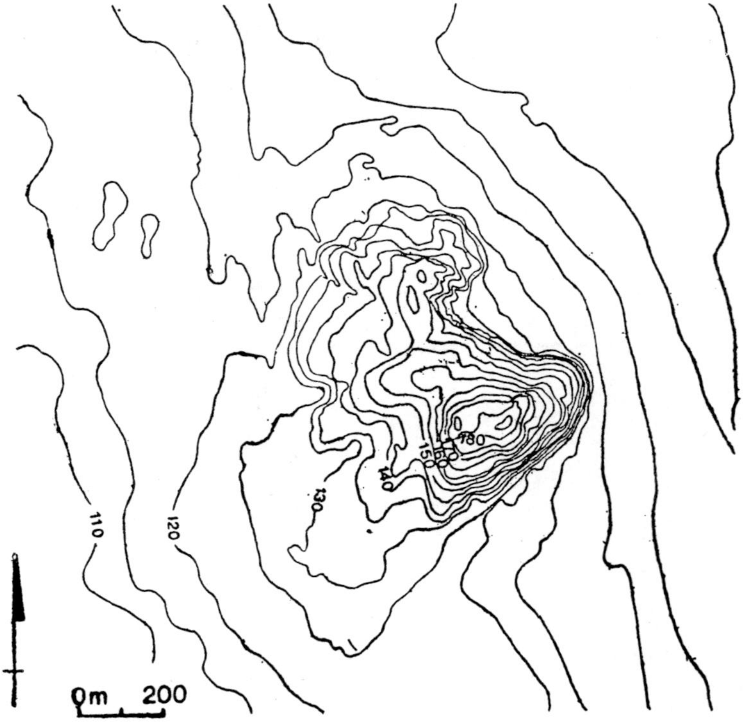

FIG. 8.10(a) : *Plans of Kokerbin, (W.A.) Carappee, (S.A.) Mt. Stirling bornhardts, (W.A.) and the Humps, (W.A.) showing Stepped form*

ought to continue (Fig. 8.10a). Thus no bornhardts ought to be older than the duration of a cycle, that is perhaps as youthful as 10 ma or as old as 33 ma, (Gilluly, 1955; Twidale, 1976b; see also Schumm, 1963). On the other hand, since they are inherently massive and resistant and particularly because of reinforcement effects (Twidale, Bourne and Smith, 1974) bornhardts may persist through several cycles in terms of the two-stage mechanism. What is the evidence?

Many bornhardts are youthful features of the landscape. In the Yosemite Valley and Domeland, California, they have only recently been exposed and continue to be revealed (though in both areas there is evidence of their having evolved beneath a

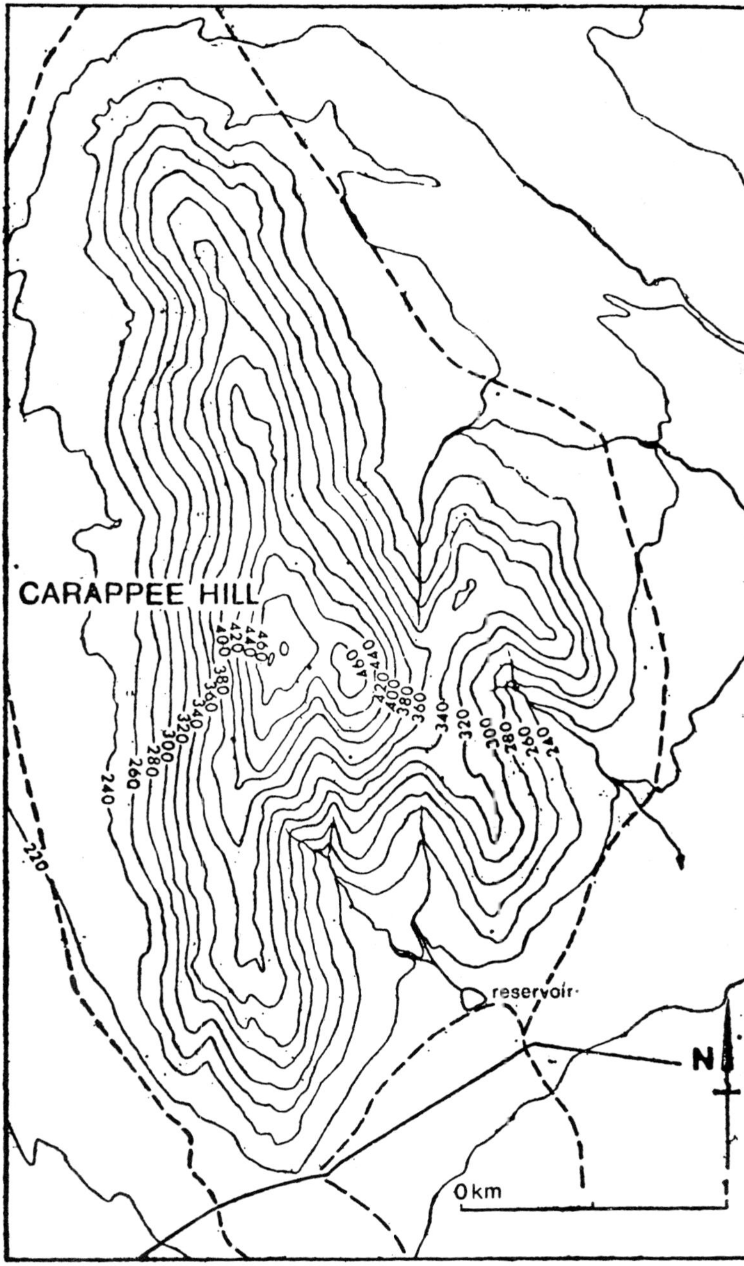

FIG. 8.10(b) : (*W.A.*) *Carapee*

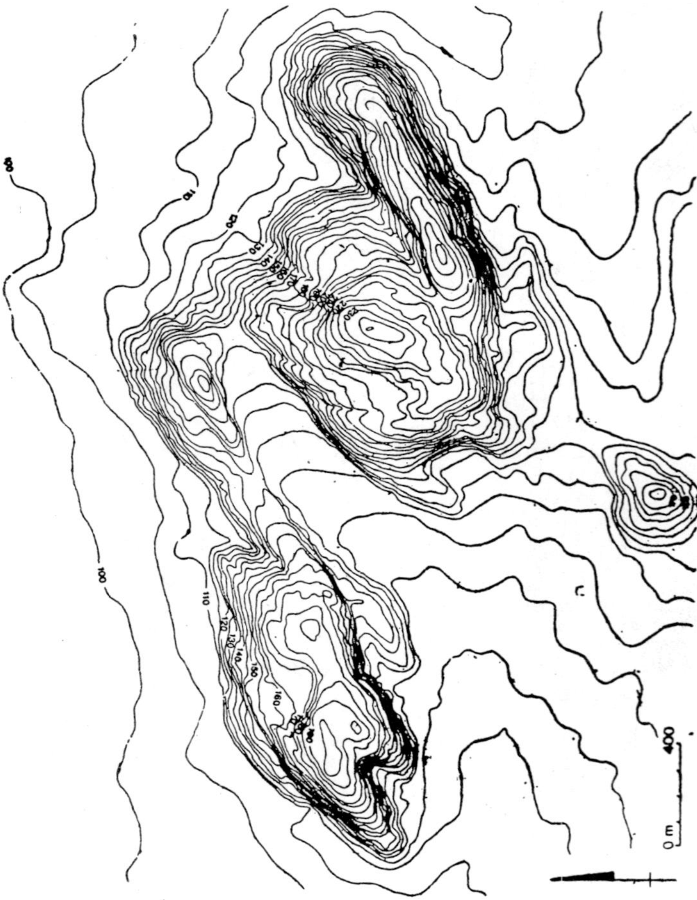

FIG. 8.10(c) (S.A.) *Mt. stirling barnhardts*

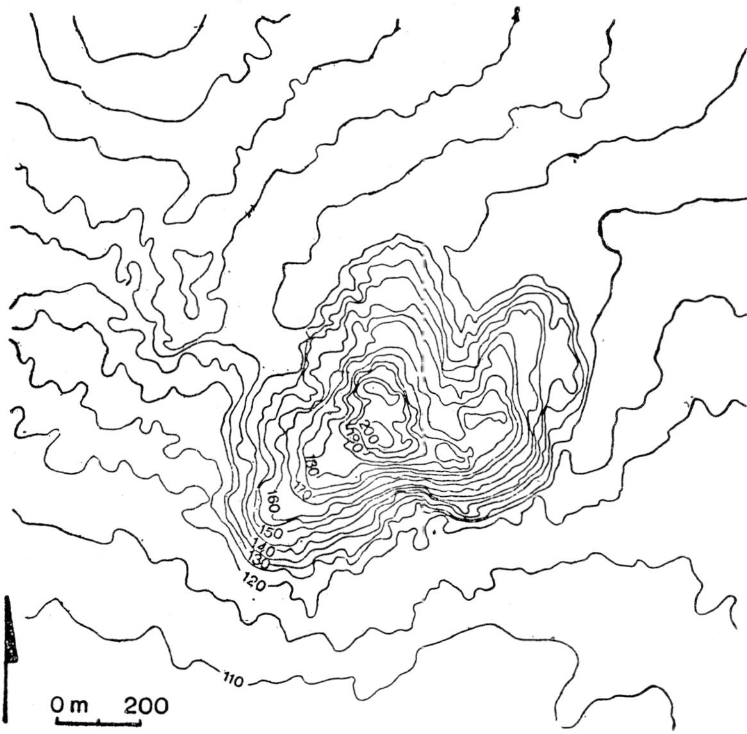

FIG. 8.10(d) : *The Humps*

palaeosurface of low relief, and breaks of slope on the flanks of some domes are suggestive of phased exposure). The same can be said of the valley of a Thousand Hills, the valleys of the Colorado Rockies, the Kamiesberge of Namaqualand (Plate 16), central Namibia, parts of eastern Transvaal (Plate 7), the Darling Ranges of Western Australia, and so on. There is no question of these forms being of any great age.

But just as clearly other bornhardts are ancient. Some have been preserved by burial and have recently been exhumed. Minor examples have been described from the Murray Gorge in South Australia where the bornhardts are of Miocene age (Twidale, 1968b, pp. 121-123; Twidale, Lindsay and Bourne, 1978). Exhumed Precambrian forms have been described from northwest Scotland (Williams, 1969), gneissic bornhardts of similar age have been observed in the Groot Kariesberg of southeastern Namibia and southern Namaqualand. But, despite their great age, these features have been preserved by burial and though interesting do not enter into the present argument.

Other bornhardts are of great age and appear never to have been buried, but rather to have been subject to exogene processes ever since exposure. The Everard and Musgrave ranges of central Australia stand above a plain that carried remnants of silcrete, a duricrust that developed in the early-mid Tertiary (Wopfner and Twidale, 1967; Wopfner, Callen and Harris, 1974; Hutton, Twidale and Milnes, 1978). These granitic residuals stood above the plain on which the silcrete formed by weathering, and so are at least 20 ma, and may be as much as 60 ma old. That the latter age is more likely suggested by the demonstrated Paleocene age of the nearby Ayers Rock and the Olgas (Twidale and Harris, 1977; Twidale, 1978c).

In Western Australia some bornhardts stand above a widespread lateritised land surface. Though there has been considerable argument it now seems clear (Fairbridge and Finkl, 1978) that this palaeosurface is of at least Eocene age so that such residuals as The Humps (Plate 17), Althorpe Peaks, and several in the Kondinin area and in the Western Pilbara are at least of that age, i.e., some 60 ma old. Correlations with duricrusts in southern South Australia where the silcrete is again early-mid Tertiary and the laterite is demonstrably older than Middle Jurassic (Daily, Twidale and Milnes, 1974) suggest that some parts of the higher inselbergs of Eyre Peninsula (Fig. 8.11) are of early Tertiary and even Mesozoic age (Twidale and Bourne, 1975a).

That some inselbergs are very old is not a new idea. Willis (1936) considered that some East African inselbergs may be of Palaeozoic age and Dresch (*in* Birot, 1958) regards some of the Saharan forms as very old. Landforms of such antiquity pose problems (Twidale, 1976b) for what is suggested is that they were features of the land surface before the granitic rocks of the Sierra Nevada for example were emplaced. In the present context however the antiquity of some inselbergs points to their having developed according to the two-stage model, and the

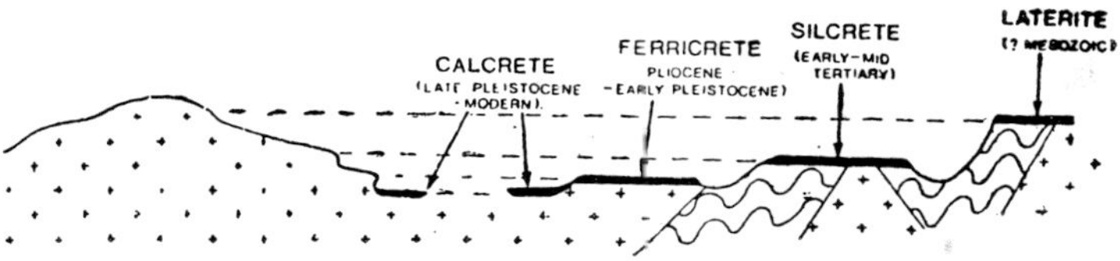

FIG. 8.11 : *Diagrammatic section through northern Eyre Peninsula showing relationship between duricrusts and flared zones on granitic residuals*

forms still being exposed are self-evidently developing according to the two-stage mechanism.

MODIFICATIONS OF BORNHARDTS

It is argued here that weathering of bornhardts, particularly subsurface weathering, produces nubbins and castle koppies (Twidale, 1980f).

Nubbins

Nubbins are boulder or block-strewn residuals. They are roughly conical, though some are flat-topped. At first sight there seems to be little or no order in the arrangement of boulders and blocks : the surface of the residuals appears to consist of a chaotic jumble. But closer examination shows that the disorder is only skin deep : at a depth of only a few metres at most the residual is massive, and moreover at many sites the core mass is seen to be domical. The boulders and blocks are merely the outer shells, or at most two shells, of bornhardts that have broken down into large more or less angular fragments.

Nubbins are developed either in wet sites or in warm humid conditions. Thus they are the characteristic granite residuals in Mount Bundey area, near Darwin, N.T., in the western Pilbara of Western Australia, in north-western Queensland (Plate 3) and in Hong Kong and the New Territories (e.g. Shatin area) all of them tropical, hot and seasonally humid (monsoonal).

But nubbins also occur in arid areas where local conditions are suitable, though it is not clear whether they merely persist, or whether they developed, in aridity. For example in the northern suburbs of Alice Springs, N.T., the granite outcrop, with many nubbins, lies in a virtually enclosed basin between sandstone ridges so that the outcrop occupies a relatively wet site regardless of regional climate. Even if, as is likely (see Kemp, 1978) the differential subsurface compartment weathering to which the residuals fundamentally are due took place in humid climates, the same arguments hold good: the site is and has long been a wet one. Again Corrobinnie Hill, though not a true nubbin is nevertheless very bouldery and is moreover bordered by a broad platform (Bourne, Twidale and Smith, 1974) both features being due to the location of the residual in the (fault-line) topographic low known as the Corrobinnie Depression. Many, though not all, of the residuals developed in the Swakop Valley, in central Namibia south of Karibib, are nubbins (Fig. 8.2). They are located in a wet site for the Swakop is the major waterway of the region but the significance of structure is again underlined, for even here bornhardts have persisted, presumably in the deep compressional cores of the antiforms, where the bedrock is sufficiently massive.

Barton (1916), Bain (1923) and Griggs (1936) have all suggested that granite is more susceptible to moisture attack than it is to weathering under dry conditions, and this is supported by field evidence. In the Witrivier area of the eastern Transvaal and the Kamiesberge of Namaqualand it is quite common to find nubbins low in the relief with domes dominating the upper levels though it is a fair to comment that there are lateral variations also, related to contrasts as between granite and gneiss. Even the same residuals change character according to elevation. Thus Rooiberg, a prominent dome in the Kariesberge, is partly gneissic and the cleavage has been exploited to give the slopes a ribbed appearance, but the exploitation of cleavage is much more pronounced below the level of shoulders that denote an old planation surface that is preserved in the region (Plate 16). Similar vertical zonation can be observed in the gneissic Reynolds Range, north of Alice Springs.

This argument implies that the disintegration of the outer

shell or shells of incipient bornhardts to produce nubbins takes place beneath the land surface under attack by moisture; and it may be so for on Tcharkuldu Hill, and on Little Wudinna Rock, both on northwestern Eyre Peninsula, on Paarlberg, near Cape Town, and elsewhere granite blocks, slightly rounded and still *in situ* can be seen in process of exposure as the soil cover is stripped away (Plate 18). At Naraku, northwest Queensland, and in the Georgetown district, some of the nubbins are flat-topped and the planate summits could be relics of old planation surfaces, in etch form; blocks and boulders located high on the residuals are pitted (Twidale and Bourne, 1976) again suggesting subsurface wheathering; the known denudation chronology of the regions is also consistent with the interpretation (Twidale, 1956, 1966). Likewise the Swakop Valley is dominated by a palaeosurface of low relief below which the gneisses could have been differentially weathered, and in the Alice Springs area there is ample evidence of old planation surfaces standing higher than the granite nubbins (see e.g. Mabbut, 1965).

But on some domes (e. g. the Everard Ranges; the Kamiesberg in Namaqualand; Enchanted Rock, central Texas) angular, unweathered remnants of such sheet structures rest in orderly arrangement on the slopes, indicating that the cubic or quadrangular blocks are *in situ* (Plate 19). The question arises as to why such arcuate slabs should split into orthogonal blocks. Though orthogonal sets and sheeting joints are the dominant fractures in granite masses the first named in broad view form a radial and tangential pattern and other radially-disposed weaknesses such as cross-joints, hard-way planes (Cloos, 1923) and joint fans developed through upthrust and arching (Balk, 1937) have been recognised. These structural planes may have been exploited by weathering to produce the ordered sets of slabs, blocks and boulders so prominent on some residuals. These little rounded quadrangular blocks and slabs are the result of subaerial splitting, possibly as a result of moisture exploitation of structural planes.

Castle Koppies

General

The genesis of castle koppies has been somewhat neglected.

Their two-stage evolution has of course been described (Linton, 1955), but their relationship with bornhardts has received scant attention. Ritchot (1975, pp. 253 *et seq.*) sees castle koppies as part of a sequential series the end product of which is the bornhardt. He argues that the marginal areas of structural domes persist as landscape features longer than do the crestal zones because the latter are in tension and therefore subject to weathering and erosion. This is correct so far as it goes but it is an over simplification of the situation and takes no account of the environments in which castle koppies occur. Godard (1977, p. 101) on the other hand considers that inselbergs of the castellated type (*inselbergs de poche*) are either large tors or the cores of inselbergs remaining after either scarp retreat or differential rock disintegration, but the discussion is carried no further.

Castle koppies are consistently smaller in area than are bornhardts, and in several areas (Gokemere, Domboshawa, both in Zimbabwe; and Waulkinna Hill, on northern Eyre Peninsula for example) there are domes with marginal developments of boulders (Fig. 8.12). This suggests that castle koppies may be bornhardts that have suffered marked marginal attack, probably due, not as Jessen (1936) surmised, to surface runoff and stream erosion, but either to pronounced marginal weathering in the subsurface, in the groundwater zone, or to aggressive exogene attack under certain conditions.

Marginal Attack

Subsurface weathering in the piedmont zone finds its most dramatic expression in flared slopes (Twidale, 1962; 1968a; 1971, pp. 90-96; 1976a, pp. 7-17) which are developed not only on granite but in other lithological environment also. They have been observed in many parts of the world though they are best developed in southern Australia.

As has been mentioned previously flared slopes are a particular form of the weathering front developed in the scarp foot zone (Twidale, 1962, 1967). They are best developed in subtropical semiarid conditions, (but also form in cold area's such as the Sierra Nevada and in true aridity) and in areas of tectonic and geomorphological stability where there has been

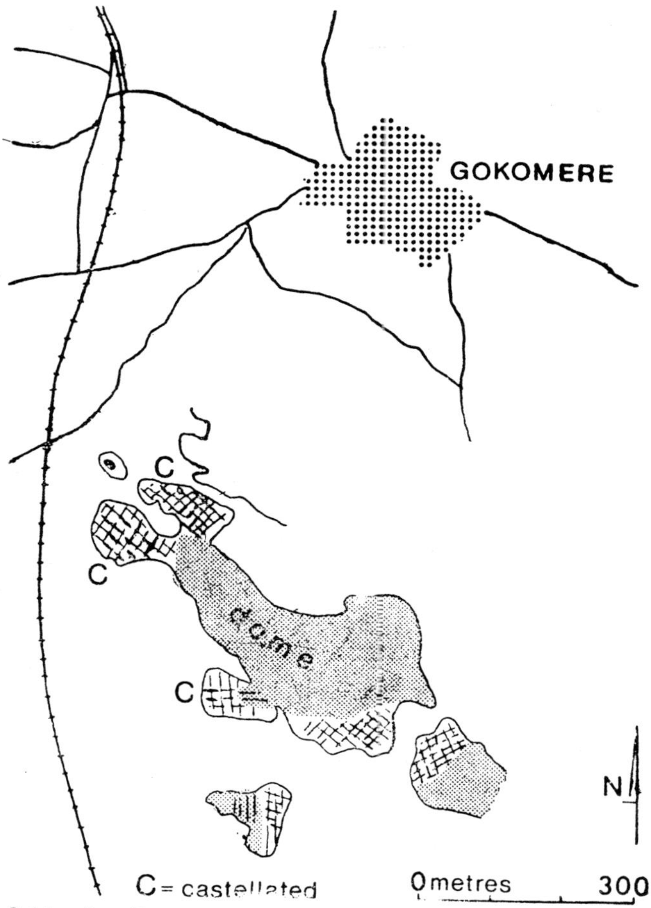

FIG. 8.12 : *Inselberg near Gokomere, Zimbabwe showing marginal disintegration of dome to form boulder-strewn nobbin areas (drawn from air photographs)*

time for pronounced marginal subsurface attack by soil moisture (Twidale, 1972).

Repeated alternations of subsurface weathering and lowering of the plains have allowed the development of several flares on some slopes (Plate 20). Some such multiple flared slopes, like that on the eastern side of Ucontitchie Hill, stand up to 30 metres above the plains and expose the truncated ends of several sheet structures. Clearly there has been a reduction in the area of the hill, and the width of the platforms exposed between the base of the flare and steep plunge of the bedrock marking the edge of the massive compartment indicates the amount of marginal trimming that has taken place. At some sites it amounts to some tens of metres.

The occurrence of flared slopes on such castellated forms as Castle Rock, near Albany Western Australia, and on the Witteklip granite-dome at Vredenberg near Saldanha, south-western Cape Province, South Africa (Scholz, 1946) (Plate 21) suggests that they, like their domical counterparts, have been exposed in phases.

Marginal attack is particularly effective in two hydrological settings. First, where groundwaters are alkaline, massive sheets that would otherwise remain intact are decayed. In some places, for instance Eyre Peninsula, the outer shells have all but been eliminated. Elsewhere and particularly where lime-stone overlies the granite, as at Remarkable Rocks on the south coast of Kangaroo Island, (Daily, Milnes, Twidale Bourne, 1979) and on the Mount Manypeaks complex east of Albany in western Australia, large residual blocks remain only on the crests of domes. That alkaline soil moisture is aggressive in the weathering of acid rocks is suggested by observations on the Brachina pediment, in the western piedmont of the Flinders Ranges in South Australia (Twidale, 1978a, 1979). The pediment mantle consists of cobbles and boulders of limestone and quart-zite set in a matrix of silt and sand. The cobbles and boulders carry a patina of lime, and lime is precipitated also in the creek beds. The limestone cobbles are marginally etched producing narrow platforms and an overall boss-and-shield effect. Many of the quartzite boulders on the other hand are completely planed off parallel to the surface as a result of alkaline soil moisture attack.

Second, the effects of marginal attack are most apparent where there are pronounced dry seasons or alternations of dry and wet periods of any sort. This produces a contrast between the moist, active weathering environments of the subsurface and the stable dry exposed rock surfaces (see e.g. Barton,

1916; Bain, 1923; Griggs, 1936; Twidale, 1962; Wahrhaftig, 1965; Twidale, Bourne and Smith, 1974).

For obvious reasons pronounced marginal attack and steepening of the flanks of massive compartments—future inselbergs—is also produced by long periods of landscape stability. It has been mentioned that flared slopes apparently are best and most widely developed in such conditions but the general point applies also. Prolonged weathering of course reduces the smaller compartments first, so that the Cottage Loaf, near Wudinna, on Eyre Peninsula, for example, is a small castle koppie set in a field of domed forms (Fig. 8.13).

Castle koppies also form in wet sites, particularly if weathering has been of long duration. For example the Devil's Marbles (Twidale, 1980a and g) is a group of large castellated blocky granite residuals located some 50 km. south of Tennant Creek in the centre of the Northern Territory, in the core of a breached anticline eroded in the Lower Protero-zoic Hatches Creek Group sediments (Fig. 8.14).

The granite residuals consist of large groups of massive angular blocks and isolated rounded boulders. Some of the blocky forms extend to plain level but others stand on what are obviously large radius domes (Pl. 22). Standing above the level of the bedrock plains are mesas of weathered granite which can be correlated with the late Cretaceous summit surface preserved on the adjacent sandstone ridges (Mabbutt, 1965, 1967; Hays 1967). Beneath the palaeosurface developed in an essentially closed basin bounded by sandstone ridges differential compart-ment weathering occurred. It was then that most of the castellated forms were incipiently developed. With the lowering of the land surface to the high plain level these compartments were left in relief.

A similar situation obtains in the Pilbara of Western Australia where there are many examples of domes in various stages of disintegration—high steep sided sheeted domes, orderly groups of blocks standing on domical plinths and low domes carrying a scatter of blocks and boulders. Few large domes survive, presumably because of the prolonged and intense weathering beneath the Hamersley Surface.

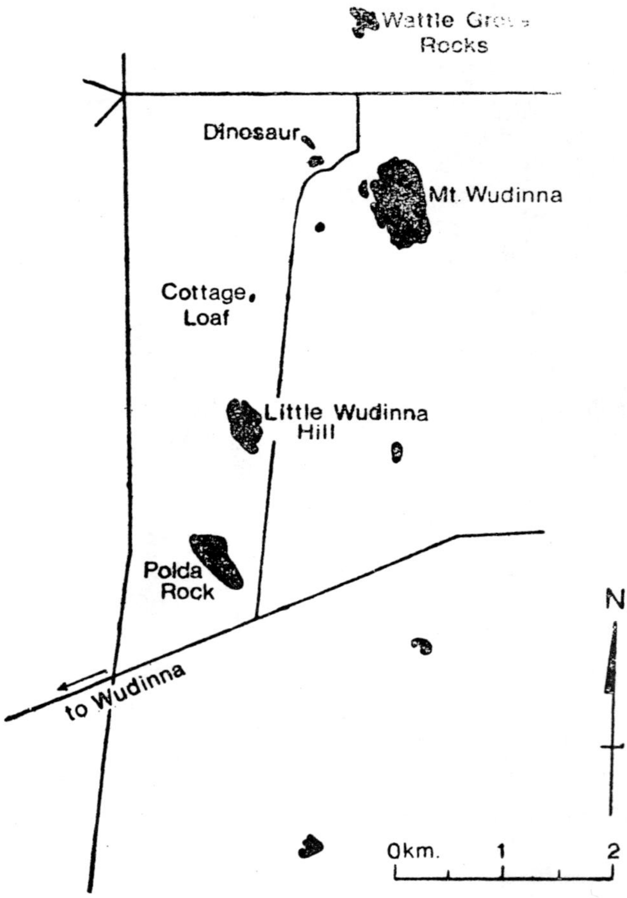

FIG. 8.13 : *Granite outcrops in the Mount Wuddinna area of north-western Eyre Peninsula, showing small areal extent of the Cottage Loaf (drawn from air photograph)*

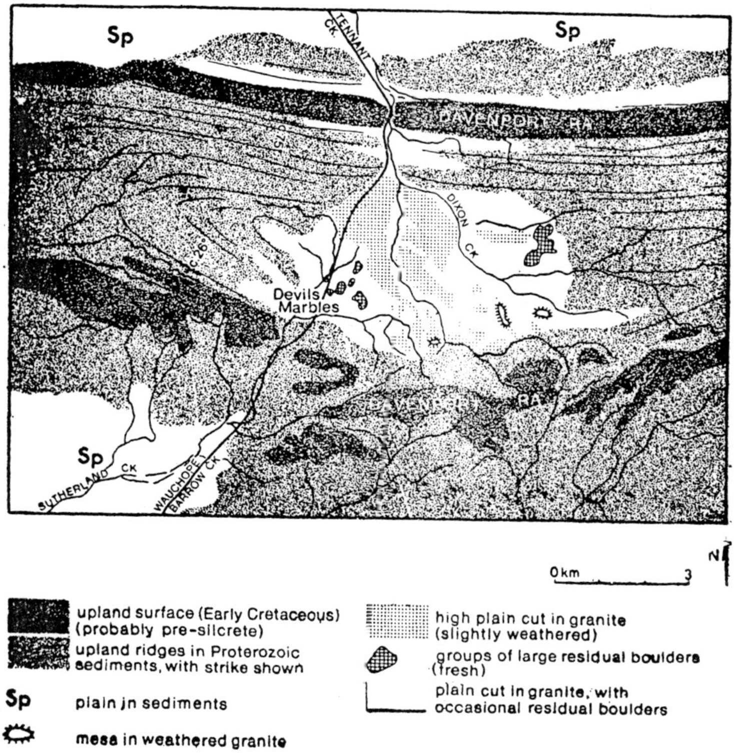

FIG. 8.14: *Morphological Map of the Devil's Marbles area, N.T.*

Nival Conditions

The buried parts of resistant rock masses are subjected to similar concentrated marginal attack in cold lands also, for soil moisture suffers alternations of freezing and thawing. But nival action on exposed residuals is probably more significant than is exogene modification in other areas. For both of these reasons, freeze-thaw action in both the subsurface and after exposure, the residuals of such areas as Dartmoor (see Linton, 1955), the Bohemian Massif (Demek, 1964), Tasmania (Caine, 1967), the high plains of the Rockies of Colorado (Eggler, Larsen and Bradley, 1969), Mongolia (Dzulynski and Kotarba, 1979), and the Pyrenees are notably angular (Plate 23). The castle koppies of the Sherman Surface in the Colorado Rockies stand in marked contrast with the domes exposed in the flanks of valleys cut into the high plain (Plate 24). The acicular forms of the Agulhas Negras (altitude 2787 m) in the Serra do Mantiqueira of Rio Province, Brazil, many come into this category, for they almost certainly were affected by frost action during Pleistocene glaciers (Plate 25).

Structural Influences

Marginal weathering attack is a significant factor in the transformation of domes to castellated forms. However it does not provide a complete explanation. In some instances sheets have all but completely been stripped away, with only a few blocky residuals remaining. Thus at the eastern end of the Manypeaks Range, east of Albany, Western Australia, the hill slopes are virtually smooth, with a few rounded boulders and sloping platforms (exposed sheets) but prominent castellated hills standing on the crests. This suggests that to some extent individual sheets above the neutral plane (Fig. 8.8) which in tension are readily weathered and eroded are attacked as entities. Once these have been eliminated the sheets in compression are exposed. They are resistant and provide a stable surface which is weathered and eroded only very slowly. This may explain why some bornhardts that are extraordinarily lacking in obvious joints (Polda Rock and Little Wudinna Hill on northwestern Eyre Peninsula; Paarl Mountain in the Cape Province, South Africa; Enchanted Rock in the Llano area of central Texas; Stone Mountain, Georgia) have essentially resisted all major weathering and erosional attack.

Though concentrated in the scarp foot zone, soil weathering is not restricted to that location. Any structural weaknesses are exploited and joints and other fractures in particular are widened by weathering in the subsurface. Thus masses that are dominated by vertical fractures in particular are especially prone to develop into castellated forms. Hence castle koppies are well developed on gneisses with vertical or near-vertical cleavage in such areas as Botswana and Zimbabwe.

Joint controlled or fault-line streams and valleys (or clefts) are a prominent feature of many granite landscapes and many of them, like those on Yarwondutta and Mount Wudinna, are

22. Large residual blocks and boulders resting on large radius dome at the Devil's Marbles, N.T.

23. Angular residual or castle koppie in the summit surface (Tosa de Gargantillar), altitude circa 2640m, in the Andorran Pyrenees

24. Angular, frost-riven castellated granite residual located on the high plain surface of the Rocky Mountains in Colorado

25. Angular granite residual in the Agulhas Negras at an altitude of 2787m in the Serra da Mantiquiera, Rio Province, Brazil

bounded by flared slopes indicative of subsurface exploitation and initiation. The cleavage planes of gneiss domes such as those of the Reynolds Ranges, north of the Macdonnell Ranges in central Australia, and of the castellated koppies of central Rhodesia, are readily exploited so that such domes are broken down into a series of angular units or castellated forms. The Shashe castle koppies consist of a complex of towers in which the strong NNW-SSE cleavage and foliation is reflected in the topographic grain (Fig. 8.5 b). However at this site there are also two or three very low outcrops which though they display the same marked structural grain as the rest of the outcrop lack the high towers of rock, and form low large radius domes, similar to those described for the Devil's Marbles.

Optimal Development

Subaerial nival attack may play a part in the shaping of castle koppies (though the castellated forms of the cold regions cited are remarkably solid and devoid of minor fretting) the massive compartments that eventually emerge are domical masses that have been modified by subsurface moisture attack, effected particularly at the margins of the masses. This takes place most readily :

1. Where the rocks have strong vertical jointing or cleavage. Hence the marked tendency for gneisses to give rise to castellated forms, as in Rhodesia, Botswana and Malawi.
2. Where there has been aggressive weathering, and especially strong subsurface weathering. This category embraces cold areas where frost action is or has been significant, areas where there is for climatic reasons a marked contrast between moist regolith and dry exposed rock surfaces, areas where soil moisture is alkaline, and areas where there has been long-continued subsurface moisture attack.

CONCLUSIONS

Granite inselbergs are of three types. Some are upstanding because they are constructed of rocks that are lithologically different (and more resistant) than those around them, and a few others are tectonic (minor horsts), but the vast majority of all types are developed on particularly massive compartments of rock.

Bornhardts are dominated by sheet structure. The residuals are formed in two stages. The first involves differential fracture-controlled subsurface weathering, resulting in the development of some zones of deep weathering and distributed amongst them as incipient inselberg relief compartments of massive rock that have been but little affected by weathering attack. In the second of the two stages erosional agents, most commonly rivers and streams, erode the land mass, the weathered compartments being lowered to baselevel, leaving the fresh-rock compartments in relief.

Bornhardts are modified by weathering to give nubbins and castle koppies. Nubbins develop in humid, particularly humid tropical, conditions, and are due to the blocky disintegration of the outer shell or shells of the domical forms. Castellated features on the other hand result from the pronounced marginal attack of bornhardts (and particularly of gneissic rocks) in conditions of especially aggressive weathering (nival action, particularly wet sites) or as a consequence of long-continued weathering.

SUGGESTED READINGS

Bain, A.D.N., 1923 : The formation of inselberge, *Geol. Mag.* V. 60, pp. 97-107.

Balk, R., 1937 : Structural behaviour of igneous rocks. *Geol. Soc. Am. Mem. 5.*

Barbier, R., 1957 : Aménagement hydroélectriques dans le Sud du Brésil, *C.R. Somm. & Bull. Soc. Géol. France*, V. 67, 2, pp. 877-892.

Barton, D.C., 1916 : Notes on the disintegration of granite in Egypt, *Jour. Geol.* V. 24, pp. 382-393.

Birot, P., 1958 : Les domes crystallins, *C.N.R.S. Mem. & Doc.* pp. 7-34.

Bornhardt, W., 1900 : *Zur Oberflächengestaltung und Geologie Deutsch Ostafrikas*, Berlin, Reimer.

Bourne, J.A., C.R, Twidale & D.M. Smith, 1974 : The corrobinnie depression, Eyre peninsula, South Australia. *Trans. Royal Soc. S. Aust.* V. 98, pp. 139-152.

Boyé, M. & P. Fritsch, 1973 : Dégagement artificiel d'un dôme crystallin au Sud-Cameroun. *Trav. & Doc. Géogr. Trop.* V. 8, pp. 33-63.

Boyé, M. & M. Seurin, 1973 : Les modalités de l'altération à la carriére d'Ebaka (Sud-Cameroun). *Trav. & Doc. Géogr Trop.* V. 8, pp. 56-94.

Brook, G.A., 1978 : A new approach to the study of inselberg landscapes, *Z. Geomorph. Supplementband* 31, pp. 138-160.

Büdel, J., 1957 : Die 'doppelten Einebnungsflashen' in den feuchten Tropen, *Z. Geomorph.* V. 1, pp. 201-228.

Caine, N., 1967 : The tors of Ben Lomond, Tasmania. *Z. Geomorph.* 11, pp. 418-429.

Chapman, C.A., 1956 : The control of jointing by topography, *Jour. Geol.*, V. 66, pp. 552-558.

Clayton, R.W., 1956 : Linear depressions (*Bergfussniederungen*) in savannah landscapes. *Geogr. Stud.*, V. 3, pp. 102-126.

Cloos, H., 1923 : Das Batholithen problem. *Fortschr. d. Gcol. u. Pal.* 1, pp. 1-80.

Cotton, C.A., 1942 : *Climatic Accidents*, Wellington, Whitcomb & Tombs.

Daily, B., A.R. Milnes, C.R. Twidale & J.A. Bourne, 1979 : Geology and geomorphology pp. 1-38 *in Natural History of Kangaroo Island* (Ed. M.J. Tyler, C.R. Twidale & J.K. Ling) Royal Soc. S. Aust., Adelaide.

Daily, B., C.R. Twidale & A.R. Milnes, 1974 : The age of the lateritised summit surface on Kangaroo Island and adjacent areas of South Australia, *Jour. Geol. Soc. Aust.* V. 21, pp. 387-392.

Dale, T.M., 1923 : The commercial granites of New England. *U.S. Geol. Surv. Bull.* 738.

Demek, J., 1964 : Castle koppies and tors in the Bohemian Highland. *Biul. periglac.* 14, pp. 195-216.

Dumanowski, B., 1960 : Comment on origin of depressions surrounding granite massifs in the eastern desert in Egypt. *Bull. Acad. Pol. Sci. 8*, pp. 305-312.

Dzulinski, ST., and A. Kotarba, 1979 : Solution pans and their bearing on the development of pediments and tors in granite. *Z. Geomorph. 23*, pp. 172-191.

Eggler, D.H., E.E. Larson and W.C. Bradley, 1969 : Granites, grusses and the Sherman erosion surface, southern Laramie Range, Colorado-Wyoming. *Am. J. Sci. 267*, pp. 510-522.

Gllbert, G.K., 1904 : Domes and dome-structures of the High Sierra. *Geol. Soc. Am. Bull.* V. 15, pp. 29-36.

Gilluly, J., 1955 : Geologic contrasts between continents and ocean basins. *Geol. Soc. Am. Spec. Paper 62*, pp. 7-18.

Godard, A., 1977 : *Pays et Paysages du Granite*, Paris, PUP.

Griggs, D.T., 1936 : The factor of fatigue in rock exfoliation. *J. Geol. 44*, pp. 783-785.

Hack, J.T., 1966 : Circular patterns and exploration in crystalline terrain, Grandfather Mountain area, North Carolina, *Geol. Soc. Am. Bull.* V. 77, pp. 975-986.

Handley, J.R.F., 1952 : The geomorphology of the Nzega area of

Tanganyika with special reference to the formation of granite tors. *C.R. Congr. Geol. Inter.* (Algiers), V. 21, pp. 201-210.

Hassenfratz, J.H., 1791 : Sur l'arrangement de plusiers gros blocs de différentes pierres que l'one observe dans les montagnes. *Ann. Chimie* V. 11, pp. 95-107.

Hays, J., 1967 : Land surfaces and laterites in the north of the Northern Territory. pp. 182-210 *in Landform Studies from Australia and New Guinea.* (ed. J.N. Jennings and J.A. Mabbutt) A.N.U. Press, Canberra.

Holmes, A., 1918 : Pre-Cambrian and associated rocks of the District of Mozambique. *Quart. J. Geol. Soc. London*, V. 74, pp. 31-97.

Howard, A.D., 1942 : Pediment passes and the pediment problem, *Jour. Geomorph.* V. 5, pp. 2-31, 95-136.

Hugo, M.L., 1970 : Inselberg – Ontwikkeling in die Pietersburgomgewing. *Tydskrif vir Aardrykskunde* V. 3, 7, pp. 715-725.

Fairbridge R.W. & C.W. Finkl, 1978 : Geomorphic analysis of the rifted cratonic margins of Western Australia. *Z. Geomorph.* 22, pp. 369-389.

Hutton, J.T., C.R. Twidale & A.R. Milnes, 1978 : Characteristics and origin of some Australian silcretes, pp. 19-39 *in Silcrete in Australia* (Ed. T. Langford-Smith), Armidale, Univ. New Engl. Press.

Jessen, O., 1936 : *Resin and Forschungen in Angola.* Reimir, Berlin.

Jones, T.R., 1859 : Notes on some granite tors. *Geologist* 2, 301-312.

Kemp, E.M., 1978 : Tertiary climatic evolution and vegetation history in the southeast Indian Ocean region. *Palaeogeog. Palaeoclim. & Palaeoecol.* 24, pp. 169-208.

King, L.C., 1949 : A theory of bornhardts, *Geogrl. Jour.* V. 112, pp. 83-87.

King, L.C., 1953 : Canons of landscape evolution, *Geol. Soc. Am. Bull.* V. 64, pp. 721-752.

King, L.C., 1962 : *Morphology of the Earth*, Edinburgh, Oliver & Boyd.

King, L.C., 1966 : The origin of bornhardts. *Z. Geomorph.* V. 10, pp. 97-98.

Lester, J.G., 1938 : Geology of the region round Stone Mountain, Georgia. *Univ. Colorado Stud.* Ser. A. V. 26, pp. 88-91.

Lewis, W.V., 1955 : pp. 483-484 *in Discussion* of Linton (1955).

Linton, D.L., 1952 : The significance of tors in glaciated lands. *Inter. Géogr. Union Proc. 8th Gen. Assemb.* 17th Inter. Cong., pp. 354-357.

Linton, D.L., 1955 : The problem of tors, *Geogrl. Jour.* V. 121, pp. 470-487.

Lister, L.A., 1976 : *The Erosion Surfaces of Rhodesia.* D. Phil. thesis, Univ. of Rhodesia.

Logan, J.R., 1849 : The rocks of Palo Ubin. *Genoots. Kunsten Wetenschappen (Batavia)* V. 22, pp. 3-43.

Logan, J.R., 1851 : Notices of the geology of the Straits of Singapore, *Quart. Jour. Geol. Soc. London*, V. 7, pp. 310-344.

Louis, H., 1959 : Beobachtungen über die Inselberge bei Hua-Hin am Golf von Siam. *Erdkunde*, V. 13, pp. 314-319.

Mabbutt, J.A., 1955 : Pediment landforms in Little Namaqualand, South Africa. *Geogrl. J.* 121, pp. 77-83.

Mabbutt, J.A., 1965 : The weathered land surface in central Australia. *Z. Geomorph. 9*, pp. 82-114.

Mabbutt, J.A., 1967 : Denudation chronology in central Australia pp. 144-181 *in Landform Studies from Australia and New Guinea* (Ed. J.N. Jennings & J.A. Mabbutt) A.N.U. Press, Canberra.

Matthes, F.E., 1930 : Geologic history of the Yosemite Valley, *U.S. Geol. Surv. Prof. Paper*, V. 160.

Mennell, F.P., 1904: Some aspects of the Matopos I. Geological and physical features. *Proc. Rhodesia Scientific Assoc.* 4, 72-76.

Oberlander, T., 1972 : Morphogenesis of granitic boulder slopes in the Mojave Desert, California. *Jour. Geol.* V. 80, pp. 1-20.

Obst, E., 1923 : Das abflusslose Rumpfschollenland im nordöstlichen Deutsch-Ostafrika. *Mitt Geogr. Ges. Hamburg* V. 35.

Ollier, C.D., 1965 : Some features of granite weathering in Australia. *Z. Geomorph.* V. 9, pp. 285-304.

Ollier, C.D., & W.G. Tuddenham, 1962 : Inselbergs of central Australia. *Z. Geomorph.* V. 5, pp. 257-276.

Price, N.J., 1966 : *Fault and Joint Development in Brittle and Semi-Brittle Rock*, London, Pergamon.

Pugh, J.C., 1956 : Fringing pediments and marginal depressions in the inselberg landscape of Nigeria. *Trans & Papers Inst. Brit. Geogr.* V. 22, pp. 15-31.

Ritchot, G., 1975 : *Essias de Géomorphologie Structurale.* Presses Universitaires Laval, Québec.

Saussure, H.B., de, 1796 : *Voyage dans les Alpes*, Nauchatel, Fauche.

Scholz, D.L., 1946 : On the younger Pre-Cambrian granite plutons of the Cape Province. *Proc. Geol. Soc. S. Afr.* 49, XXXV-LXXXII.

Schumm, S.A., 1963 : Disparity between present rates of denudation and orogeny. *U.S. Geol. Surv. Prof. Paper*, V. 454.

Selby, M.J., 1977 : Bornhardts of the Namib Desert. *Z. Geomorph.* V. 21, 1-13.

Soen, Oen Ing., 1965 : Sheeting and exfoliation in the granites of Sermasoq, South Greenland, *Med. GrΦn.* V. 179 (6).

Thomas. M.F., 1966 : Some geomorphological implications of deep weathering patterns in crystalline rocks in Nigeria, *Trans. Inst. Brit. Geogr.* V. 40, 173-193.

Thomas, M.F., 1967 : A bornhardt dome in the plains near Oyo, Western Nigeria. *Z. Geomorph*, V. 11, 239-261.

Tricart, J., 1957, Mise en point : L'évolution des versants, *L'inform. Géogr.* V. 21, pp. 103-115.

Twidale, C.R., 1956 : Chronology of denudation in north-west Queensland. *Geol. Soc. Am. Bull. 67*, pp. 867-882.

Twidale, C.R., 1960 : Some problems of slope development. *Jour. Geol. Soc. Aust.* V. 6, pp. 131-147.

Twidale, C.R., 1962 : Steepened margins of inselbergs from north-western Eyre Peninsula, South Australia, *Z. Geomorph.* V. 6, pp. 51-69.

Twidale, C.R., 1964 : Contribution to the general theory of domed inselbergs. Conclusions derived from observations in South Australia. *Trans. & Papers Inst. Brit. Geogr.* V. 34, pp. 91-113.

Twidale, C.R., 1966 : Geomorphology of the Leichhardt-Gilbert Area of North-west Queensland. C.S.I.R.O. *Land Res. Series 16.*

Twidale, C.R., 1967 : Origin of the piedmont angle, as evidenced in South Australia. *Jour. Geol.* V. 75, pp. 393-411.

Twidale, C.R., 1968a : Origin of Wave Rock, Hyden, Western Australia, *Trans. Royal Soc. S. Aust.* V. 92, pp. 115-123.

Twidale, C.R., 1968b : *Geomorphology with special reference to Australia.* Melbourne, Nelson.

Twidale, C.R., 1971 : *Structural Landforms*, Canberra, Austr. Natl. Univ. Press.

Twidale, C.R., 1972 : Flared slopes, scarp-foot weathering and the piedmont angle : comparisons between Australia, southern Africa and the western United States, *S. Afr. Geogr.* V. 4, pp. 45-52.

Twidale, C.R., 1973 : On the origin of sheet jointing, *Rock Mechanics*, V. 3, pp. 163-187.

Twidale, C.R., 1976a : *Analysis of Landforms*, Sydney, Wiley.

Twidale, C.R., 1976b : On the survival of palaeoforms, *Am. Jour. Sci.* V. 276, pp. 77-94.

Twidale, C.R., 1977 : Fragile foundations : some methodological problems in geomorphological research, *Rev. Géomorph. Dynam.* V. 26, pp. 81-95.

Twidale C.R. 1978a : On the origin of pediments in different structural settings. *Am. Jour. Sci.* V. 278, pp. 1138-1176.

Twidale, C.R., 1978b : Early explanations of granite boulders, *Rev. Geomorph. Dynam.* V. 27, 133-142.

Twidale, C.R., 1978c : On the origin of Ayers Rock, central Australia, *Z. Geomorph. Supplementband*, V. 31, 177-206.

Twidale C.R., 1979 : The character and interpretation of some pediment mantles. *Sediment. Geol.* 22, pp. 1-20.

Twidale, C.R., 1980a : The Devil's Marbles, central Australia. *Trans. Royal Soc. S. Aust.* 104. pp. 41-49.

Twidale, C.R., 1980b : The origin of bornhardts, *J. Geol. Soc. Aust.* V. 27, In press.

Twidale, C.R., 1980c : Landforms, pp. 13-41, *in Field Guide to the Flinders Ranges.* Adelaide, Rigby.

Twidale, C.R., 1980d : Origin of minor sandstone landforms. *Erdkunde.*

Twidale, C.R., 1980e, Les inselbergs en gradin et leur signification : 1'example de 1'Australia. *Ann, Geog*, In press.

Twidale, C.R., 1980f : Granite inselbergs : domed, black-strewn and Castellated. *Geogrl J.* In press.

Twidale, C.R. 1980g : Les "Boules de Diable", 1' Australie centrale, et leur signification pourle problème de la zaralité des formes granitiques. *Trav. Doc. Geogr. Trop.*

Twidale, C.R. & J.A. Bourne, 1975a : Episodic exposure of inselbergs. *Geol. Soc. Am. Bull.* V. 86, pp. 1473-1481.

Twidale, C.R. et. al. 1975b : The subsurface initiation of some minor granite landforms. *Jour. Geol. Soc. Aust.* V. 22, pp. 477-484.

Twidale, C.R. et. al. 1976 : Origin and significance of pitting on granitic rocks. *Z. Geomorph.* V. 20, 405-416.

Twidale, C.R. et. al. 1977 : Rock doughnuts. *Rev. Géomorph. Dynam.* V. 26, pp. 15-28.

Twidale, C.R, et. al, 1978 : Bornhardts. *Z. Geomorph. Supplementband* V. 31, pp. 111-137.

Twidale, C.R., J.A. Bourne, & D.M. Smith, 1974 : Reinforcement and stabilisation mechanisms in landform development. *Rev. Géomorph. Dynam.* V. 23, pp. 115-125.

Twidale, C.R. & E.M. Corbin, 1963 : Gnammas, *Rev. Géomorph. Dynam.* V. 14, pp. 1-20.

Twidale, C.R. & W.K. Harris, 1977 : The age of Ayers Rock and the Olgas, central Australia. *Trans. Royal Soc. S. Aust.* V. 101, pp. 45-50.

Twidale, C.R., J.M. Lindsay & J.A. Bourne, 1978 : Age and origin of the Murray River and Gorge in South Australia, *Proc. Royal Soc. Victoria*, V. 90, pp. 27-42.

Wahrhaftig, C., 1965 : Stepped topography of the southern Sierra Nevada California. *Geol Soc. Am. Bull.* V. 76, pp. 1165-1190.

Whitlow, J.R., 1980 : Some aspects of the geomorphology of domed inselberg terrain in Zimbabwe *S. Afri Geogr. J.* In press.

Wilhelmy, H., 1958. *Klimamorphologie der Massengesteine*, Braunschweig, Westermann.

Williams. G.E., 1969 : Characteristics and origin of a Precambrian pediment, *Jour. Geol.* V. 77. pp. 183-207.

Willis, B., 1934 : Inselbergs, *Assoc. Am. Geogr. Ann.* V. 26, pp. 123-29.

Willis, B., 1936 : East African Plateaus and Rif Valleys, in *Studies in Comparative Seismology*, Washington, D.C., Carnegie inst.

Woodard, G.D., 1955 : The stratigraphic succession in the vicinity of Mount Babbage Station, South Australia, *Trans. Royal Soc. S. Aust.* V. 78, pp. 8-17.

Wopfner, H., R.A. Callen & W.K. Harris, 1974 : The Lower Tertiary Eyre Formation of the southwestern Great Artesian Basin, *Jour. Geol. Soc. Aust.* V. 21, pp. 17-51.

Wopfner, H. & C.R. Twidale, 1967 : Geomorphological history of the Lake Eyre basin, pp. 118-142, *in Landform studies from Australia and New Guinea* (Ed. J.M. Jennings & J.A. Mabbutt) Canberra, Austr. Natl. Univ. Press.

NINE

H. S. Sharma

GEOMORPHOLOGY OF JAWAHAR LAL NEHRU UNIVERSITY, NEW CAMPUS, NEW DELHI
The Study of Quartzitic Tors

THE NEW CAMPUS of Jawahar Lal Nehru University, New Delhi, provides an excellent example for the study of morphogenic process especially weathering and surface wash and resultant features. The aim of the present paper is to provide a sound understanding of various morphogenic processes operating over a homogeneous quartzitic surface and to give a cogent account of development of quartzitic tors and slopes.

The new site of Jawahar Lal Nehru University lies southwest of Delhi metropolitan (Fig. 9.1). It occupies a nodal position on the Indian sub-continent. It is situated on the watershed dividing two mighty river systems—of the Ganga draining into the Bay of Bengal and of the Sindh falling in the Arabian Sea. The vast riverine plains of the sub-continent narrows down in this sector between the Thar-Aravalli barrier, projecting itself from the southwest and the Himalayan outliers thrusting themselves from north, thus giving Delhi Metropolitan a character of broad alluvial neck and undulating plateau in the south-southwest.

Physiographically, it is a extreme northeastern culmination of the Aravalli mountain system. The ridge which is the culminating spur of the Mewat branch of the Aravallis, constitute the most significant physiographic feature of the region. The ridge enters the territory of Delhi Metropolitan from South and extends like a "Lean but Wiry Finger" straight to the Yamuna in a north east direction. Skirting the city on the northwest and west the ridge appears like a rampart of the huge fort provided by the nature to defend the heart of India. A branch

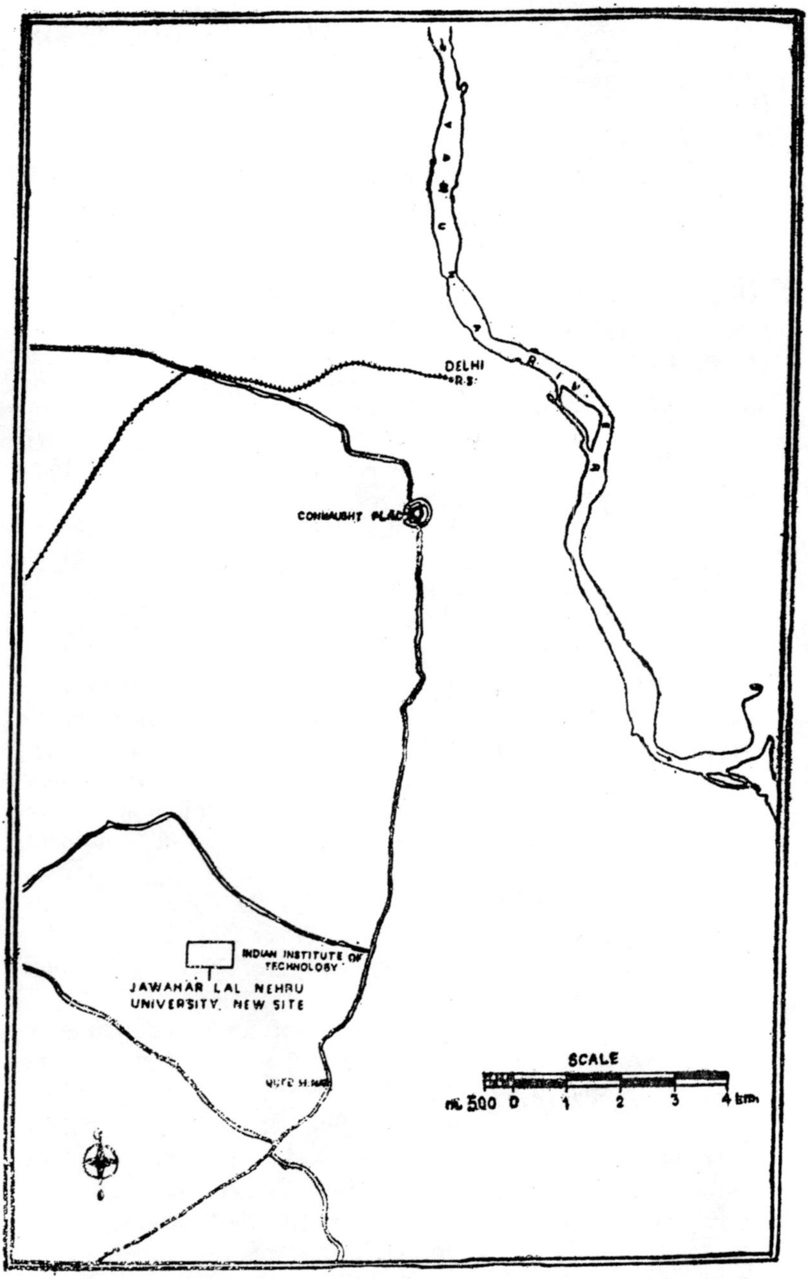

FIG. 9.1 : *Delhi Metropolitan*

of the ridge separates itself from the main trunk and extends in a north eastern direction upto Arangpur. The new site of the university lies on this branch of the ridge. It is a broad, undulated and dissected plateau of quartzites.

MORPHO-GENIC ENVIRONMENT

GEOLOGY

The rocks of the area belong to the earliest and oldest chapters of the geological history of the earth—the Pre-Cambrian and the Quarternary. The intervening records are missing here, as the region has been lying exposed to sub-aerial denudation since it rose from beneath the sea in late Pre-Cambrian times. The generalised formations met within the area are follows :

Post Tertiary — Recent and sub-recent deposits, alluvium
(Quarternary) and nodular limestone (Kankar)
——unconformity——
Delhi System — Alwar series and intrusives of pegmatites and quartz veins.

Delhi Super Group
Alwar Quartzites

Alwar quartzites are extensively exposed in the area as the elongated and discontinuous ridges which form a prominent geomorphic feature of the region. The hills, ridges and plateaus and vestiges of the major folded structure. The time worn hills and plateaus which are made up of resistant quartzites, are of special interest to the geomorphologists being the northeastern extension of the Aravalli range of Rajasthan.

In the area under investigation Alwar Quartzites form longitudinal ridges trending NNE—SSW and generally dipping 60° to 70° eastward. They form the lower group of the Delhi System and are grey to pinkish purple in colour. They are comprised predominantly of compact quartzites having immense thickness and grits and conglomerated with subordinate argillaceous bed and impure limestone. They are well bedded generally pale in colour though more or less streaked and mottled with brown red tints according to the amount of iron oxide coating the grains. Microscopically, the quartzite is usually a mosaic of interconnecting allotromorphic crystals showing little sign of original rounded grains or any secondary growth around it. Detrital pieces of muscovite and tourmaline are common. White biotite iron ore and pyrite are also found. Alwar quartzites are characterized by well developed joints and fractures. In the area four sets of joints are exhibited. In order of abundance they are :

1. Bedding joint — striking NNE to SSW and dipping 50° to 70° eastward.
2. Dip joints — striking NNW to ESE and dipping steeply up to 85° toward north and also towards south.
3. Diagonal joints— which cut across the other set of joints and dip at 20° to 40° and
4. Striking joints— dipping NNE to SSW with a dip eastwards at 20° to 35°.

These joints are open at the surface and tend to close with the increase of depth. The bedding joints are most prominent and show a varying opening from 1.5 m to 150 m and even more. The bedding joints are generally spaced at a distance of about 1 metre apart while the dip joints are spaced from 1.5 to 8 metres apart. The various types of joints present in Alwar quartzite play a dominant role in the evolution of weathering tors as quartzites break with sub-concoidal fractures and sharp splintary edges showing clearly vitreous quartz veins.

Pegmatites

In some patches thick pegmatites also occur. These rocks consist of white kaolinised felspar biotite, a little quartz and very scanty tourmaline.

Quartenary Deposits

The quarternary deposits laid on the Delhi Quartzites consist of terrestrial material such as alluvium, grit and gravel deposited by fluvial action. These deposits are of two divisions—the older division called the Pleistocene and the younger, the recent which began about 12,000 years ago since the withdrawal of the last glaciation from the Himalayas. In the present areas only older alluviums (Bangar) are found as younger alluvium (Khadar) extensively occurs in the eastern part of the Delhi ridge and north of Delhi along the course of the river Yamuna. Older alluvium in JNU campus is generally composed of poorly distributed clays, silt, sand and secondary carbonates of lime. The Kanker (lime-concretious) are generally associated with the older alluvium and occur at variable depths below the surface. The water in the older alluvium occurs in the interstitial spaces between the constituent particles of the sand. Because of the presence of lime the interstitial space is reduced leading to low water storing capacity of the formations. In Delhi region as a whole the older alluvium occupies slightly higher ground (200-250 metres) while the younger alluvium is at a lower ground (200 metres). The existence of conglomerate horizons in the Bangar at different levels (older alluvium) suggest that the sedimentation of Bangar was not continuous and fluctuations of crystal movement which resulted in the uplift of the peneplained Delhi quartzitic surface and the terracing of the Bangar itself with rest of the graded flood plain of Khadar. The presence of ravines in the Bangar itself indicate the process of degradation is going on even now.

Climate

Climatically Delhi is a part of semi-arid climate. The summer and winter are extreme as compared to most of the parts of the country. The mean minimum and mean maximum temperature are 27.6°C and 45.2°C in summer and 7.5°C and 22.2°C in winter respectively. Both daily and annual ranges of temperature are quite high. Dust storms are common during summer when day temperature in May and June commonly exceeds 40.5°C. During this period the mean wind velocity is also quite high with the value of 10.6 km/hour. The mean annual rainfall is 754 mm with the maximum 1230.1 mm and the minimum of 321.8 mm. Of the total rainfall 90 per cent occurs during the months of July to September by southwest monsoon and the rest during the winter. The rainfall in this is generally erratic and not too frequent. The mean rainfall pattern shows that it is maximum in July and August. The comparison of rainfall pattern with the evaporation rate reveals that evaporation is more than rainfall during the months of October to May.

GEOMORPHIC HISTORY

It has been pointed out earlier that the region is a northeastern prolongation of the Aravallis, therefore, the geomorphic history of it is closely related with the origin, evolution, denudation, peneplanation and rejuvenation of the Aravallis. Being also very close to the Himalayas the quarternary history of the region is associated with the various phases of the Himalayan orogenesis. The most authoritative work of Heron (1938) on the geomorphology of Rajasthan substantiated and supplemented by Depawati Sen (1972) reveal the existence of residual erosion surfaces in the Aravallis. Heron (1925, 38, 53) has recognized three erosion surfaces in this vast tract which are of Pre-cretaceous (675 mts), Mid-miocene (400 mts), and Pleistocene surface. Heron himself was not sure of the existence of later two surfaces for want of precise datable evidence like paleosols, fossils and lateritic soil covers. The hill tops of Delhi quartzites bear no soil cover. Deepawati Sen has applied the concept of dynamic equilibrium and emphasized that the existing topography of the Aravallis may have formed by non-cyclic erosion, where within a single erosion system the altitude differences produced by the Delhi Orogeny continued to exist, making the hypothesis of peneplanation and doming unnecessary to explain present day relief.

In the light of this above review the study of erosion surfaces with the help of morphometric techniques like superimposed profiles, projected profiles and altimetric frequency histogram of the Aravallis and JNU campus has also been made (Fig 9.2). Various techniques indicate the presence of two surfaces in the region, the higher surface which is represented by JNU hill, Moradabad hill, Qutub hill, Delhi University hill and other rocky flats, while the lower surface is represented by the river valleys and foothills. The field study reveals that the higher surface is essentially a planer surface and the lower one is aggradational in origin. The higher surface is a vestiges of quartzitic anticlines, thus they can be considered as flat bevelled surface. For lack of precise datable sediments on the higher surface, it is very difficult to ascertain the age of the surface. However, the existence of terrace deposits like chert and Jasper from various pockets of the region under investigation, indicate the existence of wide spread flood plain at this level which has later on uplifted in the quaternary period. The quarternary upliftment has also been indicated with the existence of Bangar and Khadar. The presence of calcium carbonates-pan in the sub-surface of the lower surface may also indicate the existence of a planer surface at higher elevation. Thus, the higher surface may be considered as a planer surface of erosional in nature and the lower one is of deposition.

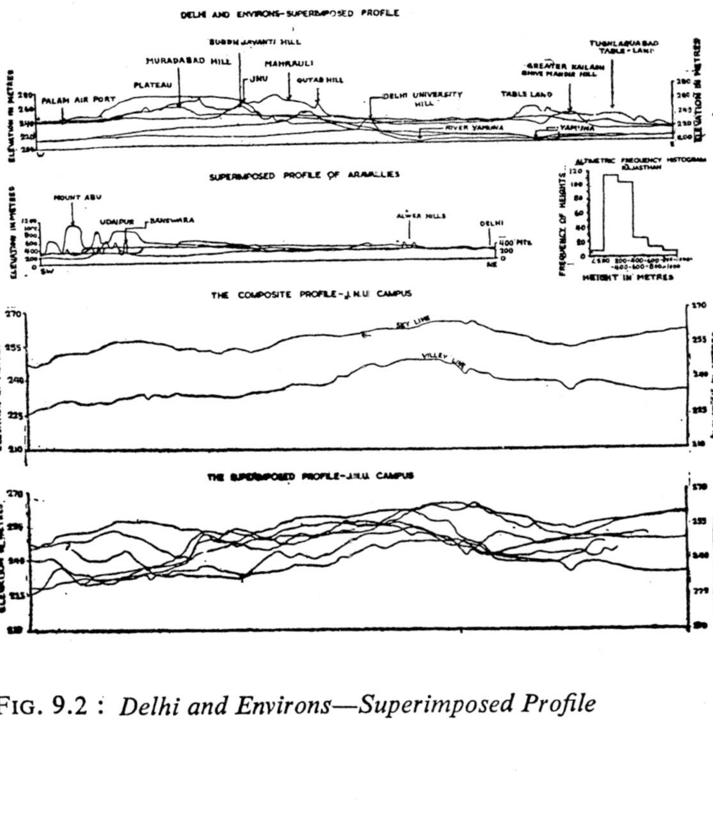

FIG. 9.2 : *Delhi and Environs—Superimposed Profile*

MORPHOGENIC PROCESSES

In accordance with the present semi-arid climatic conditions there are two exogenic processes-weathering and fluvial-have been constantly transforming the existing topography of this part of the Aravallis. The detailed description of these are given below :

WEATHERING PROCESSES

Weathering is a very effective morphogenic process in the JNU campus. Both the physical and chemical weathering processes are operating but the physical disintegration is a dominant process.

Physical Weathering Process

The arid nature of climate characterized by high range of annual and daily temperatures, scarcity of water, homogeneous quartizitic rocks attended by various sets of joints and cracks provide vulnerable sites for various forms of physical disintegration. It has been referred earlier that the quartzitic rocks are characterized by well developed bedding joints, dip joints and diogonal joints which provide vulnerable places for the attack of temperature effect. Moreover, the quartzites are white pale, grey, pale-pinkish in colour more or less mottled with brown and reddish tints due to the grains of iron oxides and oxidised pyrite, therefore, they break up with sub-concoidal features and sharp splintary edges. In the region under investigation the first state of physical disintegration is the fracturing of the rock mass with block disintegration along with joints and cracks. Later on the rectangular blocks due to presence of various minerals and their conductivity in their coefficient of expansion and in their ability to absorb heat lead to the loosening of quartzitic grains through the process of granular disintegration. Here it is significant to mention that the range of temperature and the composition of the rock play a considerable role in loosening the grain but it has also been thoroughly investigated that the water is also adding its contribution in the process. In fact temperature variation is important in the sense that it increases the permeability of the rock, so aiding the penetration of water into grains. Therefore, it is emphasized that the hypothesis that the granular disintegration is the result of temperature variation seems far from truth. As a matter of fact it should be accepted in its modified form analysing role of water in this process.

The next stage of physical disintegration is—exfoliation. It is referred above that temperature variations increase the permeability of rocks by opening of minute joints, fissures in the grains. It has also been investigated along the road cuttings that the bare quartzitic rock with steep slopes retain moisture only for a short time. But at a depth of a few centimetres and even more up to 2-3 metres water penetrates through grains and joints and persists for a longer period and initiates clay-decomposition leading to swelling of the interior section. At the same time the ferro-silicic solutions are drawn to the surface by capillary action where they precipitate to a veneer. The increase in volume resulting from clay decomposition leads to the detaching of the scale. Thus, due to this process of exfoliation accompanied by hydration the cubical blocks convert into rounded blocks or core-stones. It is thus concluded that the process of weathering in the region is very complex. Various physical and chemical processes and inter-related joints giving rise to various surface forms of weathering.

Fluvial Process

The weathered material of quartzitic masses is carried from higher surface to lower one by various terms of fluvial processes. There are no perennial streams evidenced in the region but it can be deducted through the alignment of quartzitic ridges and surface slope that most of the small rivulets flow between elongated ridges. The rivulets of this part are part of the Mahroli Basin (Sen, 1976). In this region all waters generally collect through main drains, link drains and small rivulets and discharge into the Yamuna. The abandoned clay-pits generally become reservoir and water persists in them for a pretty long time in a year. At the foothills of Parthasarthi hill the clay-pits have become a reservoir. The most significant amongst them is towards east which is of a sufficient dimension and catchment (Fig 9.3). The water from adjoining ridges rushes down and collects in the form of an elongated lake.

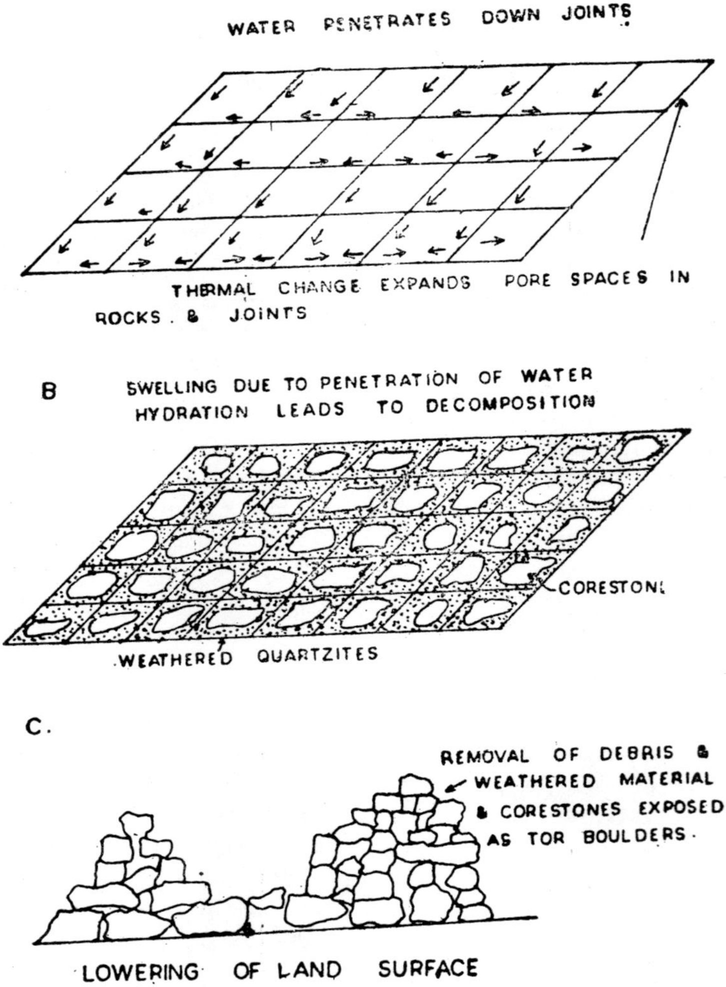

FIG. 9.3 : *The two-stage development of Quartzitic Tors*

Surface Wash

Amongst various fluvial processes surface wash is very significant and effective morphogenic process in this part. It effects individual fragments of detritus and works at a speed, i.e. infinitely greater than the speed of rock disintegration and decomposition. During rainy season whatever material is disintegrated and decomposed in the form of grain and clay is being washed from higher area to the lower. It carries only material size of the grain and clay. For material size of gravel it is less efficient. Therefore, surface wash is a selective method of transport but is a significant agent of differential erosion. In the region it is still relatively efficient on a slope of 1°.

Pedogenic Processes

The soils of the area are product of various pedogenic processes. The soils comprise of very deep, light brown colour. Textually these are clay, clay loam and loam with single structure. The thickness of the soil is very less on the summit plains but it is thicker in the valley hills. Three horizons are developing in the soils of plain. The presence of lime concretion is a common feature of lower horizon. The surface soils are high and a few are medium in texture and the fineness increases with depth. The surface soils comprise mostly of ferrogenous lime quartzites, grit and schistose rock. The epidotes and other calcium bearing minerals have the tendency to be replaced by the ferro-magnesium silicates causing the formation of calcium carbonate and separation of iron oxides. The relative order of high minerals in the soils are—quartz > orthoclase feldspar > plagioclase feldspar > microline feldspar > mica. The upper soil layers are relatively more weathered as indicated by the lesser presence of resistant minerals like quartz and zircon.

MORPHOGENIC FORM

It has been emphasized that the existing form like quartzitic tors, planar surfaces and slope is an outcome of various morphogenic processes studied earlier.

DEVELOPMENT OF TORS

The present area provides a unique example of development of quartzitic tors. Tors are essentially a pile of rounded blocks in equilibrium on a substratum. They may form at depth by joint controlled weathering within the regolith or they may form sub-areally by differential weathering and erosion of structures (Wahrhaflig, 1965). Both arid and semi-arid and periglacial climates have been claimed to be specially favourable for tors (Thomas, 1974). A regional landscape of schist tors in New Zealand and Australian has been described by Australian geomorphologist Twidale (1976). Tors are of interest to geomorphologists because they indicate that structurally controlled subsurface weathering may pre-condition in shape of a landscape long before the sub-areal surface is lowered by erosion to expose the differentially weathered terrain.

In the region under investigation numerous tors are exhibited throughout the surface. The Parthsarthi tor is an excellent example and most significant amongst them. It seems that these rounded tors are derived from cubes delimited by joints. It is evidenced due to the presence of clay pits that the development of tors is produced by crypto-decomposition in a more humid climate than that of the present and more favourable to clay decomposition.

After taking into cognizance various evidences and analysing various morphogenic processes it has been found that the origin of tors on resistant quartzitic rocks is a complex process. Observations along road cuttings and the clay-pits in pegmatitic intrusions reveal that tors are well developed on areas where the joint blocks are reasonably large and that the intervening flats have formed where the joints are close together. Tors occur where the joint-blocks are in some cases delineated by orthogonal joints and therefore cubic in shape. They are few metres in diameter. There marginal areas are weathered but the centre of each block remains untouched and is occupied by a corestone of fresh quartzitic shape like a sphere or a triaxial ellipsoid. The rounded shape derives from the greater susceptibility to granular disintegration and exfoliations of the corners and edges of cubic and quadrangle blocks where more surfaces are exposed to rain water.

Subsequently, due to removal of weathered material the land-surface is lowered. But the corestone are too massive to be transported and so remain strewn over the landscape as boulders. Some remain perched blocks. Thus it can be suggested that first there is a direct relationship between joint-spacing and landform and second that tors in many areas have developed in two stages, differential sub-surface weathering followed by differential erosion under semi-arid morphogenic environment. Thus it can be concluded that quartzitic tors develops in two stages has also been suggested by Twidale (1976) :

— Differential sub-surface weathering controlled by joint pattern, and
— Lowering of base level and differential erosion with removal of weather quartzites thereby exposing the corestones as tors boulders (Fig. 9.3).

General agreement now has been reached that tors are truly the result of semi-arid morphogenic processes. With high year round temperatures and heavy seasonal rainfall, chemical weathering extends to a great depths. Masses of fresh rock that are bypassed by the deep weathering are eventually exhumed as tors. Thus it can now be concluded that structural control and climate combine to produce one of the more striking feature like tors.

Slopes

Slope analysis forms one of the most significant aspect of geomorphology. The regional slope analysis of JNU campus has been done applying Wentworth's method. It has been observed that the slope values ranges from 0.15° to 1.43°. Maximum slope above 1° is in the southern part. In the northwestern part of the slope is also less than 0.51°.

The frequency histogram of slope (Fig. 9.4) values has been prepared. It indicates that the maximum frequency of slope ranges from 0.30° to 1.0° out of 160 samples 107 falls within the limit of 0.30° to 1.0°. Minimum frequency is above 1·30° i.e. only 4. The frequency between 1.00° and 1.30° is moderately high, i.e., 30.

The slopes of the tors and elongated quartzitic ridges moderately high between 30 to 35 degrees from the horizontal plain. It appears that due to the physical disintegration undergoing process of granular disintegration and exfoliation and surface wash the slope retreat is evidenced. In terms of slope elements there are only three elements namely convex (upper part), vertical (rocky face) and immediately below it is a pediment surface.

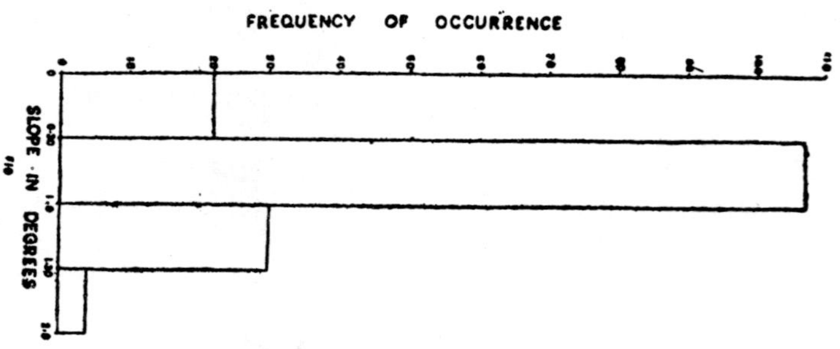

FIG. 9.4 : *Slope Histogram of J.N.U. Campus, New Delhi*

LAND UNITS

A detailed geomorphological map (Fig. 9.5) has been prepared with exhibits three broad land units in the area under investigation. These land units are characterized by varied morphological characters.

Planar Surface

It is a widespread land unit which is characterised by flat undulated surface dotted with quartzitic tors (both developed and developing). The surface slope ranges between 0.30° to 1.00°. The surface soils larger is very thin mostly covered by disintegrated quartz grains. It is a planar surface which has been produced by continuous differential weathering and erosion. In some pockets there are flood plain deposits like chert and Jasper which indicate the existence of widespread flood plain supporting the existence of a planar surface.

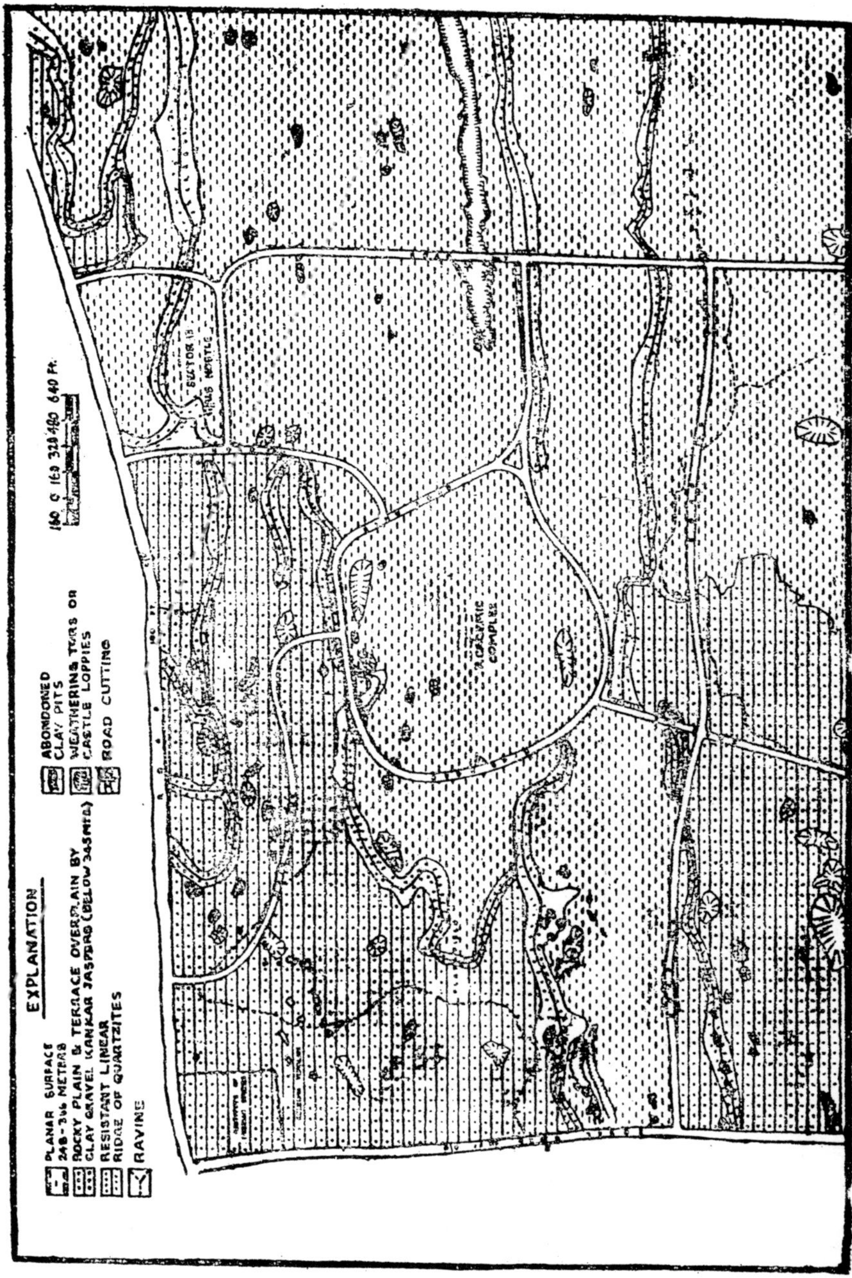
EXPLANATION
PLANAR SURFACE
ROCKY PLAIN & TERRACE OVERPLAIN BY
RESISTANT LINEAR
RIDGE OF QUARTZITES
RAVINE
ABONDONED
CLAY PITS
WEATHERING TORS OR
CASTLE LOPPIES
ROAD CUTTING
ACADEMIC
COMPLEX

Resistant Linear Ridges

An eroded anticlines in the form of linear ridges also form a conspicuous feature. These ridges are made of quartzites and run from southwest to northeast. They have also been detached and isolated due to continuous weathering and erosion and have converted into tors and isolated hillocks.

Aggradational Plain

It is usually an inter-ridges plain formed of clay, silt and other flood plain deposits. The sub-surface is characterized by the presence of calcium carbonate pan which has been exposed to surface at some place due to renewed erosion. They also exhibit development of gullies which are of 2 to 3 metres in depth.

CONCLUSION AND RECOMMENDATION

In the foregoing pages various aspects of geomorphology of the new campus of Jawahar Lal Nehru University have been discussed. The history of geomorphological research bears the testimony of its important applications in the investigation, evaluation and classification of land as a resource. The foregoing analysis of landforms of JNU campus has been done in order to provide a sound understanding of the various morphogenic processes and form so as to make use in planning of the university on the basis of this study. The following recommendations may be made :

1. It has been studied that the Kankar layer is generally associated with the older alluvium. Therefore, the water in the older alluvium occurs in the interstitial spaces between the constituent particles of the sand. Because of the presence of lime the space is reduced leading to low storing capacity of the deposits. Hence the older alluviums are generally poor water bearing formations.
2. It has also been studied that the quartzites are characterized by various sets of joints therefore in such rocks the ground water occurs along the joints and fissures which circulate them.

3. The physico-chemical weathering has extended well below the surface up to a depth of 5-6 metres and solid quartzitic rectangular blocks have weathered into corestones. Therefore, while laying foundations of skyscrapers this fact must be taken into cognizance.
4. The development of an artificial reservoir in abandoned clay-pit in Purwanchal area is a good proposition in order to maintain a healthy ecosystem in the campus.

SUGGESTED READINGS

Heron, A.M., 1925 : Geology of N.E. Rajputana, *Memoir*, Geological Survey of India, Vol. XLV, pp. 1-128.

Heron, A.M., 1938 : The physiography of Rajputana: Presidential Address, *Proceedings of the Indian Science Congress*, 25th session, Calcutta, pp. 119-132.

Heron, A.M., 1953 : Geology of central Rajputana, *Memoir*, Geological Survey of India, Vol. LXXIX, pp. 1—389.

Sen, D., 1972 : Geomorphology of the Aravalli Range, Rajasthan and a reinterpretation of residual erosion surfaces. *Geographical Review of India*, Vol. 34, No. 3, pp. 262-286.

Thomas, M.F., 1974 : *Tropical geomorphology* : John Wiley & Sons, Inc. New York (a Halsted Press Book). pp. 332 (In UK by the Macmillan Press Limited).

Twidale, C.R., 1976 : *Analysis of landforms*: Joint Wiley & Sons, Australasis Printing Limited, Sydney, pp. 572.

Wahrhafty, C. 1965 : Stepped topography of the southern Sierra Nevada, California: *Geological Society of America Bulletin*, Vol. 176, pp. 1165-89.

TEN

C. R. Twidale

GEOMORPHOLOGICAL SIGNIFICANCE OF DURICRUSTS, WITH SPECIAL REFERENCE TO AUSTRALIA

INTRODUCTION

DURICRUSTS are hard surficial encrustations widely developed in tropical and subtropical lands, but preserved also in many extratropical regions. In Australia for example five mineralogically distinct and separate types are known, and others are known from other parts of the world. But whatever their composition they are geomorphologically significant in that they form caprocks; because they disintegrate to form the stony carapaces variously known as gibber, hamada, serir, desert armour in different parts of the world; because they stabilise the land surface; and because at the regional level they can be used as morphostratigraphic markers that indicate either the relative or the stratigraphic ages or age ranges of landforms and landscapes. Here, Australian examples are used to illustrate these general points.

DURICRUSTS

Several major types of duricrust are described together with various variants and minor occurrences.

Laterite is a highly weathered material rich in secondary oxides of iron and alumina or both (see for example Prescott and Pendleton, 1952; Alexander and Cady, 1962; Sivarajasingham, Alexander, Cady and Cline, 1962; Maignien, 1966). It is virtually devoid of bases and primary silicates, although in some samples there are large amounts of quartz and kaolinite. Laterite is soft when wet but hardens on exposure and drying. This is indeed the origin of the term laterite for it is used as a building material in southern India : this was observed by the early Europeans (see e.g., Buchanan, 1807) who, calling upon their classical educations coined the term laterite from the Latin *later*=a brick.

In Australia (see e.g., Prider, 1966) the typical laterite profile consists of a sandy A-horizon, 6-7 m thick but more frequently missing as a result of erosion of the loose friable material. This is underlain by a ferruginous B-horizon up to 5 m thick. It is this ferruginous zone that most commonly is exposed in the landscape. It is commonly pisolitic and vesicular and in many places displays a rudimentary stratification. It has a shiny brown or brownish purple exterior but inside the pisoliths are white, brown, yellow and red, and most commonly display several of these colours. Beneath the iron-rich layer is a zone, up to 30 m thick, consisting essentially of kaolinised clay. White, with splashes of yellow, red and brown the clay contains irregular, usually small, masses of chalcedony. Many workers subdivide the kaolinised zone into an upper mottled and lower pallid zone, but the two are frequently indistinguishable, patchily distributed or occur in reverse order, with the mottled material below the pallid. Below the Kaolinised material is the weathered bedrock.

Bauxite is similar to laterite. It is rich in alumina and is the result of intense desilication. *Ferricrete* is a ferruginous iron encrustation lacking the profile development of true laterites. There are obvious terminological difficulties for Goudie (1973) has suggested, logically but with utter disregard for historical precedent, that the various duricrusts be named according to their dominant chemistry and with the suffix—crete to denote hardness. Thus laterite becomes ferricrete, bauxite alcrete, and so on. In these two particular instances the present writer prefers to recall that there is not one logic but many, and to be reminded of bricks being made from laterite, and of bauxite occurring at Les Baux, in France. Moreover, there is the practical advantage of then retaining the term ferricrete to denote ferruginous carapaces that differ morphologically and genetically from laterite.

Many hypotheses have been offered in explanation of laterite (see Maignien, 1966). Some early workers observing their vesicular structure, thought they were of volcanic origin. Now however, though there are controversies as to detail it is agreed that laterites are weathering profiles. Both laterite and bauxite are presently forming in the humid tropics and particularly in such monsoon areas as southern India, Malaysia and Indonesia. They are formed in regions where high temperatures and rainfall and abundant products of vegetational decay induce intense desilication. The precipitation of iron oxides is a matter of debate but is generally considered to be optimised by a fluctuating water table, though as is so often the case, the feature may evolve in different ways at different sites.

Paton and Williams (1973) have suggested that laterite development is not restricted to the humid tropics and have pointed to contemporary ferruginous accumulations in temperate areas. Nonetheless it is fair to comment that the extratropical developments in question lack the full profile development characteristic of laterite and that they are of limited areal extent. They are best regarded as local ferricrete accumulations. The vast sheets of true laterite seem only to be developing and to have evolved in the humid tropics; where they are found in semiarid regions or in such extratropical regions as Tasmania and South Australia, climatic change is implied.

Silcrete is a silica-rich rock consisting of 95 per cent or more of quartz (see Frankel and Kent, 1937; Hutton, Twidale and Milnes, 1978). It is characterised by the presence of porphyroclasts of quartz set in a quartzose (hence crystalline) matrix. In hand specimen silcrete is grey or yellow-brown with patches of red and dark brown; it has a vitreous sheen and a good concoidal fracture which led to its widespread use in the manufacture of artefacts by Australian aborigines. The great bulk of silcrete occurs in thick sheets, of considerable extent, and with rounded pebbles, cobbles and boulders, mainly of quartz but in some instances of silcrete, included in the rock mass. The siliceous material is frequently underlain by kaolinised bedrock but there is no necessary genetic connection between the two; the chemistries of the silcrete and of the substrate are such that in some cases at any rate the two cannot be genetically related. An extreme case is of silcrete overlying limestone near Quorn in the southern Flinders Ranges of South Australia.

Some silcretes, minor in amount but of considerable interest occur as skins developed on blocks and boulders, usually of quartzite. They are high in such resistates as silicon, titanium niobium, zirconium, (containing for example up to 25 per cent TiO_2); aluminium, potassium and iron are virtually absent. They differ from the sheet silcrete mineralogically, chemically and genetically (see Hutton, Twidale, Milnes and Rosser, 1972; Hutton, Twidale and Milnes, 1978). Some sheet silcretes restricted to the southwest of Lake Eyre and north of Lake Torrens have an opaline matrix (Wopfner and Twidale, 1967).

The major occurrences of silcrete are in the form of quite extensive plains and plateaux and it has been assumed that they are remnants of a once-contiguous sheet. This is not necessarily so and indeed is unlikely. The sheets are characterised by the presence of fragments of alluvial origin. The situation is seen in miniature in the Beda Valley, southern Arcoona Plateau, South Australia (Twidale, Shepherd and Thompson, 1970) where the sheet silcrete with pebbles is restricted to the valley floor (Fig. 10.1). Clearly this and the other more extensive sheets represent old valley floors, and their present disposition is the result of relief inversion on a regional scale (cf. Miller, 1937). Silcrete does not appear to be forming at present anywhere in the world : it is a truly relic accumulation.

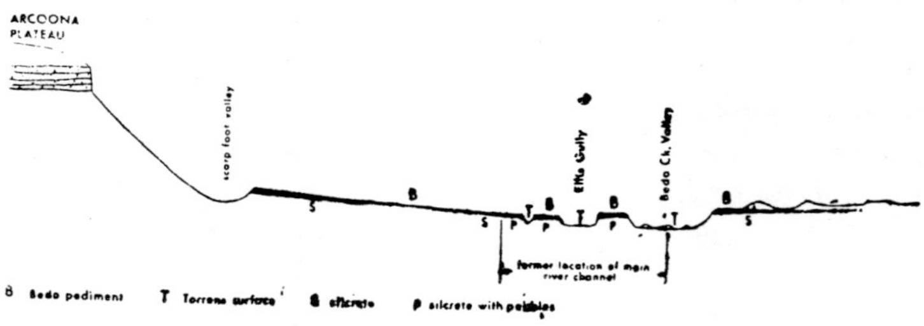

FIG. 10.1 : *Diagrammatic cross-section through the Beda Valley and Camp Hill, a remnant of the quartzite-capped Arcoona plateau (after Twidale, Shepherd and Thompson, 1970)*

Several theories, not mutually exclusive, have been proposed in explanation of silcrete. Some authors have suggested short-distance translocation, followed by concentration and precipitation of silica (Jack, 1915). Others consider that alkaline solutions aid in the coagulation of colloidal solutions of silica involved a short distance transport (Frankel and Kent, 1937). Long distance translocation has been used by Stephens (1964, 1970), who with respect to the Lake Eyre Basin has suggested the evaporation in the arid interior caused silica precipitation. Precipitation from silica rich gels in lacustrine conditions (Öpik, 1954) and concentration by primitive plants (Lovering, 1959) are other mechanisms that have been proposed. Gradual cyrstallisation of silica on stable surfaces of low relief over very long periods of geological time, probably aided by plants, is yet another suggestion (Hutton, Twidale and Milnes, 1978).

Calcrete, also known as kankar, kunkar and caliche, is a pedogenic accumulation rich in calcium carbonate. It commonly contains 80 per cent or more of $CaCO_3$ with minor amounts of quartz, dolomite and clays. Calcrete takes various forms, which can be viewed as an evolutionary sequence (Netterberg, 1971). Some calcrete consists of isolated nodules of concentric structure, and these obviously grow and coalesce so that the rock consists of a mass of lime nodules cemented together. Continuation of the process leads to the formation of a solid sheet or hardpan of lime. Further accumulation can lead to a space problem and the buckling or arching of the surficial material: pseudoanticlines that are geometrically anomalous with respect to the country rock (see Price, 1925; Jennings and Sweeting, 1961). Some calcrete has a honeycomb structure and in many deposits old roots and other structures are preserved.

Calcrete is distinguished from travertine which is lime precipitated from springs and stream waters and also from the so-called valley calcretes of West Australian geologists (see Mann and Horwitz, 1979) which are of commercial interest because of their associated uraniferous (carnotite) deposits and which seemingly accumulate from below groundwaters but along stream lines.

Calcrete is of arid as well as semiarid provenance, though it is better and more widely developed in the latter than in the former. For example, in both Western Australia and South

Australia calcrete is prominent (and is still forming) in southerly semiarid areas (in the west, south of the Menzies line—see Butt, Horwitz and Mann, 1977). The sheet (nodular, honeycomb, hardpan) forms either through the precipitation or carbonate because of variations in pore pressure, through capillary rise and evaporation, through water table oscillations, or because of organic activity.

In addition to these major occurrences *gypcrete* (Wopfner and Twidale, 1967) has accumulated over considerable areas to the west and southwest of Lake Eyre. It consists of 2-3 m of crystalline (butterfly twinned) gypsum underlain by gypsiferous silts.

DURICRUSTS AS CAPROCKS

Whatever their condition when wet, duricrusts become hard when dry, so that once dissected and left above the local or regional water table these materials come to form resistant cappings and to plateau, mesa and butte forms, depending on the degree of dissection. Gypcrete and calcrete are especially emphemeral because of their physical softness and chemical reactivity. They persist only in hot arid conditions. Laterite and silcrete however survive longer (indeed very long periods in some instances) and in a wide range of climates, though their preservation is again aided by heat and aridity. The sandy A-horizon of laterites is usually stripped away to expose the pisolitic ferriguous meterial, though in some areas, such as the Gilberton Plateau of northwest Queensland (Twidale, 1966) it is preserved beneath the eucalypt woodland. Most laterites and silcretes are essentially flat-lying, so that plateau forms predominate (Plate 1). Elsewhere they are folded, and in still other areas gently warped though it is in some instances difficult to distinguish between such distortions and the original relief. But where they are folded they again act as resistant strata and form ridges, the precise morphology of which depends on the disposition of the duricrust.

Thus, folded silcrete occurs in southwest Queensland, and adjacent areas of South Australia and N.S.W. (Wopfner, 1960). The anticlines and domes have been breached, and the silcrete forms cappings to the cuestas on the flanks of the structures

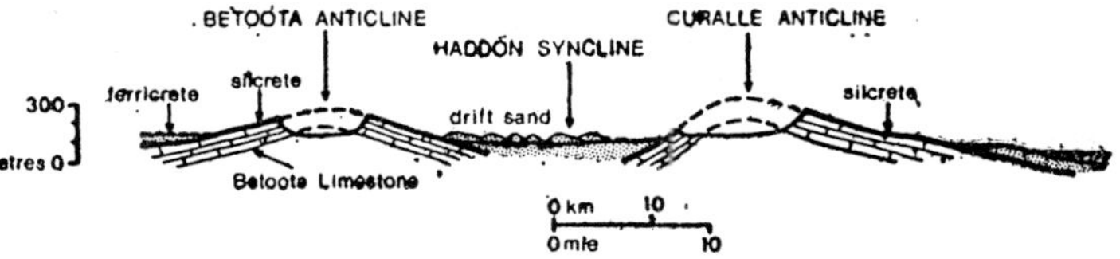

FIG. 10.2 : *Diagrammatic section through the folded silcrete involved in the Betoota and Curalle anticlines, Southwest Queensland (after Wopfner, 1960).*

(Fig. 10.2). Similar features are developed on folded silcrete near Oodnadatta (Plate 2) while in the northeastern piedmont of the Flinders the tighter folds developed there in relation to the Paralana Fault Zone form low hogbacks (Fig. 10.3).

Where dissected, as in the Ashburton Valley of Western Australia (Plate 3), and in the lower parts of the Lake Eyre Basin, calcrete also gives rise to plateaus, mesas and buttes, but elsewhere, as in the western Murray Basin, it has a stabilising effect, preserving against weathering and erosion a high plain cut in Pliocene calcarenite and of Pliocene-early Pleistocene age (Twidale, Lindsay and Bourne, 1978). Again, on Yorke and Eyre peninsulas the later Pleistocene planation surface is effectively protected by a veneer of calcrete (Twidale, Bourne and Smith, 1974, 1976).

To refer to gypcrete as a caprock may seem absurd in view of the physical softness of the constituent mineral (gypsum has a physical hardness of 2 on Moh's scale) and the solubility of hydrous calcium sulphate. But it must be remembered that though powdered (flour) gypsum is fairly soluble, the crystalline form is not readily dissolved so that, in an arid climate such as that which obtains in central Australia, the meterial is more resistant than would at first seem the case. The massive gypcrete stratum exposed in the cliffs on the western side of Lake Eyre (Plate 4) is certainly more resistant to weathering and erosion than the gypsiferous silts that underlie it, and it here forms a caprock and holds up precipitous cliffs.

Even where the duricrust is fragmented it retains a protective function, for the blocks, in the form of gibber, form a carpet which tends to be durable and hence to become more

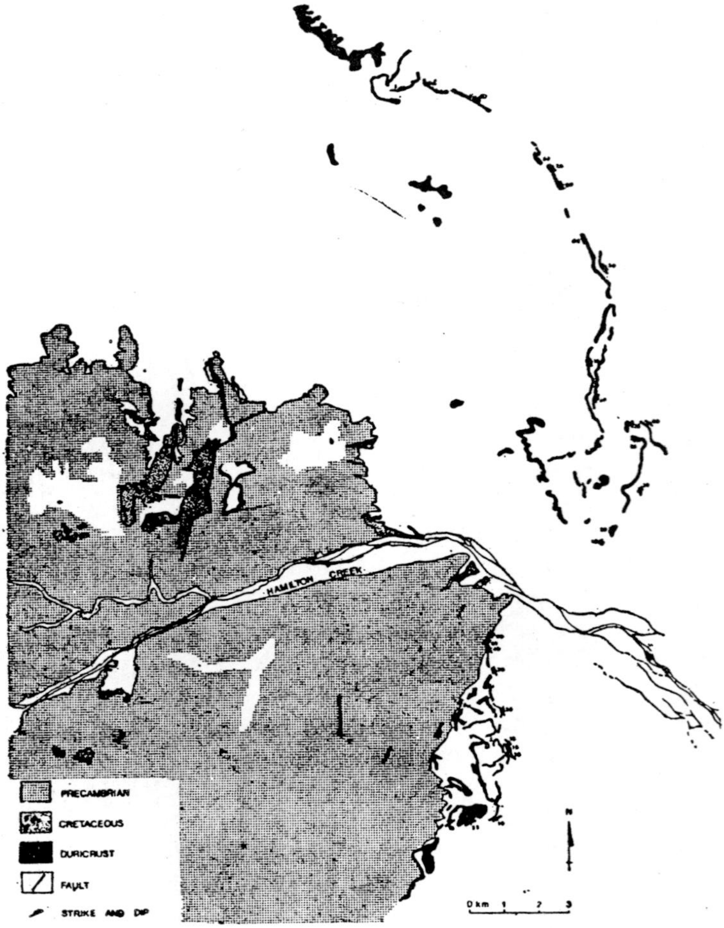

FIG. 10.3 : *Folded silcrete in the northeastern piedmont of the Flinders Ranges, South Australia (after Canpana et al. 1961*)

concentrated until it forms a secondary encrustation (Plate 5). Silcrete and calcrete, in particular lend themselves to such gibber development. Sturt's Stony Desert, the angular stones of which played such havoc with his horses' hooves during his explorations of 1844, is carpeted by stones "coated with oxide of iron", composed of "indurated or compact quartz" and in the shape of "shivered fragments" (Sturt, 1849); that is, by fragments of silcrete derived from the dissection of the original duricrust capping still preserved in adjacent areas.

DURICRUSTS AS A STABILISING FACTOR

Duricrusts are comparatively hard and stable. They thus tend to preserve the landforms and landscapes on which they are developed and that would otherwise be destroyed by weathering and erosion.

The Pleistocene peneplain that occupies large areas of central and northern Eyre Peninsula, in South Australia, is well preserved despite later baselevel lowering largely by virtue of the widely developed calcrete encrustation. It is likely to remain essentially unchanged until more humid climates prevail for only then will the lime be dissolved. Even then there is likely to be a prolonged period of karstic development rather than the direct modification (dissection) of the land surface.

Similarly the widespread remnants of laterite and silcrete preserved in central and northern Australia, for example, allow a partial reconstruction to be made of the landscapes at the time of their formation. Allowance has to be made for the sandy A-horizon in the case of most laterites (for only rarely it is still *in situ* and over wide areas it is the ferruginous B-horizon that is exposed) and for relief inversion in the case of many silcrete formations (for the sheets of silcrete accumulated in broad valley floors), but though incomplete these reconstructions indicate the general nature of the topographic conditions under which the duricrusts formed.

AGE OF THE DURICRUSTS

A distinction must be made between the laterite (and bauxite) of southern Australia and that of northern regions. Laterite forms a capping to the Lincoln Uplands, the Mount Lofty Ranges, and Kangaroo Island, all in the 'Gulfs region' of South Australia (Fig. 10.4a). The regional morphology of the region is determined by faulting (Benson, 1911; Fenner, 1930, 1931; Twidale, 1976) and it is clear from the stratigraphic evidence (Glaessner, 1953) that the most recent phase of major faulting, and the faulting responsible for the uplift of the present Mount Lofty Ranges, for example, occured in the latest Cretaceous or earliest Eocene. As the lateritised summit or high plain surface was uplifted by this faulting, the laterite obviously developed during the later Mesozoic.

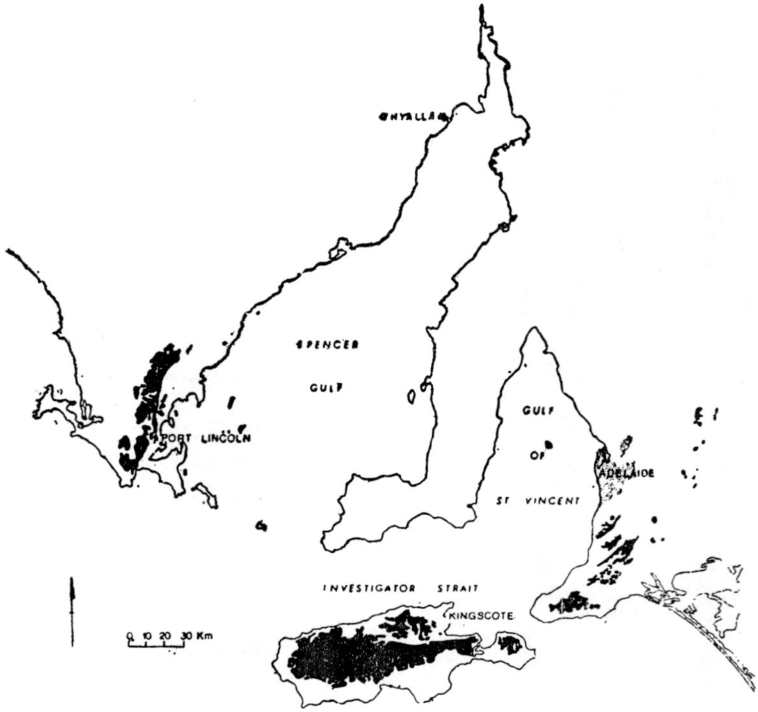

FIG. 10.4(a) : *Laterite plateaux and High Plains in the Gulfs regions of South Australia*

(From map sheets Lincoln, Kingstone, Barker, Geol. Surv. Aust.)

Some of it at least is of much greater antiquity. The laterite capping the plateau a high plain that occupies the major part of Kangaroo Island is developed on strata of Cambrian and Permian age. The latter are of glacigene origin and consist of

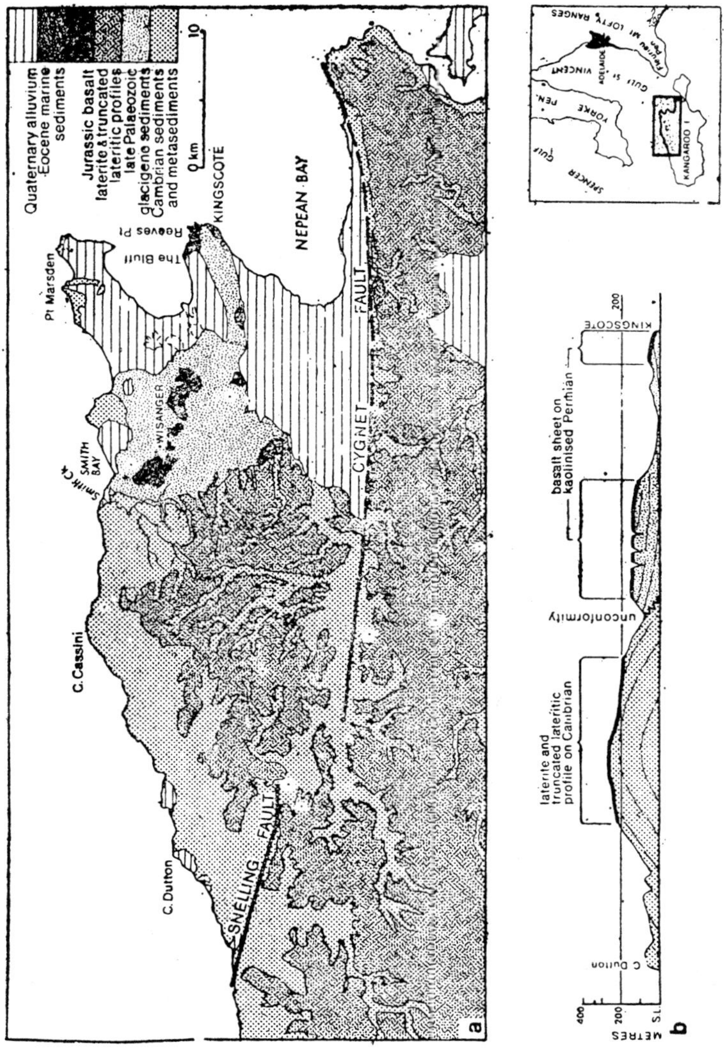

FIG. 10.4(b) : *Laterite and bedrock geology of part of northern Kangaroo Island (After Daily, Twidale and Milnes, 1974)*

sands and clays that are readily eroded. West of Kingscote, basaltic lavas were extruded on a plain cut in mottled zone materials developed on these glacigene materials (Fig. 10.4b), though to the east, near Panneshaw, the basalt overlies the ferriginous zone of the laterite. The laterite is thus clearly older than the basalt, but younger than the Permian strata on which it is developed. Radiometric datings show that the basalt is of Middle Jurassic age (Wellman, 1971) and the 'best' age for the laterite, taking into account both geological and palaeoclimatic data is Triassic (Daily, Twidale and Milnes, 1974).

In Western Australia, the laterite and bauxite of the Darling Ranges also appears to be of considerable antiquity. It has generally been considered to be of Pliocene of even Pleistocene age (Prider, 1966), but various general considerations, both geological and geomorphological, have caused that conclusion to be questioned, and more recent stratigraphic and tectonic analyses suggest that lateritisation in the area extended through the later Mesozoic and the Eocene (Fairbridge and Finkl, 1978).

Thus in southern Australia the lateritised land surfaces are of considerable antiquity. In this they are not alone for palaeosurfaces of low relief of Cretaceous age, but lacking any duricrust development, have been reported from such areas as the uplands of eastern Victoria (Hills, 1975), the Arcoona Plateau (Twidale, Shepherd and Thompson, 1970), the Cleve Hills and Gawler Ranges (Twidale, Bourne and Smith, 1976), and in various of the ranges of Central Australia (Mabbutt, 1967; Twidale, 1980a). The prominent summit surface developed in the Hamersley Ranges primarily on Precambrian banded iron formation is also probably of Cretaceous age.

In Northern Australia the laterite has a considerable age range but appears to have begun during the Cretaceous (Hays, 1967), and lasted until the Miocene. Certainly Cretaceous and early Tertiary strata have been lateritised but rocks of later Cainozoic, certainly Pleistocene and possibly Pliocene, age, have not. There are from time to time suggestions that lateritisation and bauxite formation are still active but the weight of evidence favour an early-middle Tertiary age for the process and the resultant rock.

Silcrete has a similar age range. Though Jurassic silcretes have been reported (Wopfner, 1978) and strong silicification was taking place in Central Australia during the Cretaceous (Wopfner, Callen and Harris, 1974). The main period of silcrete development was again the early-middle Tertiary (Wopfner and Twidale, 1967). The process had ceased by the Pliocene for the Etadunna formation of the Miocene or later age is not affected.

In Western Australia silicification commenced in the Oligocene (Fairbridge and Finkl, 1978). The date of its termination is not known with precision but it predates the Quaternary. Near Mount Magnet and elsewhere silcrete forms infillings in joints in the laterite.

The northern laterite and the silcrete have the same age-range, and both developed under humid tropical conditions (Ludbrook, 1969; Twidale and Harris, 1977; Kemp, 1978); but whereas the laterite mainly occupies regions of exoteric drainage, the silcrete occurs in basins of interior drainage (Fig. 10.5). In the south the silcrete is younger than the laterite, and consistently occupies lower topographic positions. Again, however, major elements of the present landscape are of considerable antiquity.

Ferricrete, gypcrete and calcrete are much younger than the laterite and silcrete. Ferricrete is developed in valley floors and basins in the Mount Lofty Ranges and on plains eroded in folded silcrete in southwest Queensland (Fig. 10.2) and on northeast Eyre Peninsula (Twidale, Bourne and Smith, 1976). Thus it clearly postdates the Miocene and stratigraphic evidence from Yorke Peninsula (Horwitz and Daily, 1958) suggests that the iron encrustations, both buried and exposed, of that and other areas is of Plio-Pleistocene age. Certainly this is consistent with all the evidence so far adduced (see also Horwitz, 1960).

The gypcrete is developed on a Pleistocene land surface is associated with Pleistocene fossiliferous sediments (Wopfner and Twidale, 1967) and predates the later Pleistocene but no greater precision is at present possible. Calcrete is still forming and has been developing for several tens of thousands of years and possibly longer. Again no great precision is at present possible.

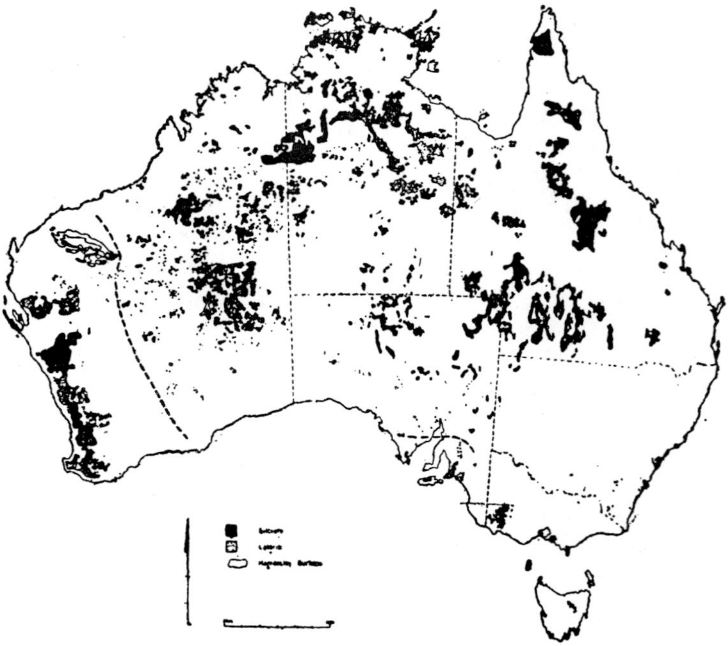

FIG. 10.5 . *Distribution of laterite and silcrete in Australia*

In Western Australia the lateritised surfaces are in west, and in south Australia those in the south of the dashed line are of Mesozoic age. The remainder and the silcrete are of early-middle Tertiary age.

Quite apart from the surfaces on which these various duricrusts are preserved being dated by their presence: a maximum age is given for valleys and plains eroded below them. Thus the plains eroded below the level of the silcrete plateaux of Central Australia are clearly late Cainozoic forms. General thought it may be, a maximum age is thus defined. It is also possible to deduce rates of erosion from such data.

Again, bornhardts can be dated with reference to duricrust. In Western Australia both in the western Pilbara and in the so-called Wheat Belt of the southeast, several of the granite residuals stand above the laterite and are at least as old as the duricrust in those areas. In the Wheat Belt for instance the Humps (Plate 6), Althorpe Peaks, and several residuals east of

1. (a) Silcrete mesa and plateau, central Australia (CSIRO)

1. (b) Butte eroded in granite capped by laterite south of Cloncurry, northwest Queensland (CSIRO)

2. (a) Ridges capped by folded silcrete : Southwest Queensland

2. (b) Oodnadatta area of northern South Australia

3. Calcrete mesa in the Ashburton valley, northwest of western Australia

4. Western fault-scarp of Lake Eyre salina showing costal cliffs capped by gypcrete

5. Silcrete gibber or hamada near Kamaran Downs, southwestern Queensland

6. The Humps, a complex of granite domes or bornhardts in the south-west of Western Australia, with laterite exposed in a shallow quarry (foreground) at its base

Kondinin are of at least early Tertiary age. On the other hand such bornhardts as Mount Stirling and Kokerbin Hill are exposed in valleys cut below the laterite surface and are just as obviously of late Cainozoic age.

On Eyre Peninsula many of the granite inselbergs are stepped, indicating phased exposure through the episodic lowering of the surrounding plains (Twidale and Bourne, 1975). Moreover, the various steps can be correlated with duricrusts of various types (laterite, silcrete, ferricrete, calcrete), preserved either on the adjacent plains or in the uplands to the east. If these correlations are correct the crests of such high residuals as Mount Wudinna and Carappee Hill are of Mesozoic age (Twidale and Bourne, 1975; Twidale, 1980b).

CONCLUSION

Duricrusts have not only a direct influence on landform development in the tropics and subtropics, but also provide information on the environments in which landscapes evolved and the age or age-ranges of their development.

SUGGESTED READINGS

Alexander, L.T. & J.G. Cady, 1962 : Genesis and hardening of laterite in soils. *U.S. Dept. Agric. Tech. Bull.* 1282.

Benson, W.N., 1911 : Note descriptive of a stereogram of the Mt. Lofty Ranges, South Australia. *Trans. R. Soc. S. Austr.* 35, pp. 108-111.

Buchanan, F., 1807 : *A Journey from Malebar through the Countries of Mysore, Canara and Malabar*. Vol. 2., East India Co., London.

Butt, C.M., R.C. Horwitz & A.W. Mann, 1977 : Uranium occurrences in calcrete and associated sediments in Western Australia. *C.S.I.R.O. (Aust) Div. Min. Rept. FP 16.*

Daily, B., C.R. Twidale & A.R. Milnes, 1974 : The age of the lateritized summit surface on Kangaroo Island and adjacent areas of South Australia. *J. Geol. Soc. Aust.* 21, pp. 387-392.

Fairbridge, R.W. & C.W. Finkl, 1978 : Geomorphic analysis of the rifted cratonic margins of Western Australia. *Z. Geomorph.* 22, pp. 369-389.

Fenner, C., 1930 : The major structural and physiographic features of South Australia. *Trans. R. Soc. S. Aust.* 54, pp. 1-36.

Fenner, C., 1931 : *South Australia : A Geographical Study*, Whitcombe & Tombs, Melbourne.

Frankel, J.J. & L.E. Kent, 1937 : Grahamstown surface quartzites (Silcretes) *Trans. Geol. Soc. S. Afr.* 40, pp. 1-42.

Glaessner, M.F., 1953 : Some problems of tertiary geology in southern Australia. *J.R. Soc. N.S.W.* 87 (2), pp. 31-45.

Goudie, A., 1973 : *Duricrusts in Tropical and Subtropical Landscapes*, Clarendon Press, Oxford.

Hays, J., 1967 : Land surfaces and laterites in the north of the northern territory. Ch. 9 *in Landform Studies from Australia and New Guinea.* (Ed. J.A. Jennings & J.A. Mabbutt) A.N.U. Press, Canberra.

Hills, E.S., 1975 : *Physiography of Victoria.* (2nd Ed.), Whitcomb & Tombs, Melbourne.

Horwitz, R.C., 1960 : Géologie de la région de Mt compass, (feuille Milang), Australie méridionele. *Eclogae géol. Helv.* 53, 211-263.

Horwitz, R.C. & B. Daily, 1958 : Yorke Peninsula. Chap. 3 *in Geology of south Australia.* (Ed. M.F. Glaessner & L.W. Parkin) Melbourne.

Hutton, J.T., C.R. Twidale & A. R. Milnes, 1978 : Characteristics and origin of some Australian silcretes Ch. 3 *in Silcrete in Australia.* (Ed. T. Langford-Smith) Univ. New Engl. Press, Armidale.

Hutton, J.T., C. R. Twidale, A.R. Milnes & H. Rosser, 1972 : Composition and genesis of silcretes and silcrete skins from the Beda Valley, southern Arcoona Plateau, South Australia. *J. Geol. Soc. Aust.* 19, pp. 31-39.

Jack, R.L., 1915 : The geology and prospects of the region to the south of the Musgrave Ranges and the geology of the western portion of the Great Artesian Basin. *Geol. Surv. S. Aust. Bull.* 5, pp. 7-54.

Jennings, J.N. & M.M. Sweeting, 1961 : Caliche pseudoanticlines in the Fitzroy Basin. Western Australia. *Amer. J. Sci.* 259, pp. 635-639.

Kemp, E.M., 1978 : Tertiary climatic evolution and vegetation history in the southeast Indian Ocean. *Palaeogeog. Palaeoclim. & Palaeoecol.* 24, pp. 169-208.

Lovering, T.S., 1959 : Geological significance of accumulator plants in rock weathering. *Bull. Geol. Soc. Amer.* 20, pp. 781-800.

Ludbrook, N.H., 1969 : Tertiary Period *in Handbook of South Australian Geology*, (Ed. L.W. Parkin) Geol. Surv. S. Aust., Adelaide.

Mabbutt, J.A., 1967 : Denudation chronology in central Australia. Ch. 8 in *Landform Studies from Australia and New Guinea.* (Ed. J.N. Jennings and J. A. Mabbutt) A.N.U. Press, Canberra.

Maignien, R., 1966 : Review of research on laterites. *Nat. Resour. Res.* IV, UNESCO.

Mann, A.W. & R.C. Horwitz, 1979 : Groundwater calcrete deposits in Australia : some observations from Western Australia. *J. Geol. Soc. Aust.* 26, pp. 293-303.

Miller, R.P., 1937 : Drainage lines in bas-relief, *J. Geol.* 45, pp. 432-438.

Netterberg, F., 1971 : Calcrete in road construction. *Nat. Inst. Road. Res.* (*Pretoria*) *Bull.* 10.

Opik, A.A., 1954 : Mesozoic plant-bearing sandstones of the Camooweal region and the origin of freshwater quartzite. *Bur. Min. Resour.* (*Australia*) *Tech. Rep.*

Paton, T.R. & M.A.J. Williams, 1973 : The concept of laterite. *Amer. Assoc. Geogr. Ann.* 62, pp. 42-56.

Prescott, J.A. & R.L. Pendleton, 1952 : Laterite and laterite soils. *Comm. Bur. Soil. Sci. Tech. Comm.* 47.

Price, W.A., 1925 : Caliche and pseudo-anticlines. *Bull. Assoc. Am. J. Sci.* 9, pp. 1009-1017

Prider, R., 1966 : The lateritized surface of Western Australia. *Austr. J. Sci.* 28, pp. 443-451.

Sivarajasingham, S., L.T. Alexander, J.G. Cady & M.G. Cline, 1962 : Laterite. *Adv. Agron.* 14.

Stephens, C.G., 1964 : Silcretes of central Australia. *Nature* (London) 203, p. 1407.

Stephens, C.G., 1970 : Laterite and silcrete in Australia. *Geoderma* 5, pp. 5-52.

Sturt, C., 1849 : *Narrative of an Expedition into central Australia.* Boone, London.

Twidale, C.R., 1966 : Geomorphology of the Leichhardt-Gilbert Area of northwest Queensland. *Land. Res. Series* 16, (C.S.I.R.O. Aust.).

Twidale, C.R., 1976 : Geomorphological evolution. Ch. 2 in *Natural History of the Adelaide Region.* (Ed. C.R. Twidale and M.J. Tyler) R. Soc. Aust. Adelaide.

Twidale, C.R., 1980a : The Devil's Marbles, central Australia. *Trans. R. Soc. S. Aust.* In press.

Twidale, C.R., 1980b : Les inselbergs en gradin et leur signification : l'example d'Australie. *Ann. Géogr.* In press.

Twidale, C.R. & J.A. Bourne, 1975 : Episodic exposure of inselbergs. *Geol. Soc. Amer. Bull.* 86, pp. 1473-1481.

Twidale, C.R., J.A., Bourne & D.M. Smith, 1974 : Reinforcement and stabilisation mechanisms in landform development. *Rev. Géomorph. Dynam.* 23, pp. 115-125.

Twidale, C.R., J.A. Bourne & D.M. Smith, 1976 : Age and origin of palaeo-surfaces on Eyre Peninsula and in the southern Gawler Ranges, South Australia. *Z. Geomorph.* 20, pp. 28-55.

Twidale, C.R. & W.K. Harris, 1977 : The age of Ayers Rock and the Olgas, central Australia. *Trans. R. Soc. S. Aust.* 101, pp. 45-50.

Twidale, C.R., J.M. Lindsay & J.A. Bourne, 1978 : Age and origin of the Murray River and Gorge in South Australia, *Royal, Soc. Vict. Proc.* 90, pp. 27-42.

Twidale, C.R., J.A. Shepherd & R.M. Thomson, 1970 : Geomorphology of the southern part of the Arcoona Plateau and of the Tent Hill region west and north of Port Augusta, South Australia. *Trans. R. Soc. S. Aust.* 94, pp. 55-67.

Wellman, P.W., 1971 : *The age and palaeomagnetism of the Australian Cainozoic Volcanic rocks.* Unpubl. Ph.D. thesis, A.N.U. Press, Canberra.

Wopfner, H., 1960 : On some structural development in the great Australian Artesian Basin. *Trans. R. Soc. S. Aust.* 83, pp. 179-194.

Wopfner, H.. 1978 : Silcretes of northern South Australia and adjacent regions, Ch. 7 *in Silcrete in Australia.* (Ed. T. Langford-Smith) Univ. N. Engl. Press, Armidale.

Wopfner, H., R.A. Callen & W.K. Harris, 1974 : The Lower Tertiary Eyre Formation of the southwestern Great Artesian Basin. *J. Geol. Soc. Aust.* 21, pp. 17-51.

Wopfner, H. & C.R. Twidale, 1967 : Geomorphological history of the Lake Eyre Basin, Ch. 7 *in Landform Studies from Australia and New Guinea.* (Ed. J.N. Jennings and J.A. Mabbutt) A.N.U. Press, Canberra.

SECTION TWO

KARST—PROCESS AND LAND FORMS

INTRODUCTION

In certain regions of the world solution becomes the dominant process in landform development with resulting production of a unique type of topography to which the name Karst has been applied. The word Karst is a comprehensive term applied to limestone or dolomite areas that possess a topography peculiar to and dependent upon underground solution and the diversion of surface waters to underground roots. A vast amount of data has now been accumulated concerning the denudation of the Karst areas; further we are now aware that Karst denudation is not a simple processes involving just one world wide variation of a single climatic variable, but a complex problem involving the techniques of many disciplines. At the same time some of our basic assumptions are being called into question.

There are some geomorphologists who feel that the isolation of Karst denudation as a separate phenomenon is unreal. We should view Karst denudation in total denudational context not in isolation.

There has been a less emphasis on the mechanical erosion in Karst regions. This is of course a long dispute in Karst studies particularly in fluvial Karstic areas. Newsom (1971) in his work of the fluvial Karst areas of Mandips in southern Britain has stated that only 0.10 per cent of the total solution figures in that area comes from corrosion of Cave passages and that by far the greater part of the denudation is by mechanical erosion.

The problem of collection of data for our studies in Karst denudation also deserves mention. The measurements of Karst denudation have been dominated since 1950s by the use of indirect methods relying basically on the analysis of various Karst waters in terms of Ca and Mg hardness. Thus in recent years much attention has been paid to the different type of waters in a Karst environment. Numerous problems are associated with the analysis of Karst waters, many of these are of highly chemical nature and are probably better dealt with by the Commission for the study of physio-chemical methods.

It is desirable that we should evolve reasonably quick, cheap and reliable methods for ascertaining calcium and magnesium hardness without too much apparatus.

One of the most important factors affecting limestone solubility is the variability of the amount of carbon dioxide present. The importance of vegetation and soil cover in affecting limestone corrosion is also well known. Carbon dioxide production under soils causes corrosion and in some tropical areas is regarded as the most important way in which limestone is attacked; in fact, one idea for the formation of cockpits, uses this theory only (Smith, Drew, Atkinson, 1972). From recent works the effects of vegetation on solution and hardening of limestones may also be significant. It is well known that limestones become case hardened with exposure; the nature of case hardening has so far received little attention.

Climate is above all a very important factor contributing to Karst denudation system. As a matter of fact we need much more information on different kinds of rainstorms—the relationship of limestone corrosion to the direction of rain bearing winds, the effect of wetting and drying upon the rock of heat and cold, the effect of different kinds of waterflow on limestone. All of these will affect corrosion of rocks in a variable way. The nature and variability of limestone themselves is also very significant in Karst denudation system.

Ivan Gams in his study, "Karst, Karst denudation, chemical erosion—climate" has studied the relationship between Karst denudation and climate. He has emphasized the role of climate in the Karst distribution system. Chemical erosion is mainly controlled by run off, surface Karst depressions which are typical of the Karst in extra tropical zone are an effect of process combination of solution and deep denudation.

In the last three decades when the study of processes and morphometry have appeared to dominate geomorphology, a better understanding of cave formation has been achieved by approaches, than by classical methods of field observation, followed by inductive and deductive reasoning. Straightformed exploration and mapping of caves have been the greatest source of food for thought. These have been pressed further and farther a field from the early centres of speleology in Europe and N. America.

J.N. Jennings in his paper "the problem of cavern formations" has systematically discussed the processes in limestone cave formation and evolution. He has reviewed the progress that has been done on the evolution of cave formations. He has also emphasized the hydro-dynamics and factors in cave evolution.

The paper by Marushvil has discussed the features of mountain Karst of the southern part of the USSR-Georgia. The paper discusses general characteristics Georgian Karst and then discusses Karst forms like Karst funnels and sink-holes, Kettle holes and poljes and caves. According to him, Georgia Karst is a relatively young formation and on the basis of paleontological and archaeological evidences of caves he has dated them as early pleistocene. Developments—of tiers of Karst caves has been considered peculiar and rhythmic geomorphological process.

REFERENCES

Newsom, M.D., 1971 : A model of subterranean limestone erosion in the British Isles based on Hydrology. *Trans. Inst. Brit. Geographers*, No. 54, pp. 55-70.

Smith, D.I., D.P. Drew, and T.C. Atkinson, 1972 : Hypothesis of karst landform development in Jamaica. *Trcns. Cave Res. Group*. Vol. 14, pp. 159-173.

ELEVEN

Ivan Gams

KARST—KARST DENUDATION—CHEMICAL EROSION—CLIMATE

IN SPITE of the copious literature on this topic it still remains a question how and in which manner the climate controls the karst denudation.* Here are examined in particular the roles of two climatic factors, temperature and precipitation.

Recent research trends to confirm the old thesis that the karst denudation in the warmer and humid climates is higher, due to higher biological activity in the soil and higher partial pressure of CO_2 in the soil air. In the Northern Sedimentary (Limestone) Alps in Austria (Maurin-Zötl, 1964), in Northwestern Yugoslavia (Gams, 1967), in Northern America (Harmon et al., 1975) in Venezuela (Gascoyne, 1978) a positive correlation between the total hardness of karst waters and their temperature was confirmed. This correlation was explained by a more rapid solutional process (Bögli, 1971), a higher production of CO_2 in the soil due to the higher soil temperature (Harmon et al., 1975) and a greater production of organic acids (Eraso, 1975). Gerstenhauer (1972) established a high correlation coefficient (0, 90) between the soil temperature and Pco_2.

* In English terms karst denudation, chemical erosion, corrosion intensity and solution rate are sometimes used in the same sense although national differences exist. In the Commission on karst denudation at the International Speleological Union karst denudation is mostly used (s. discussion in the book Karst process and relevant karst forms, Ljubljana 1976, p. 41).

But there exists also another correlation : higher precipitation and higher runoff means also lower total hardness. In the mountains *both the temperature* is lower and the precipitation (runoff) is higher than in the lowland. This was established in the Northwestern Dinaric Karst and in Alps (Gams, 1976). For this reason the correlation between the total hardness and water temperature is of local value and in the arid cold and hot deserts the total hardness is highest or higher than in warm humid tropics (s. Bleahu, 1974, pp. 118-119). Therefore, the dependence of the total hardness from the water temperature is only apparent and not genetically conditioned (Gams, 1980).

Much less discussed in recent time is the role of the precipitation for the amount of karst denudation. Nearly all who have studied this correlation from the quantitative aspect confirm the existence of a close negative correlation of this factor (Gams, 1967, 1972, Habie, 1968, Balacz, 1971, Jennings, 1972, Kotarba, 1973, Priesnitz, 1974, Pulina, 1974, Douglas, 1968, Nicod, 1976, Kunaver, 1978, Julian et al., 1978, Lang, 1977). Acc. to Priesnitz the correlation coefficient in the world wide scale is 0, 743.

Runoff stands thus in negative correlation to the total hardness and in positive correlation to the karst denudation. For a better understanding of this double effect let us consider the diagram showing the relation between the discharge of the proper karst rivers in province of South France, in Leban (Julian et al., 1978) and two rivers in Slovenia (Kolbezen, 1977; Gams, 1980—Fig. 11.1). Here the weight of dissolved carbonates in transport is plotted against discharge. In the river basins there exist differences in lithology, hydrographical properties, soil, vegetation, landuse and relief. In spite of this the correlation line is nearly direct.

Special studies have disclosed differences in water hardness due to different limestone in ratio 1 : 4, (Sweeting 1973, 1976), resp. 1 : 5.6 (Trutgill, 1976). Differences in water hardness due to different soil types have been found in ratio 1 : 100 (Groom-Ede, 1972) resp. 1 : 10 (Trutgill, 1976). But in case of these rivers discharge remains the most decisivc factor. Karst denudation which corresponds only to the runoff is called normal corrosion (Gams, 1980).

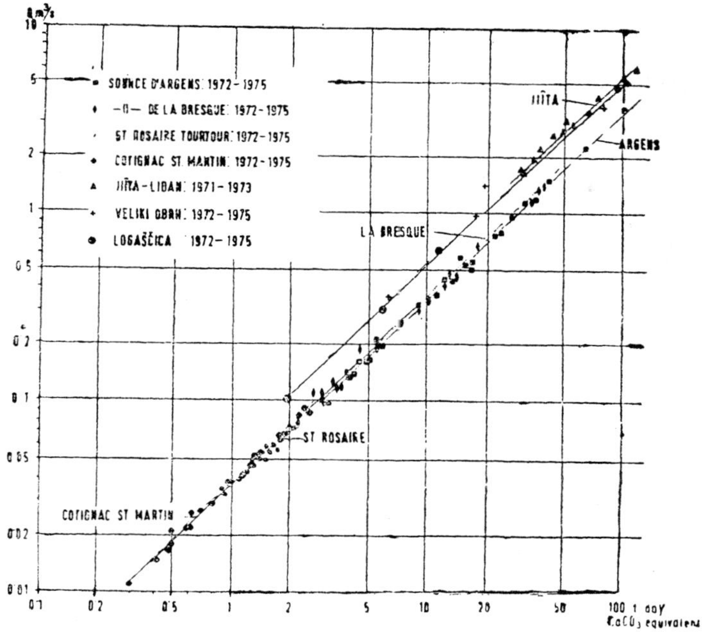

FIG. 11.1 : *Relation between the discharge of the proper karst river in Province of South France*

In the karst springs in the Dinaric Karst discharge reacts with delay of only one to two days after heavy rain and with only slightly change in total hardness. This is in temperate low Dinaric Karst and in karst in France usually above 180 mg $CaCO_3+MgCO_3$/1 eqv. Sweeting (1972) believes that the uniform hardness between 250 and 300 ppm $CaCO_3+MgCO_3$ exists everywhere in the world. Few changes in the total hardness after rain was registered also at the percolating water in the Cave of Postojna. From January to June an increase of discharge results there in a slightly diminishing hardness and in the second half of the year in an increase of hardness (Gams, 1967). But in laboratory testings with stable Pco_2 in the water and in the air after two days of exposure of different limestones to the aggressive water only up to 100 ppm $CaCO_3$ was obtained (Markovicz, 1972; Rock-Ek, 1973).

These results and the double effect of the runoff can be explained if we consider that there is a great quantity of retention (non-gravitational) water. It exists in the soil and in the bedrock in form of hygroscopic and capillary water. It is attached to the soil and rock particles and rock surface by absorption, cohesion, capillary attraction preventing gravitational flow. Visualized as a layer, retention water in proper karst is many metres thick. New rain water forces a part of the retention water to flow gravitationally, especially in its lower strata. Both replacement and mixing of new and old water occurs. The less of annual (seasonal) precipitation, the older the retention water body and the higher is total hardness. In identical conditions elsewhere, the total hardness is an indicator of the age of karst water body. In arid zone the relation of retention water to annual precipitation is in favour of the first one and water is older and harder.

Only few measurements of the age of karst water are known to have been made so far. The construction of the tunnel under the Schneealp karst in the Austrian Alps showed a part of water to be 5.5 years old (Bauer, 1969). In Northern Israel the discharge reacts to precipitation with a delay of 2 years (Gerson, 1974) and retention water is probable still older. Technically speaking, the retention water is a mixture of water from different ages due to different speed of percolation in different conduit systems inside of the karst body and inside the drainage area as well between the drainage basins.

On the other hand, the speed of solution is a function of ion concentration (Markovicz, Lohinovicz, 1972). It declines with the time of contact of water with carbonates. Therefore, the total karst denudation is higher in instances of lower total hardness of karst water and higher precipitation (runoff).

The total hardness of karst rivers studied in the world vary in span 1 : 5 and the discharge 1 : 40 (Priesnitz, 1974).

The relations treated so far in the present article are of value only in proper karst, developed also on the surface in special karst forms. These surface karst forms are at their most impressive in the belt passing Southern Europe and Asia along the Alpide mountain system. In SE Asia and in N. America there are two belts, one in humid temperate and the other in the humid tropic climates. (also Kosak, 1952). There the runoff is the highest. These belts have been limited by a kind of climatic threshold.

Threshold denotes the existence of zones of conditions where the topographical form or processes occur or cease. In a recent study D.C. Ford (p. 360) stated : "climate thresholds and limits prove to be elusive, perhaps illusory, except in the broadest terms". He stresses the lithology and time as the most important factors.

Almost whenever the studying the dependence of karst and climate the question of the definition of karst arises. The name karst is of Indo-European origin and its original meaning was stone resp. stony land (Gams, 1973). But in the modern geomorphology it is a denomination for land with surface karst forms in karst rocks. Recently the special Commission for terminology at the Int. Speleological Union (s. also Trimmel, 1965) and also the Yugoslav terminology (Gams, 1972, ed.) the following definition have adopted : karst is land where underground water circulation prevail due to fissured rock, where solution is efficient and where characteristic surface forms *can* occur. Underground (endokarst) water circulation and corrosion are essential and surface karst forms are not obligatory, acc. to these definitions. In this sense karst was termed also by Sweeting (1972, p. 5) : "The sinking of water and its circulation underground is the essence of the karst processes. This process is dominated by a chemical (solution) activity and true karst land forms result largely from the action of one erosion process, namely solution".

It is obvious after many decades of postware research that the amount of chemical erosion of carbonates is not necessarily an indicator of the karstification process. If the drainage of the precipitated water is completely underground, then erosion on the surface is absent. But there is still active denudation of soil and stony regolith along the slopes. Inside of permeable carbonate sediments there exist regions without surface drainage and without karst depression forms. These areas are more numerous in more arid climate and in easy weathered sediments (marls, chalk, sand, dolomites). In the climate with frost occurrence the denudation is more intensive and karst depressions are rare. Density of dolines is a function of slope gradient (Maksimovic 1963, p. 127), as on steeper slope more active denudation occurs. The higher the runoff is, the higher is the solution rate; the harder the rock, more steep are slopes where dolines can be still found. The depth of the depressions is an indicator of the relation between the local corrosion at the bottom and of the accumulation of the weathered material from the slope which hinders or stops the deepening of depression.

If precipitation and evapotranspiration are balanced in the soil, the horizont of $CaCO_3$ accumulation occurs. In arid climate it comes to the soil surface. It also hinders the solutional deepening of the depressions (s. Fig. 11.2 —Gams, 1976). For all the solutional surface forms the humid climate is therefore at favourable condition. Sweeting 1972) suggests a rainfall threshold of 250-300 mm/yr for limestone karst, Verstappen 1500 mm/yr for polygonal karst, Balazs (1971) 1200 mm/yr and 18°C for the tower karst of South China.

But climatic conditions (Fig. 11.3) are only a frame in which surface karst forms can occur. The other conditions are proper lithology (this means also more resistant rock for weathering processes) and piezometric water level below surface. Under very favourable related conditions, surface karst occurs also outside of this climatic threshold. But there is no climatic threshold for the endokarst and for the development of the caves which are controlled by two main factors mostly, solution of water and resistivity of bedrock to collapsing (cave-in). Above underground cavaties roof can collapse irrespectively of the climate and in this way the collapse dolines came into being. There are also contact karst forms. Contact karst is karst in transition from surface to underground river pattern. There, long caves and collapse dolines above the karst plains, blind valleys, some kind of dry valleys and border (contact) poljes occur. In classical Karst contact karst features are at most fascinated forms (Gams, 1974).

In the world river transport $CaCO_3$ and $MgCO_3$ take 59 per cent of the dissolved material (Ishikava, 1980). The dissolved carbonates derived only partially from the surface karst areas where corrosion is a form-decisive process. Acc. to Balacz (1977), these areas take about 4 per cent of the earth surface. The rest is deriving from the areas built of carbonate sediments but without surface karst forms. This type of endokarst is in N. Dinaric Karst developed mostly in the Triassic dolomites. There the karst

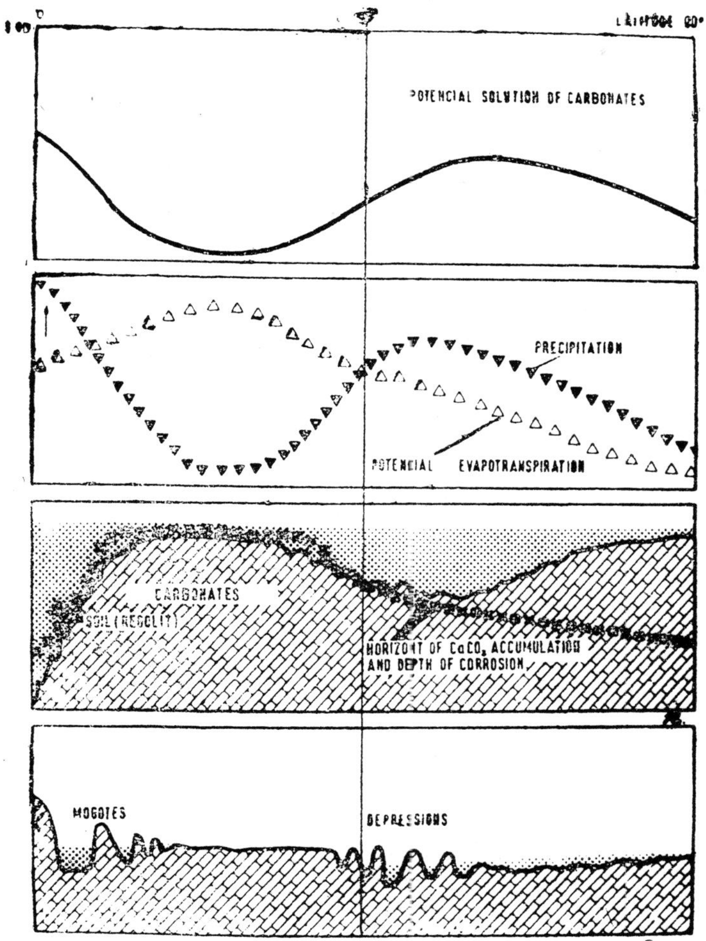

FIG. 11.2 : *Solutional deepening of the depressions*

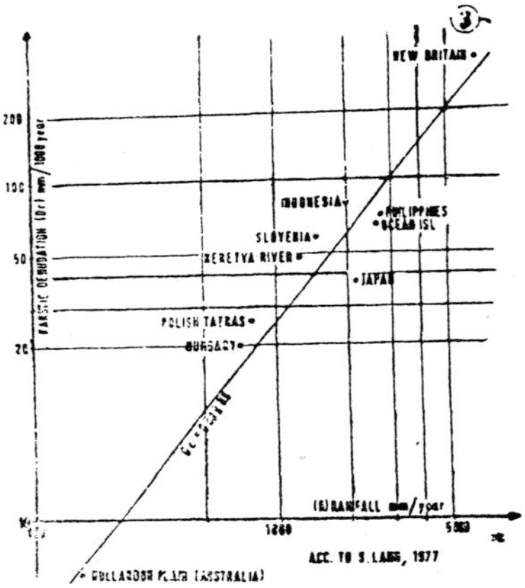

FIG. 11.3 : *Climatic conditions*

denudation is nearly the same as in the neighbouring river basins built of limestone. The total hardness of the rivers draining dolomite is usually higher but more reduced at high discharge (Gams, 1972). There in endokarst the relief is fluvial with crests and valleys and only karst springs testify the predominant underground drainage. This karst type is called fluviokarst (s. also Roglic, 1974). On the slope there denudation is the form decisive process and in the bottom of the valley the erosion of running water although the most efficient process of mass waste is solution proven by measurements of river transport. Chemical erosion is the most active quantitative process also in the drainage areas in low and hilly relief built of unpermeable, partially carbonatic sediments (of Tertiary age mostly). There the corrosion takes place only inside the soil and directly below it. Nevertheless, there the corrosion as non-form-decisive process stands behind the proper karst for 1/4 to 1/2 only and runoff is still the most important factor for it. This is shown also in the (Fig. 11.4), where drainage areas in northwestern Dinarides in Yugoslavia are shown.

Taking into account the mentioned 59 per cent of the world river transport in form of dissolved $CaCO_3$ and $MgCO_3$ as well as the parcentage of carbonate in the world (acc. to Maksimovic, 1963—44 per cent) further research of chemical erosion in the

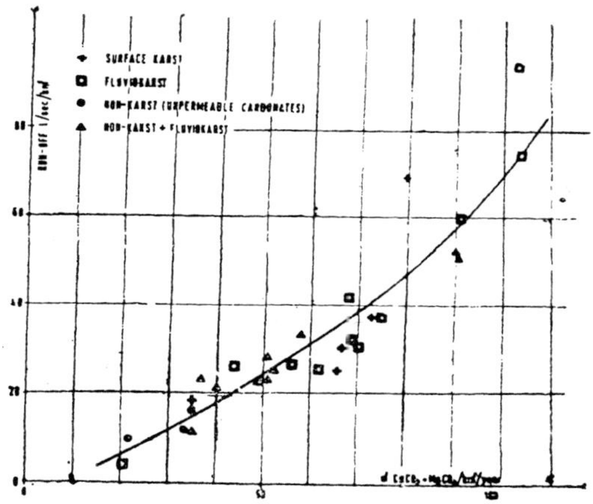

FIG. 11.4 : *Corrosion as non-form-decisive process stands behind the proper Karst*

world is necessary, even if we take into account that the corrosion is mostly not a form—decisive process.

CONCLUSIONS

Karst denudation resp. chemical erosion is mainly controlled by runoff. Surface karst depressions which are typical of the karst in extratropical zone are on effect of process combination of solution and weak denudation. The last mentioned process is dependent of lithology, relief and climate. The most close dependence on climatically controlled process is shown by dolines, the loosest dependence the endokarst phenomena and contact karst.

Climatic threshold is a large frame where lithology influences the denudation processes and controls the development of surface forms. Corrosion is form-decisive process only in a small part of the areas built of carbonates. As the form non-decisive process it is not yet recognized in the geomorphology. Many forms especially at the contact of concarbonatic and non-carbonatic terrains can be thus explained.

SUGGESTED READINGS

Balazs, D., 1971 : Relief types of karst areas. *Symp. Karst Morphogenesis, I.G.U.*, Hungary.

Balazs, D., 1977 : The geographical distribution of Karst Areas. *Proc. 7th Int. Speleological Congress Sheffield*, England.

Bauer, F., 1969 : Karsthydrologische untersuchungen im schneealpenstollen in den steirisch-niederösterreichischen Kalkalpen. *Steirische Beiträge zur Hydrologie*. Graz.

Bleahu, D.M., 1974 : *Morfologia carstica*. Bucuresti.

Bögli, A., 1971 : Karstdenudation-das Ausmass des korrosiven Kalkabtrags. *Regio Basiliensis, XII/2*, Basel.

Douglas, I., 1968 : Some hydrologic factors in the denudation of Limestone Terrains. *Z.F. Geomorpologie*, N.H., Bd. 12, p. 3.

Eraso, A., 1975 : The function on physical-chemical factors in the process of karstification. *Int. Symposium of Physical Chemistry in Karst*. Granada.

Ford, D.C., : Threshold and limit effects in Karst geomorphology (Separate).

Gams, I., 1967 : Faktorji in dinamika korozije na karbonatnih kameninah slovenskega dinarskega in alpskega krasa. *Geografski vestnik* XXXVIII, 1966, Ljubljana.

Gams, I., 1972 : Effect of Runoff on corrosion intensity in the northwestern dinaric karst. *The Transactions of the Cave Research Group of Great Britain. V. 14*, p. *2*.

Gams, I., 1973 : The evolution of the slovene words kras and dolina into international terms until the end of 19th Century. In : *Slovene Karst terminology* (ed. Gams I.), Ljubljana.

Gams, I., 1974 : Kras. Ljubljana.

Gams, I., 1976a : Variations of total hardness of karst waters in relation to discharge (Case Studies in Slovenia, Yugoslavia). In : *Karst Processes and Relevant landforms* (ed. Gams I.). Ljubljana.

Gams, I., 1976b : Chemical erosion as a landscape-forming climatozonal phenomenon. *Int. Geographical Congress-Abstracts, 1*, Moskva.

Gams, I., 1980 : Retention water in Karst : its role in influencing total water hardness and denudation rate. *Proc. Symposium on Karst Denudation I.S.U. Aix en Provence-Marseille-Nimes.*

Gascoyne, M., 1978 : Hydrology and solution chemistry in northern Venezuelan Karst. *Bol. Soc. Venezolana Espel.*, 9(17).

Gerson, R., 1974 : Karst processes of the eastern upper Galilee, Northern Israel. *Journal of Hydrology*, 21.

Gerstenhauer, A., 1972 : Der einfluss des CO_2 Gehaltes der Bodenluft auf die Kalklösung. *Erdkunde*, XXVI.

Harmon, R., et al., 1975 : Regional hydrochemistry of North American carbonate terrains. *Water Resources Research, 11.*

Groom, G.E., Ede, D.P., 1972 : Laboratory simulation of limestone solution. Transac. *Cave Research Group of Great Britain. V. 14, 2.*

Habic, P., 1968 : Kraski svet med idrijco in Vipavo, Ljubljana.

Keller, R., 1963 : Der einfluss von boden, Gestein und Klima auf den Chemismus des Grundwassers. *Int. Ass. Scientific Hydrology. Publ. No 64.* Barkely.

Ichikawa, M., et al., 1980 : Inland water and water resources in Japan. In : *Geography of Japan.* Teikoku-Shein, Tokyo.

Jennings, J.N., 1972 : Observations at the blue waterhole, March 1965-April 1969, and Limestone Solution on Cooleman Plain, N.S.W. Helictite.

Julian, M., Martin, J., Nicod, J., 1978 : Les Karst mediterraneen. *Mediterranée, Nr. 1-2.*

Kolbezen, M., 1977 : Discharge and corrosion intensity in the drainage basins of the Karst River Ljubljanica. *In : Karst Processes and Relevant Karstforms.* Ljubljana.

Kosak, H., 1952 : Die Verbreitung der Karst-und pseudokarsterscheinungen über der Erde. *Pett. Geogr. Mitt. 96*, Gotha.

Kotarba, A., 1972 : Powierzshniowa denudacja chemiczna w wapienno-dolomitowych Tatrach Zachodych. *Prace geograficzne, Nr. 96 Inst. geografii PAN.* Wroclaw.

Kunaver, J., 1978 : Intenzivnost zakrasevanja in njegovi ucinki v Zahodnih Julijskih Alpah-Kaninsko pogorje. *Geografski vestnik*, L, Ljubljana.

Lang, S., 1977: Relationship between world-wide Karst denudation (Corrosion) and Precipitation. *Proc. 7th Int. Speleological Congress* Sheffield, England.

Maksimovic, G.A., 1963 : Osnovi Karstovedenia. T 1, Perm.

Markowicz-Lohinowicz, 1972 : Aktywnosé chemiczna wód krasowych. *Speleologia, VII. 1-2*, Warszawa.

Maurin, V., Zötl, J., 1964 : Karsthydrologische untersuchungen im toten gebirge, *Oesterreichische Wasserwirtschaft*, 16, 5/6.

Nicod, J., 1976 : Relation débit-teneur et débit-erosion dans les karst méditerranées. Actes du 99 congrés nat. Soc. savantes. Sect. de Géographie. *Etudes géographiques sur le montagnes.* Paris.

Picknett, P.G., 1977 : Rejuvenation of aggressiveness in calcium carbonate solutions by means of magnesium carbonate. *Proc. 7th Int. Speleological Congress*, Sheffield, England.

Picknett, R.G., 1977 : Foreign substances and calcite solubility in carbonated water. *Proc. 7th Int. Speleological Congress*, Sheffield, England.

Priesnitz, K., 1974 : Lösungsraten und ihre geomorphologische relevanz. Abh. Akad. Wiss. in Göttingen. Sonderabdruck : *Geomorphologische Prozesse und Prozesskombinationen in der Gegenwart unter verschiedenen Klimabedingungen.* Göttingen.

Roques, H., Ek., C., 1973 : Etude experimentale de la dissolution des calcaires par un eau chargée de CO_2. *Ann. Spéléologie, 28-4.*

Roglic, J., 1974 : The Delimitations and morphological types of the dinaric Karst. *Nase jame VII, 1-2*, Ljubljana.

Sweeting, M.M., 1972 : *Karst Landforms.* Macmillan, London.

Sweeting, M.M., 1973 : Some comments on the lithological basis of Karst landforms variations. *Int. Speleology, I, subs.* In : Geology of soluble rocks. Olomouc.

Sweeting, M.M., 1976 : Recent problems in the Karst denudation research in the world. In : *Karst Processes and Relevant Landforms* (ed. Gams I.,), Ljubljana.

Trimmel, H., 1965 : Speläologisches fachwörterbuch. 3. *Int. Kongress für Speläologie.* Wien-Obertraun-Salzburg. Wien.

Trutgill, S.T., 1976 : The role of soil cover in limestone weathering, cockpit country, Jamaica. *Proc. 7th Int. Speleological Congress* Sheffield, England.

Trutgill, S.T., 1976 : The marine erosion of Limestone on Aldabra Atoll, Indian Ocean. *Z. Geomorphologie, Supp. Bd. 26.*

TWELVE

J. N. Jennings

THE PROBLEM OF CAVERN FORMATION

IN THE LAST three decades when process study and morphometry have appeared to dominate geomorphology, a better understanding of cave formation has been achieved, less by these approaches, than by the classical methods of field observation, followed by inductive and deductive reasoning. Straightforward exploration and mapping of caves have been the greatest source of food for thought. These have been pressed further and farther afield from the early centres of speleology in Europe and N. America, and pursued with much improved techniques, including the new avenue of cave diving. In this circumstance one should yield tribute to William Morris Davis, whose classic paper on limestone caves of 1930, when he was 70 years old, employing just these basic methods, gave fresh orientation and vital stimulus to this field of study at that time in the English speaking world.

Of course, there is not one problem of cave formation but many if for no other reason but that caves in different kinds of rock involve various processes and undergo diverse evolutions. Extensive cave systems have been found in gypsum, including the world's third and fourth longest, and also in lavas, into the tens of kilometres, but the most elaborate and extensive ones are still found in carbonate rocks, especially limestone. The world's longest cave, the Flint Mammoth Cave system in Kentucky with over 300 km of passage, and the world's deepest, the Gouffre Jean Barnard in the French Alps 1402 m in depth, are formed in pure limestone.

In these recent decades, the most significant change in discussion about limestone caves has been a shift away from controversy, arising because opposing standpoints were monogenetic and maintained that a particular mechanism or a given locus of activity is generally dominant, to recognition that at different times and places one of a range of mechanisms or combinations of mechanisms takes on supremacy (cf. Halliday 1960; White and Longyear 1962; Ford, 1965).

PROCESSES IN LIMESTONE CAVE GENESIS AND EVOLUTION

Despite what has been said above, it is logical to begin by considering the relevant processes, in which also advances have been made in this period.

Corrosion

The single most important process, the chemical removal of limestone, is a most complex matter (Picknett, Bray and Stenner, 1976) which can be treated here only in a crude, non-mathematical manner. Simple physical solution is normally the initial phase yet it can only bring about a concentration of 13-15 mg/*l* according to temperature. Natural concentrations in cave waters are much higher than this, commonly reaching between 100 and 200 mg/*l*, and this excess is due to carbonation. The carbon dioxide, which mildly acidifies the water for this, dominantly is provided by the life and decay processes of plants, macroscopic and microscopic (including bacteria), in vegetation, its litter and the soil. The atmosphere provides much less. The transport of organic matter to the depths of caves allows for a renewal of CO_2 by decay at considerable remove from the surface (James, 1977).

The maximum amount of $CaCO_3$ that can be dissolved—the saturation equilibrium—varies directly with the content of CO_2 (its partial pressure Pco_2) in the water and inversely with the temperature in accord with Henry's law. However the first effect ranges one hundredfold whereas the second has only a threefold range over cave temperatures. Therefore, the indirect influence of temperature in promoting organic processes to provide CO_2 overrides its opposite effect through saturation equilibrium.

There are other controls on saturation equilibrium—ion pairing as with magnesium which has a complex effect (Picknett, 1972), fortunately minimal when magnesium is about 15 per cent of calcium, a common ratio in karst; ion strength such as the presence of much NaCl in sea water increasing the saturation equilibrium for calcium; presence of metal ions such as copper which reduces the saturation amount, to mention only some.

It is fortunate, therefore, for our attempts to understand caves and karst that natural waters are much more commonly undersaturated than saturated so that the kinetics of the corrosion process governs the situation mainly, both in time and space, namely the rate at which limestone coming into solution in a thin skin of water as the rock surface is carried away into the main stream that removes it altogether. This mass transfer varies

(i) directly with the difference between the solute concentration in the main waterbody and the saturation concentration reached at the interface (which is controlled by the factors mentioned above),
(ii) directly with the diffusivity of the solute, itself varying directly with temperature,
(iii) inversely with the viscosity of the solute but this in turn varies inversely with temperature,
(iv) directly with the velocity of the water,
(v) with the nature of the flow, being greater with turbulent than with laminar flow, and
(vi) with the nature of the rock surface, whether it is rough or smooth, in open channel or in pipe.

Also where kinetics are in control, the rate of chemical reaction assumes importance and this doubles with every 10° C increase in temperature according to Arrhenius' law.

What generalities can be drawn from all this? Two points of great importance need stress. The prime factor is the amount of liquid water available; how much water is available matters much more than how much limestone can go into solution per unit volume. Water availability depends on the effective precipitation (P-ET) providing the autogenic water supply from the atmosphere above (Williams and Dowling, 1979) and on the magnitude of allogenic water inputs from surrounding rocks. With small karst areas, which often are recessive in relief, this last can be of great importance (cf. Jennings, 1977a). The second factor is biological activity; the more plants and microbial growth, the more carbon dioxide is available; and this in turn depends on precipitation (P-E this time) and temperatures above those minimal for plant growth.

Two other kinds of corrosion need notice. The weathering of pyrites found in some limestones but more significantly in interbedded shales, produces the strong acid, sulphuric acid, which attacks the limestone powerfully. Some caves owe much to this process but it is of local importance only. Some organic acids, such as fulvic and crenic acids, also act strongly on limestone and these are produced in more widespread fashion, in forest litter, for example, but more in association with bogs, swamps, marshes and mangroves. Unfortunately for our understanding of their role, it is hard to disassociate the effects of organic acids from that of carbon dioxide produced at the same time from organic matter (Trudgill, 1979) and from sulphuric acid generated by sulphur bacteria in peat (Bray and O'Reilly, 1974). Their effects may therefore be restricted to a subcutaneous zone, though they have been thought to be significant in the formation of large blind shafts developed in limestone beneath sandstone covers with peaty soils. Few would think there is as yet evidence to support Jakucs' claim (1977) that in tropical rainforest karst organic acids are as important as carbon dioxide in limestone corrosion.

That large caves have been created by corrosion at great depth, commonly in a water-filled condition from their detailed morphology, presents a problem of process because of the exhaustion of carbon dioxide with its surface sources. Again the rare exception has ready explanation—there are caves formed by hydrothermal waters from below such as in Zbrasovska Aragonite Cave at Hranice in Moravia; the hot water contains much CO_2 but there are other corrosive constituents. The general occurrence has engaged the attention of Bögli (i.a. 1964) particularly and in his view the main factor is *mixing*, *corrosion*, which depends on the exponential relationship between $CaCo_3$ saturation equilibrium and Pco_2 (Fig. 12.1). When two flows of water, both saturated but at different partial

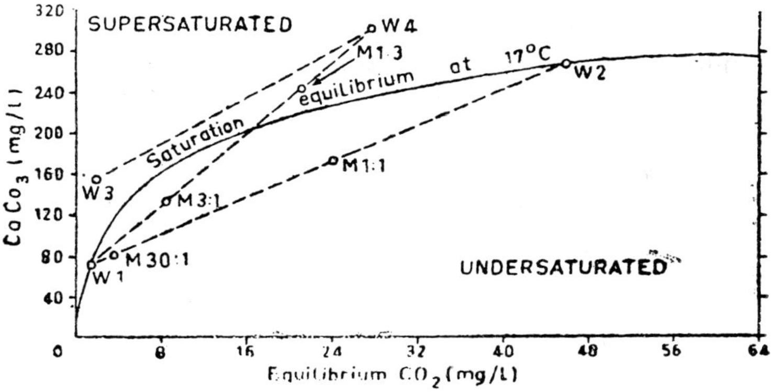

FIG. 12.1 : *Mixing Corrosion* (*after Bögli and Thrailkill*)

Mixing of saturated waters W_1 and W_2 in any proportions yields aggressive water but M 30:1 (30 parts W_1 of low Pco_2 with 1 part W_2 of high Pco_2) is more aggressive than M 1:1, an equal mixture of the two. No mixture of supersaturated waters, W_3 and W_4, will become aggressive. M 1:3 (1 part saturated W_1 with 3 parts supersaturated W_4) remains supersaturated but M 3:1 (3 parts W_1 with 1 part W_4) becomes aggressive.

pressures of carbon dioxide, mix, the mixture is undersaturated and capable of fresh solution of limestone. This can be result of the meeting of downwardly infiltrating autogenic water, saturated at high concentration, with allogenic cave stream water, saturated after its underground transit but at lower concentration. Bögli pointed also to *cooling corrosion*; thus in summer time, water heated at the surface may be cooled as it penetrates underground because cave walls are conservative of heat. This cooling will render previously saturated water aggressive once more. Another source of renewed aggresivity lies in the mixing of waters with different magnesium carbonate contents (Picknett, 1972). Another localised effect is where stream currents carry air bubbles into water-filled passages when in descent they are compressed, increasing the partial pressure of carbon dioxide. This can produce a central roof channel (Bögli, 1978). Finally one can point again to the transport of organic matter into cave depths, whereby decay produces carbon dioxide. This proceeds faster in the presence of oxygen but some micro-organisms digest organic matter anaerobically so this action can take place in water-filled cavities.

From this discussion it is evident that rates of cave corrosional enlargement will vary tremendously in place and over time and that these will be best obtained by determining the time it has taken whole caves or parts of caves to form, i.e. by speleochronology. Measurements of present day rates of enlargement are valuable, nevertheless, but few are available. Those of the longest term—over a few years—are from caves in Clare, western Ireland, where High and Hanna (1970) used the micro-erosion meter to determine floor lowering in inflow caves of 0.25 mm/year and in outflow caves of 0.17 mm/year.

Corrasion

In early cave study, mechanical stream work was over estimated because many of the features associated with it—waterfalls, rapids, gorges, meander under-cuts, plunge pools, swirlholes (i.e., fluvial 'potholes' or rockmills) were prominent in the active river caves most frequently explored then. Later it suffered prolonged neglect in favour of the less obvious but more distinctive chemical attack. Nevertheless all aspects of mechanical erosion come into play in caves—impact, cavitation, abrasion, attrition—and as for the most part there is no difference between their underground and their surface operation, they need little discussion here, except to point out that since caves fill up like pipes, water moves uphill under hydraulic head as well as downhill under gravity so this mechanical work can be carried to roofs as well as walls and beds.

Calcite is a soft mineral and clasts of limestone wear out quickly through attrition, apart from their susceptibility to solution. Therefore, there can be a dearth of large tools in cave rivers, though the fine insolubles will constitute some abrasive material. There are, however, sometimes chert and flint nodules in limestones to yield up and cave breakdown may maintain a supply of limestone blocks. Where there are allogenic inputs from neighbouring insoluble rocks, no lack may develop if the caves are of no great length; thus at Cooleman Plain, New South Wales, igneous gravels run right through the systems.

In recent years, when much effort has gone into measuring

rates of chemical denudation in karst, only a few attempts have been made similarly to quantify corrasion. The most notable are those of Newson (1971) working in the Mendip caves in England. Elaborating a lead by the famous French cave explorer, Pierre Chevalier, Newson exposed tablets of limestone in cave streams caged in nylon of varying mesh, the finest mesh excluding any corrasion but permitting corrosion, the intermediate allowing sand and silt to abrade, whilst the the coarsest exposed the rock to gravel. Much greater weight loss was suffered by the tablets in the two meshes allowing mechanical attack. Jennings (1977b) performed the same exercise over longer term in River Cave Cooleman Plain, New South Wales, and determined that the tablets lost about as much from mechanical as from chemical attack.

A more comprehensive approach is to measure the solid load as well as the solute load emerging from a cave and this Newson (1971) did for cave springs at Cheddar and Burrington in the Mendip. He found that corrosion was overall more important than corrasion; only in floods was more solid load carried than dissolved. Nevertheless the importance of floods in caves must not be underestimated; an exceptional storm in the Mendip in 1968 made drastic changes to some of the caves; in G.B. Cave speleologists familiar with it were amazed by the changes wrought.

Cave Breakdown

By cave breakdown the caver understands the breaking away of large bedrock fragments from roof and walls, especially common when passages and chambers get large. It largely corresponds with the dry rockfall and dry rockslide categories of surface mass movements. This is not solely an effect of gravity, aided by the solutional freeing of blocks, and a response to lithostatic load, but also a yielding to tectonic stresses which modern measurements reveal are widespread. Except at depth rarely reached in caves, limestone yields elastically, not plastically, i.e., it fractures. It seems possible that unequal disposition of tectonic stresses may favour development of caves along certain belts within a limestone plateau (Renault, 1970). More patently unloading of pressure by surface erosion along valleysides allows release of tectonic stress in the form of sheeting joints more or less parallel to the slopes (Fig. 12.2). This can lead to the preferential development of caves roughly parallel to valleysides, e.g., Wyanbene Cave, New South Wales. Likewise, when a cave approaches a valleyside more or less at right angles, it is liable to deflection for the same reason and break up into several passages, e.g., Murray Cave, New South Wales.

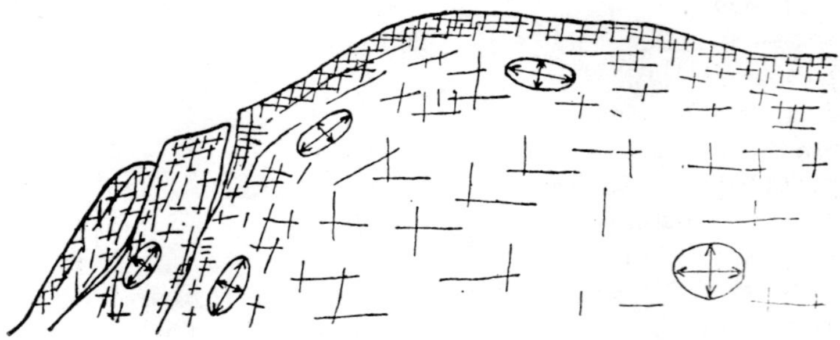

FIG. 12.2 : *Offloading of Valleyside by erosion to cause sheeting joints (after Renaults). Distribution of stresses indicated by ellipsoids)*

Loss of support of roof and walls when a cave ceases to be filled with water is commonly cited as a time when cave breakdown will be particularly likely (Davis, 1930).

Within a given cave, breakdown resulting from release of stress at its surfaces frequently produces barrel-arched roofs and domes. This may be the net result of many little fractures along joints and bedding planes, e.g., in the Nullarbor plain caves in Australia, or at greater depth through onion-skin sheeting parallel to the cave surfaces, e.g., the Gross Badlhöhle in Austria (Trimmel, 1968).

Cave breakdown cannot enlarge a cave except in conjunction with a stream removing some of its products because rockpiles occupy greater volume than the space the undisturbed bedrock occupied; otherwise the blocks would accumulate to the roof preventing further such action. However, when fallen rock is removed by water transport and erosion, breakdown may stope upwards until the roof falls in, creating a collapse doline at the surface. This can procced to the total destruction of a cave;

indeed some limestone gorges are of this origin, e.g., the Rak valley, Slovenia, though the majority are of surface origin.

Weathering

The equable climate of most caves minimises many kinds of rock weathering there but they do not escape all. In cold climates frost shattering may penetrate appreciable distances into caves (as well as creating shallow caves, e.g., in the Nahanni karst, Northwest Territories, Canada); enterances and nearby chambers are enlarged by this mechanism, e.g., in Eisrie senwelt Cave, Austria.

More widely reaching is salt crystallisation. Gypsum in particular shatters limestone as it crystallises in pores or cracks on rock surfaces. This occurs on its largest scale in dry climates where there can be more of the evaporation necessary for $CaSO_4$ saturation and here halite crystallisation can also be important. This kind of weathering is well expressed in Nullarbor Plain caves, where barrel-arched roofs and domes on a medium scale, along with large honeycomb weathering and tafoni, are produced in this way. The most remarkable feature is The Dune 9 m high in Mullamullang Cave, built of detritus from salt crystallisation.

Many other kinds of cave weathering, including biological, but with lesser import, occur.

Subsidence

This term is more commonly applied to a complex process of mass movement affecting surface sediment and soil in karst but there is no substantial difference between this and similar movement down solution enlarged fissures into caves. Flowing, sliding and slumping of mixed materials from clay to limestone blocks into caves is an important mechanism there as Pengelly (1864) showed for Devonshire caves at an early stage in cave science, though it has received inadequate attention since.

Deposition

Fluvial deposition of clastic sediments from gravel to clay sizes is an important mechanism in cave development also, affecting the course and pattern of bedrock enlargement (Renault, 1967-8). Here this must be neglected in favour of the more distinctively spelean process of calcite deposition of the most varied nature and dimensions (White, 1976). The formation of stalactites, stalagmites and other speleothems is still attributed to evaporation and this is true in cave entrances and also farther into caves in dry climates (Jennings and Sweeting, 1966). However, the dominant cause of super-saturation is diffusion of CO_2 from drips and water films into the cave atmosphere. Soil water in equilibrium with soil air rich in CO_2 passes into a closed system of water-filled, narrow cracks in the bedrock; there much limestone goes into solution, often approaching saturation; then on entering cave spaces, where the partial pressure of CO_2 commonly departs little from that of the open atmosphere, restoration of equilibrium between water and air requires this diffusion.

This happens in active free surface stream passages but the resulting speleothems are liable to destruction, both mechanically and chemically in floods. In passages abandoned by streams, accumulation may continue unhindered to the point of complete choking of passages.

HYDRODYNAMIC DOMAINS

Although there is interplay between these cave processes, far more than has been indicated above, the most important role is played by water and so the different conditions for water to operate underground are of critical importance. In this there is a big difference between the vadose and the phreatic condition; in the latter, voids are permanently filled with water whereas, in the former, air and water occupy them in a manner varying complexly in time and space. After wet weather or in wet seasons, the amount of water stored underground increases and the phreas invades the lower part of the vadose zone; the resulting transitional zone is of great speleogenetic significance. The oscillating interface between the two zones is called the watertable in hydrology; there is, however, serious liability of misunderstanding if this term is applied to karst regardlessly. The same applies to the alternative term, piezometric surface. *Water rest level* is a neutral term which can be used in karst hydrology without prejudging answers to questions that need careful resolution, answers which may indeed condemn the use of these other terms in specific karst contexts.

Vadose Seepage

At the surface, water infiltrates into the bedrock, either directly in alpine, polar and arid climates (and in anthropogenically denuded terrain) or through the medium of soil and superficial deposits. If the limestone has much intergranular or primary porosity, percolation can take place areally. Such primary permeability is found, for example, in dune limestones, which are largely joint free, but even here preferential seepage, perhaps under root guidance, soon develops, and soil-filled solution pipes begin to diversify the soil-rock interface. More commonly, limestones have little primary porosity; joints (and bedding planes where the rocks are dipping) concentrate downward seepage along lines or at points such as joint intersections (secondary permeability). In consequence fissures and shafts (karst 'potholes') are formed both in bare and covered karst (Fig. 12.3). With the latter they are generally filled with debris from the surface but as dolines develop which concentrate water at their low points by overland flow, throughflow, etc., the enlarged joint may be kept open.

Initially flow will be laminar, in the tight joint filling the narrow space or down the two faces when they widen. As the routes enlarge and inputs more concentrated into them, turbulent flow will take over. Since CO_2 from the atmosphere or from soil air dissolved in the water will be progressively used up in corrosion, the vertical features will close up downwards. This should still apply even if organic matter is washed in which renews the water's capacity to dissolve the rock.

Nevertheless many shafts and fissures enlarge downwards so it has been argued that these were developed from below upwards by water which percolated through a tight joint before it began to enlarge its routeway (Maucci, 1960). Developing upwards, a shaft may narrow in this direction, the lower the wall, the longer it has suffered corrosion. In support of this concept is the occurrence of many blind shafts, which reach upwards near to the surface but don't break through to it. The presence of a cave below to act as a starting point for vadose seepage shafts makes this mechanism more feasible.

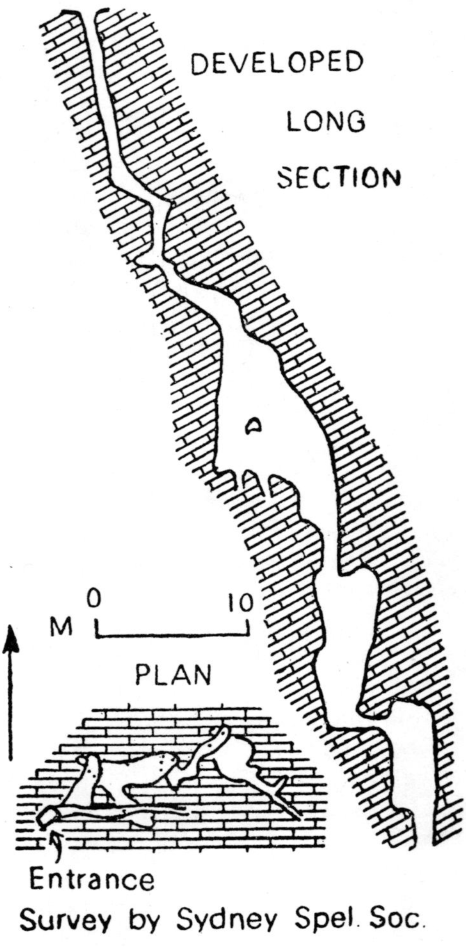

FIG. 12.3 : *Putrid Pit, Bungonia caves, New South wales open shaft cave in joint planes due to Vadose seepage solution*

The Nothephreas

A cave is usually defined as an underground cavity penetrable by human beings. Though segregating our modes of investigation, this anthropomorphic limit has no natural significance and generally there is much more interconnecting void smaller than greater than it in limestone. A natural limit is to be found in a diameter of about 5 mm for circular tubes, below which flow is laminar but above which flow may be either laminar or turbulent (White and Longyear, 1962; Howard, 1964). Here is also a chemical threshold, a critical level of aggressivity beyond which corrosion proceeds much faster, about 7 times according to White and Longyear.

Though some limestones of high primary porosity may provide tiny interconnecting tubes permitting laminar water movement through them, this is not the case with most and neither tectonic activity nor erosional unloading will provide open, interconnecting joints to permit it. The beginning of cave development therefore, is attributed to a state of water saturation, in which slow laminar motion gradually creates interconnecting systems of tiny tubes that fork and rejoin many times. These frequently develop in planes of weakness and become bedding plane and joint anastomoses (Bretz, 1942). In rocks of high primary porosity, they may develop in all directions, e.g., White Wells Cave, Nullarbor Plain. Insolubles may accumulate on the lower side of a plane of weakness, inhibiting solution there so that half-tubes develop rather than tubes. Earth tides may play a big part by pumping water through tight planes of weakness to initiate through movement (Davies, 1966); joints have been shown to dilate and contract with appropriate periods for this in caves in Hungary and the United States.

When tubes pass the critical size, turbulent flow may take over but if velocities remain small through lack of hydraulic head, further enlargement may not change its style drastically. The result can be spongework—irregular interconnecting cavities, which are either symmetrical (Fig. 12.4) or if systematically asymmetrical this reflects structural control. The Boneyard in Carlsbad Caverns, New Mexico, provides an example in large dimensions. Alternatively network mazes of joint and bedding plane controlled passages may develop, e.g., Cameron Cave, Missouri. Rock pendants are another cave form characteristic of these hydrodynamic conditions. These smoothly moulded projections from roof or upper walls can be of large dimensions, e.g., in Ryan Imperial Cave, Chillagoe, Queensland.

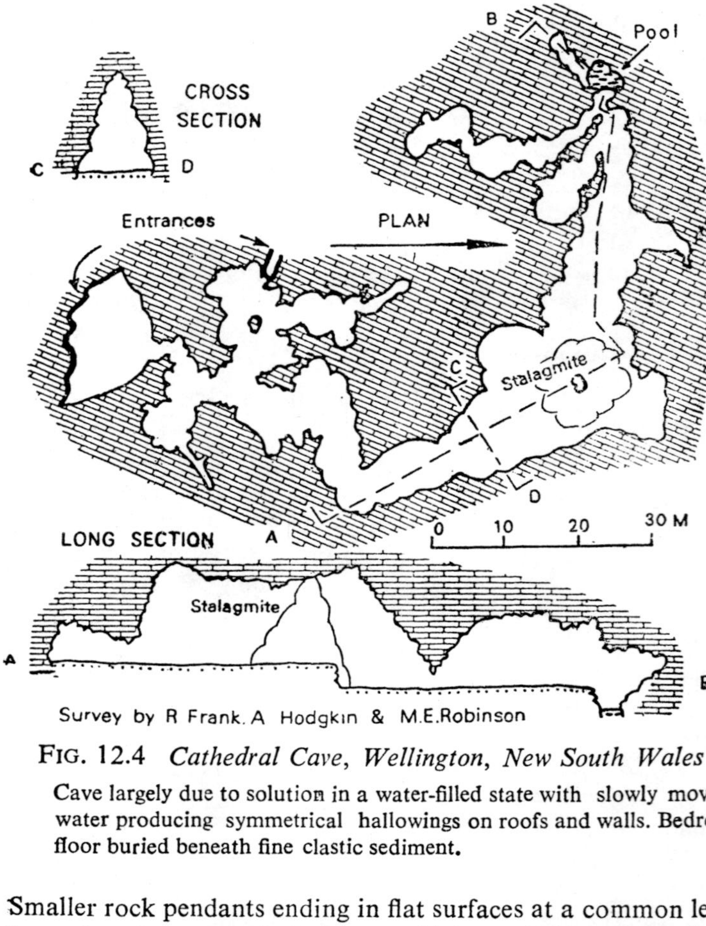

FIG. 12.4 *Cathedral Cave, Wellington, New South Wales*

Cave largely due to solution in a water-filled state with slowly moving water producing symmetrical hallowings on roofs and walls. Bedrock floor buried beneath fine clastic sediment.

Smaller rock pendants ending in flat surfaces at a common level have found two explanations, which probably imply form convergence rather than contradiction. In the one the flat surfaces belong to one side of a plane of weakness, the other side of which has been completely removed by breakdown. In the other, aggradation has confined slow flow against the roof of a passage and this has resulted in anastomosing phreatic solution deep enough to give pendants (Bretz, 1942). This is one kind of paragenetic cave development of Renault (1967-8).

Caves developed in this domain can become large yet with no connection of penetrable size to the surface; the magnificent

Ochtinska Aragonite Cave in Slovakia was encountered in mining.

Slow flow characteristic of this domain of cave development corresponds in large measure to Darcy flow through sands and gravels and is to be measured in a few metres per day, whereas with the following domains it is rather a matter of hundreds of metres per day. It is the diffuse flow of White (1969). The domain is also called the deep phreatic zone but it can occur near to the surface and be of narrow vertical dimension, e.g., in dune limestone caves near Augusta, Western Australia. Glennie (1958) argued that phreatic should be restricted to this zone and not extended to others where pipe flow with much larger Reynolds Numbers takes place, since phreatic has Darcy flow connotations. His plea has gone unheaded. Therefore, the present author (1977) has put forward the term *nothephreatic* for it.

The Dynamic Phreas

Slow flow permits the co-existence of numerous interweaving routes but if there is sufficient hydraulic head to drive water through at speed, a few or a single route will grow at the expense of the others (Ewers, 1966). It is a self-promoting effect; as a tube gets larger, the frictional loss of energy against the cave surface becomes smaller in relation to the total loss and velocity increases; as velocity increases, mass transfer of solute from that surface is enhanced, enlarging the passage further and velocity also. In the roof of the Grand Arch, Jenolan Caves, New South Wales, a large solution half-tube can be seen meandering through a large area of roof pendants, all in a bedding plane; the development of this passage must have reduced movement through the neighbouring pendants greatly.

In this way pressure conduits come to play a greater and greater role in the water movement and so in cave development. Circular or elliptical in cross-section, according to whether planes of weakness in the rock exert a strong influence or not, but with solutional attack on all surfaces, these conduits may rise and fall because they are under pressure head. Both in plan and profile, they are liable to structural control however; Ford (1965) pointed out, for example, how in steeply dipping beds, Mendip caves such as Swildon's Hole have former pressure conduits repeatedly alternating between longer, gently inclined downdip tubes in bedding planes and shorter steep chimneys in joints.

Pressure tubes characteristically have smoothed surfaces but these may alternatively carry current markings of the scallop type; these are asymmetrical hollows with a crescentic steeper upcurrent side and a gentler downcurrent one ending in a point because of intersection with neighbouring scallops. If these are on lower walls and floor only, they may be due to later vadose flow modelling but when found on high walls and roofs, it is evident that they developed during dynamic phreatic flow. Similarly although solution is dominant at all times, velocities are frequently great enough to carry abrasive materials through, even above sand size, and corrasion can contribute to their formation.

Fluvial deposition takes place in them, influencing the further course of enlargement. Whether cave breakdown plays a similar role in this domain is a matter to which little attention has been paid. In some deep caves of the Nullarbor Plain massive breakdown extends into phreatic parts but this has so far been attributed to periods of lowered water rest levels during Pleistocene glacial low sea levels (Lowry and Jennings, 1974).

It has already been emphasized that that at the top of the phreas there are passages alternatively subject to vadose flow at lower discharges and to dynamic phreatic flow at higher discharges. This, together with the top part of the phreas proper, is regarded as a zone of particularly effective speleogenesis by many investigators (Sweeting, 1950; Davies, 1953; Moore, 1960). Velocities are likely to be great here because routeways to the outflow points (the *Vorfluter*), a German term misunderstood by some English language workers (Bögli, 1978) are the most direct and involve least undulation. There is also more likely to be undersaturation and most opportunities for the renewal of corrosive power in the ways already discussed. More or less horizontal caves at and just below the level of the springs are the product, e.g., the caves of the R. Ourthe, Belgium (Ek, 1961). Flat solution roofs are typical, e.g., the Punchbowl, Signature and Dogleg Caves system at Wee Jasper, New South Wales (Fig. 12.5). The caves developed in this transitional zone between vadose and dynamic phreatic zones have been variously named: 'water-table stream' (Swinnerton, 1932) has the disadvantage already mentioned that many karst hydrologies do not justify this term at all; 'shallow phreatic' (White, 1960) has the disadvantage that such development can occur at great depth beneath the surface, it is only the vertical amplitude of the action which is intended; 'epiphreatic' (Glennie, 1958) has the appropriate eytmological derivation with no misleading overtones.

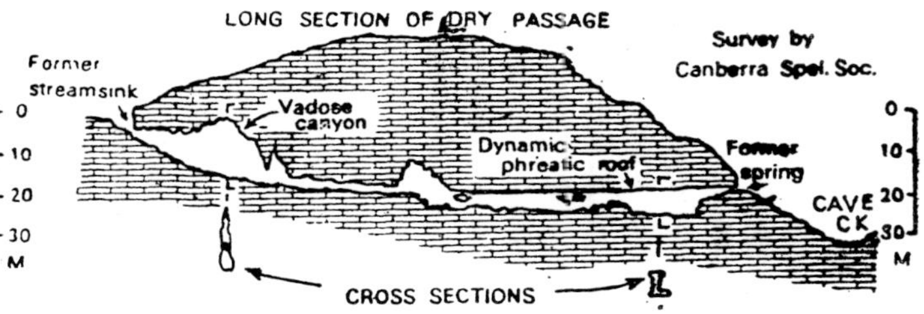

FIG. 12.5 : *Barber Cave*

Inactive passage of Barber Cave, Cooleman Plain, New South Wales. Upper part dominantly a vadose flow canyon; lower part an epiphreatic cave modified by vadose flow.

Despite the importance in many cave areas of the epiphreatic type of cave, there would have been much less controversy between the protaganists of vadose and phreatic cave development if other kinds of dynamic phreatic action did not also loom large in the overall picture of cave evolution. Ford and Ewers (1978) present a penetrating analysis in this direction, recognising three states between the nothephreatic and the epiphreatic (Fig. 12.6).

The Bathyphreatic

Davis (1930) was the first to point to the possibility of caves developing to great depth because groundwater has the capacity to descend under pressure head to considerable depth and rise back much of the way to emerge at the surface. For this he was relying on Darcy law hydrology, which applied to primary permeability, whereas most karst depends mainly on secondary permeability in pipe flow obeying different laws. Nevertheless caves which follow such courses are now known, partly through the boldness of cave divers. Thus the Fontaine de la Vaucluse

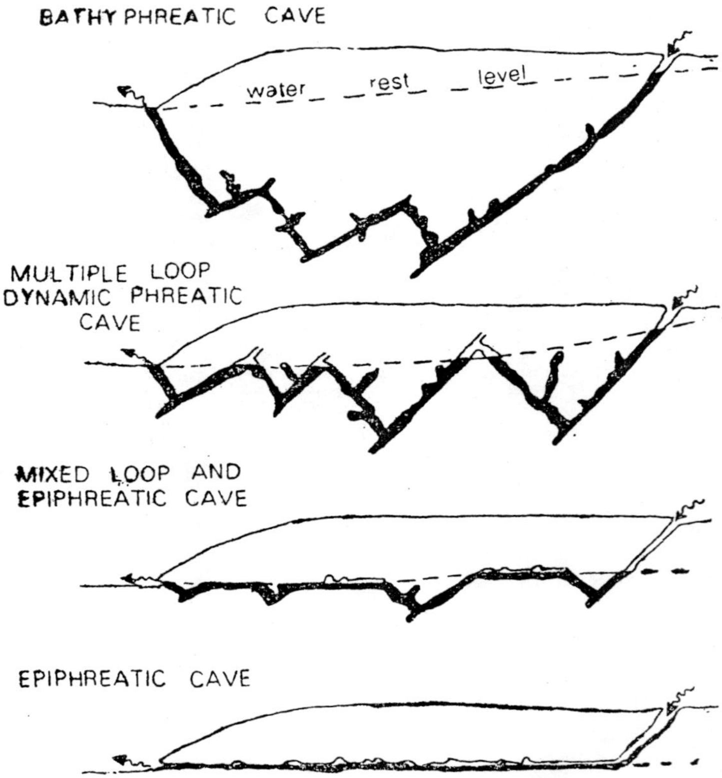

FIG. 12.6 : *Four States of dynamic phreatic cave (after Ford and Ewers)*

in southern France rises up a smooth pressure tube of some 15 m diameter from a water depth of more than 100 m. In the Sierra de El Abra in eastern Mexico, such a system but now relict loops downwards more than 300 m (Fish, 1978). It is not sufficient to have a great thickness of pure limestone, mechanically strong and of great depth, in high effective precipitation, for such bathyphreatic action. Ford and Ewers maintain that widely spaced planes of bedrock weakness are also necessary, though what joints or bedding planes there are may be extensive. The El Abra caves are in massive reef limestones. Steeply dipping beds also favour this kind of hydrology, provided the tectonics involved have not increased joint frequency unduly.

The Multiple Loop Phreatic

With greater frequency of planes of weakness, a sequence of loops develop below water rest levels. Ford and Ewers cite the Hölloch in Switzerland, the second longest cave in the world. Many but still substantial loops make up this system.

The Mixed Loop and Epiphreatic

Ford's study (1965) of Swildon's Hole in the Mendip provides the type case for this condition. Some horizontal passages form, linking the lower parts of phreatic loops; these are called bypass tubes.

Vadose Flow

The enlargement of phreatic caves will entail a lowering of water rest levels (assuming a uniform water supply) so that a vadose zone is created. The gathering of seepage waters to form underground streams is not the only source of autogenic vadose flow; as closed depressions grow at the surface so will autogenic streamsinks appear. However vadose flows will also derive substantially from surrounding and inlying impervious rocks—the allogenic contribution; these streamsinks often accompany blind valleys and poljes, many of which are partly walled and/or floored by impervious rocks.

Vadose flow operates only under gravity and so there is a strong vertical component in this circulation; nevertheless in most circumstances the horizontal component is the greater. With vadose streams corrasion can be actually as well as visually important as discussed above so erosional forms characteristic of surface rivers come to the fore. River canyons are cut in the floors of former pressure tubes and these may exhibit sequences of curved channel incuts in their sides. Ford and Ewers (1978) term caves where vadose streams inherit and modify phreatic passages in this sort of way drawdown vadose caves; such is G.B. Cave in the Mendip (Ford, 1964). Vadose canyons will entrench the tops of phreatic loops as drawdown occurs, helping to convert caves in the multiple loop state into the mixed state described above.

To be distinguished from them are the invasion vadose caves of Malott (1937) where vadose flow has developed along small routeways due to vadose seepage and early phreatic solution. In these, vadose characteristics dominate and mainly vertical caves are the most likely to reach this state.

A vadose cave necessarily assumes a branchwork plan, resembling a surface stream system in plan, though its active passages may comprise only a part of an inherited phreatic network. In addition, since a cave stream is confined and has no flood plain over which to dissipate flood discharges, it may well maintain in intermittent action higher level passages as flood overflows and so a three-dimensional braiding system persists. Therefore, there can be no easy distinction between phreatic and vadose nature of caves by simple examination of cave plans.

In the vadose caves, breakdown, subsidence and weathering come fully into action. Also deposition of their products and those of corrosion (speleothems) and of corrasion not only contribute directly to the cave morphology but also modify further bedrock excavation in ways too manifold to venture upon here (see Renault, 1967-8).

FACTORS IN CAVE EVOLUTION

To identify the main spelean processes and the domains in which they operate is only to note the ground rules for cave evolutions which have all the complexities of games of chess. It is impossible here to do more than indicate some of the factors at work to produce the kaleidoscopic variety of the underground.

The tectonostatic geological framework (Tricart, 1947), on which the active agents operate, conditions them in turn to a degree far greater than previous discussion has suggested, despite much reference to the nature and density of planes of weakness. Fissuration, as the French call this, is crucial to the development of most limestone caves and innumerable writings have been devoted to the relative roles of joints and bedding planes (e.g, Bõgli 1969). It must suffice here to refer to Ford (1970, 1971a) who has attempted the highest level of generalisation in this regard. The role of faults is conflicting, sometimes their influence is small. However, at Bungonia Caves, New

South Wales, more detailed mapping (James and others, 1978) revealed previously unknown faults, which gave answers to unresolved speleogenetic issues and disproved some earlier interpretations (Jennings and others, 1972).

In the discussion so far there has generally been the implicit assumption that limestone is the bedrock, with or without soils and superficial deposit as cover. This is by no means always the situation in karst as the example of the huge Flint Mammoth Caves system reminds us with its overlying sandstone (and locally shale) several tens of metres thick. These overlying strata have localised much of the water input near and at their margins and saved the cave system from more dismemberment than it has suffered, almost divided as it is into the Flint Ridge, Mammoth Cave Ridge and Joppa Ridge sectors (Quinlan, 1970). Palmer (1975) also argues that diffuse input from the sandstone has resulted in a maze of solution passages.

Effects of this subjacent condition are inevitably greater when the rocks are folded and hydrostatic pressure forces water down the limestone deep under an aquiclude. Such an artesian condition is incapable of being explored and resort must be had to drained systems relict from this special hydrodynamic domain. At depth it becomes nothephreatic and White (1969) describes the product as three—dimensional network caves inclined down the beds. In so far as much limestone is impermeable, i.e., it lacks primary permeability, many limestone caves may take on something of an artesian character in their phreatic sections (Glennie, 1954), provided there is substantial dip. Ford and Ewers (1978) point to the contrast between cave systems in flat-lying strata where epiphreatic systems predominate and those in steeply-dipping beds where the bathyphreatic and multiple loop systems are more frequent. They go on, however, to consider that, in tight, isoclinal folds, high joint frequency may be induced so that here again the epiphreatic predominates. This is probably one of the factors causing the poverty of New South Wales karsts in deeply looping caves (Jennings, 1977a). It is relevant to note that Cailleux and Tricart (1956) found in a statistical analysis of dips that there are three modes corresponding with these three categories of geological structure.

The geological frameworks in which karst processes operate have their tectonodynamic aspects as well; earth movements take place during the time that caves evolve. The caves of Mount Hoyo in eastern Zaire are of dynamic phreatic nature, though inactive and opening out some 300 m up a fault scarp. The fault has moved since the caves formed and there is only modest vadose modification of the phreatic character. Brook (1976) maintains that the Finim Tel caves, including the great Selminum Tel, in Papua New Guinea, formed as the high Hindenburg Wall hinged up along a fault in the late Pliocene and the Pleistocene, tilting back caves. Thus the higher relict caves now have gradients directed down the former input points.

General epeirogenic uplift and depression affect cave development more widely but their effects resemble those of other factors such as glacioeustatic sea level change, which ranged from a few tens of metres above sea level to more than 100 m below repeatedly in the Upper Pleistocene. The lowerings permitted cave development below sea level in coral reefs, e.g., in the Bashmas (Dill, 1977) and in aeolian calcarenite, e.g., in the Bahamas (Bretz, 1960), which was subsequently drowned in the Holocene transgression. It is also thought that many of the submarine springs off the high limestone coasts of Yugoslavia, Greece and Turkey represent the mouths, not simply of bathyphreatic caves adjusted to present sea level, but of other kinds of cave developed when sea level was low.

High relative sea level can interfere in the evolution of caves also; thus the sea has pushed gravels into Kairimu Cave at about 100 m A.S.L. in the North Island of New Zealand and into Wet Neck Cave at about 60 m A.S.L. at Paturau in the South Island. To what extent these are due to interglacial high sea level and to epeirogenic uplift is as yet uncertain.

The commonest factor intervening in the evolution of caves is rejuvenation of drainage as a result of a negative movement of base level, whether this be sea level or a local temporary base level. Valley incision increases hydraulic head, energising underground water circulation. At Cooleman Plain, New South Wales, there are only two caves to be attributed to a phase of planation of the limestone prior to the cutting of gorges, to which all the larger caves are related (Jennings, 1967). Incision incurs the substitution of a vadose regime for a phreatic one of one kind or another. Frequently it leads to multi-level caves, each level of which may exhibit that substitution. Such a sequence is most plainly exhibited where epiphreatic caves develop at each outflow level; the nine levels of the Demänovskè Caves in the Low Tatra Mountains of Slovakia have been related to the terraces of the Vah R. and its tributaries. Another possibility is the abandonment of caves entirely, even in a nothephreatic state. The Cathedral Cave at Wellington, New South Wales (Fig. 12.3), is not quite so completely unmodified from this condition as is the Ochtinska Aragonite Cave in Slovakia. The Hölloch in Switzerland consists of a sequence of abandoned multiple loop cave passages and one active such level (Fig. 12.7).

With flat-lying beds sequences of levels in caves may be a simple expression of structure; the controversy about the Flint Mammoth Cave system is opposite here because both structural and rejuvenation explanations have been preferred for its five levels mapped as long ago as 1908. The last detailed study, however, by Miotke and Palmer (1972) firmly maintains that the main passages cut at a low angle across beds of varying resistance and are related to Green R. terraces going back well into the Pleistocene, dating now supported by palaeomagnetic determinations from the cave sediments by Schmidt.

Positive movement of base level leads to the filling up and even burial of caves with sediment so that our knowledge of such is likely to be less than for any other kind. Only when their contents are economically valuable are they likely to be excavated, though sometimes they are also microtomed by the quarrying of the surrounding limestone. Caves beneath a deep Tertiary valley fill were partially excavated in search of gold in the Canadian Lead at Gulgong in New South Wales (Jones, 1940). This alluviation probably has its origin in epeirogenic movement but more commonly the filling of caves is associated with Pleistocene climatic change acting directly, not through sea level change. Baldocks Cave at Mole Creak, Tasmania, is partly filled with gravels washed from glaciers on the plateau above (Jennings, 1967). The paragenesis of Renault (1967-8), whereby caves are stoped upwards by phreatic solution at the

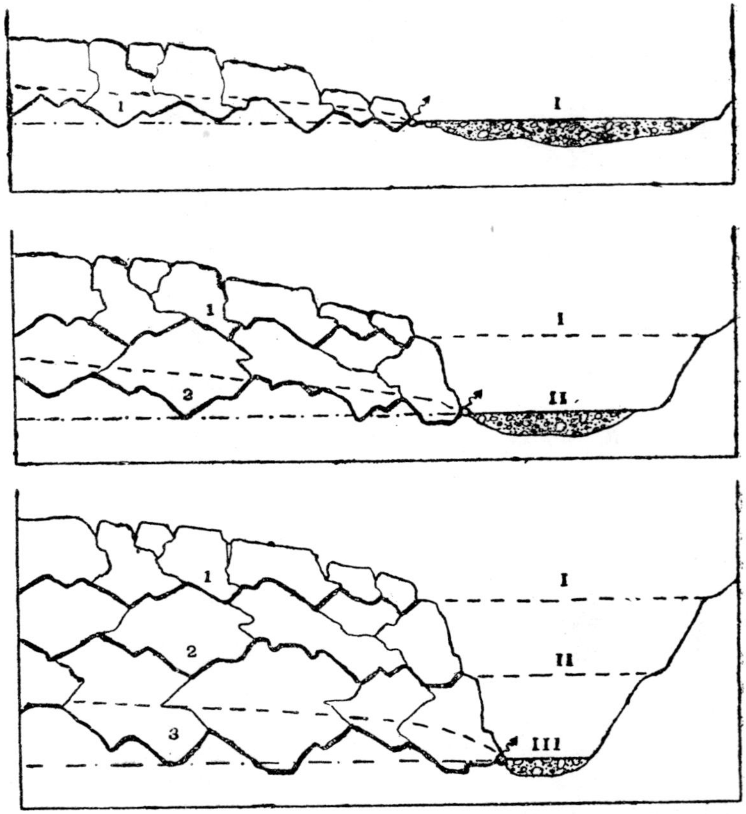

FIG. 12.7 : *Development of an Alpine Akarst cave of the Holloch Type (after Bogli)*

I, II, III,=Valley floor and outflow levels.

1, 2, 3=Cave evolution levels of multiple loop dynamic phreatic type.

top of accumulating fills, is favoured in circumstances such as these.

Even without filling in caves, glaciofluvial and periglacial fluvial sedimentation may intervene in cave development; the big development of caves in the divide between the Mole Creek and Lobster Rivulet valleys in Tasmania is attributed by Jennings and Sweeting (1959) to the decanting of water from the flanks of gravel fans of this origin into the limestone to form invasion vadose caves. Glacial meltwater at the base of the Columbia Icefield in the Canadian Rockies is feeding the active part of Castleguard Cave and its inactive upper level ends against a plug of glacier ice (Ford, 1971b). Permafrost inhibits cave formation within its zone and restricts it in the thin seasonally thawing layer above it but phreatic action still proceeds beneath it where geothermal heat keep water liquid. The presence of gaps in the permafrost (taliks) in Central Siberia allows phreatic water to burst up in large springs (Popov and other, 1972); this pattern is bound to exert an influence on the pattern of cave development below.

Glacial erosion has alternated with karst action in many karst areas in the Pleistocene, making it hard for example to weigh their relative roles in the production of closed basins as in the Mt. Owen Plateau in the South Island of New Zealand, which are part glacial erosion basins and part dolines. Caves lead off from these depressions so their history is involved as well. In the Craven karst of England, the recognition of glacial valley deepening, together with the exploration of new caves, has lead to significant revision of ideas about the cave development in general. Sweeting (1950) considered the main speleogenetic theme was in terms of successive levels of epiphreatic/vadose cave development in accord with Tertiary-Pleistocene fluvial valley deepening but in the light of the later knowledge, Waltham (1970) thinks that the valley deepening between the two main levels of caves about 90 m apart was due to glacier erosion in a glaciation earlier than the Weichselian.

Cases of climatic change affecting caves could be multiplied but the question remains whether there are any substantive differences between caves in different climates or whether it is a matter of the same processes operating in the same domains. This was the tenor of an earlier glance by the present author (1968) at this problem, since the differences mainly concerned deposits near the entrances and scarcely reached into the waterfilled parts, which are vital in cave development. Though their examples range widely in the world, Ford and Ewers (1978) did not bring in any climatic factors for their important generalisations. The question is, of course, beset by the perpetual problem of geomorphology that the factors involved are so manifold that the available array of cases may not be sufficient to disentangle them. Thus, whilst recognising many effects of geological structure, Lowry and Jennings (1974) attribute a dearth of caves in the Nullarbor Plain to a history of low precipitation since the area emerged from the sea in the middle of the Miocene. On the other hand, williams.(1978) thinks that this dearth may reflect lithology and relief rather than climate. To reverse the argument, one can point to a great difference in cave development in Oligocene-Miocene limestones not very much disturbed tectonically between those on the perhumid west coast of the South Island of New Zealand (Nile River; Paturau) and those on the much drier east coast (Craigmore; Mt. Cookson). Effective precipitation may well exert a detectable control despite the complications of time and structure.

More difficult altogether is the question of temperature, though its *net* effect on corrosion must be a positive one. The fact that the longest and deepest known caves lie in mid-latitudes may well not be an artefact of intensity of application of Western science but one has a subjective impression that caves of large volume are proportionately more common in hot humid climates with such examples to mention as the cave near Koripobi in Bougainville, Solomon Islands (Parker, 1970) and the Clearwater Cave in the Mulu in Sarawak (Brook and Waltham, 1978). Ford and Ewers (1978) have similarly ventured a notion that epiphreatic and mixed systems will be relatively more numerous than bathyphreatic and multiple loop systems in the tropics, though they link this also with young limestones of high primary permeability and joint frequency. The Mulu Caves limestones have low porosity however (Wilford, 1964). It may be some time before the intensity of cave exploration and mapping in the tropics will match that in the mid-latitudes to permit such speculations to be put to statistical test.

It will be as well, however, to end on lesser considerations more readily elucidated. Not all vicissitudes in cave evolution are due to large external factors. Thus, the progress of cave preparation—in the sense of initial enlargement of planes of weakness by vadose seepage and nothephreatic action—alters by itself the parameters for cave development. If density of fissuration in this sense is critical to the hydrodynamic domains to be established, it follows as Ford and Ewers (1978) argue that there can well be change in cave style as time progresses from bathyphreatic caves with early fissure infrequency, through multiple loops to epiphreatic caves when fissuration is dense. Again, there occur at a given level in a cave variations in style from vadose to phreatic. This association may be due to purely local factors such as big rockfalls and inputs of allogenic gravels, which block flow along roomy vadose passages and bring about mazes of even nothephreatic kind through deflection of it. Palmer (1975) proposes the first pattern occurs in Skull Cave, New York, and the second in Big Brush Creek Cave, Utah.

CONCLUSION

It has been necessary to confine this essay to limestone caves, though much of its content applies also to dolomite caves and some to gypsum caves. However, this restriction is unfortunate in a book intended primarily for use in India since much of India was part of Gondwanaland. This applies also to Australia and carbonate rocks are of limited extent in these ancient lands compared with countries built more of Mesozoic and Cainozoic rocks by the Alpine-Himalayan orogeny. Consequently more attention proportionately has been given by Australian speleologists to lave tunnels, caves in laterite break-aways, weathering caves in sandstone and other rocks, and sea caves than is the case elsewhere. It is true that there is much limestone in the Himalayas so that India has an advantage over Australia in this respect. Preliminary cave exploration suggests, however, that some of the Himalayan limestones have behaved plastically rather than in brittle fashion and so joints, which promote cave development, are not numerous in them (Waltham, 1972). So as speleology grows in India one can expect a greater interest in non-limestone caves than is at present common in European and N. American nations.

SUGGESTED READINGS

Bögli, A., 1964: Mischungskorrosion—ein Beitrag zur Verkarstungs-problem. *Erdkunde* 18, pp. 83-92.

Bögli, A., 1969 : Neue anschauungen über die rolle von schichtfugen und Klüften in der Karsthydrographischen Entwicklung. *Geologische Rundschau* 58, pp. 395-408.

Bögli, A., 1978 : *Karsthydrographie und Physische Speläologie*. Springer, Berlin.

Bray, L.J., and O'Reilly, P.M., 1974 : Preliminary oxidation studies in some waters from the Ogof Frynnon Ddu system; Breconshire. *Transactions of the British Cave Research Association* 1, pp. 75-84.

Bretz, J.H., 1942 : Vadose and phreatic features of limestone caverns. *Journal of Geology* 50, pp. 675-811.

Bretz, J.H., 1960 : Origin of Bermuda caves. *Bulletin of the National Speleological Society* 22, pp. 19-22.

Brook, D., 1976 : The karst and cave development of Finim Tel. *Transactions of the British Cave Research Association* 3, pp. 183-91.

Brook, D B., and Waltham, A.C. 1978 : *Caves of Mulu*. Royal Geographical Society, London.

Burke, A.R., and Bird, P.F. 1966 : A new mechanism for the formation of vertical shafts in Carboniferous Limestone. *Nature, London* 210, pp. 831-32.

Cailleux, A., and Tricart, J. 1956 : Le probléme de la classification des faits géomorphologiques. *Annales de Géographie*, 65, pp. 162-85.

Davis, W.E., 1960 : Origin of caves in folded limestone. *Bulletin of the National Speleological Society* 22, pp. 5-18.

Davis, S.N., 1966 : Initiation of groundwater flow in jointed limestone. *Bulletin of the National Speleological Society* 28, pp. 111-19.

Davis, W.M., 1930 : Origin of limestone caverns. *Bulletin of the Geological Society of America* 41, pp. 475-628.

Dill, R.F., 1977 : The Blue Holes – geologically significant submerged sink holes and caves off British Honduras and Andros, Bahama Islands. *Proceedings of the Third International Coral Reef Symposium, Miami, Florida* 2 Geology, pp. 238-42.

Droppa, A., 1966 : The correlation of some horizontal caves with river terraces. *Studies in Speleology* 1, pp. 186-92.

Ek, C., 1960-61 : Condiuts souterrains en relation avec les terrasses fluviales. *Annales de la Société Géologique de Belgique* 84, pp. 313-40.

Ewers, R.D., 1966 : Bedding-plane anastomoses and their relation to cavern passages. *Bulletin of the National Speleological Society* 28, pp. 133-40.

Fish, J.E., 1977 : Karst hydrogeology and geomorphology of the Sierra de El Abra and the Valles-San Luis Potosi region, Mexico. Unpublished Ph.D. thesis, McMaster University.

Ford, D.C., 1964 : On the geomorphic history of G.B. Cave, Charterhouse-on-Mendip, Somerset. *Proceedings of the University of Bristol Speleological Society* 10, pp. 149-88.

Ford, D.C., 1965 : The origin of limestone caverns : a model from the central Mendip Hills, England. *Bulletin of the National Speleological Society* 27, pp. 109-32.

Ford, D.C., 1971a : Geologic structure and a new explanation of limestone cavern genesis. *Transactions of the Cave Research Group of Great Britain* 13, pp. 81-94.

Ford, D.C., 1971b : Alpine karst in the Mt Castleguard-Columbia Icefield area, Canadian Rocky Mountains. *Arctic and Alpine Research* 3, pp. 239-52.

Ford, D.C., and Ewers, R.D., 1978 : The development of cave systems in dimensions of length and depth. *Canadian Journal of Earth Sciences* 15, pp. 1783-98.

Glennie, E.A., 1954 : Artesian flow and cave formation. *Transactions of Cave Research Group of Great Britain* 3, pp. 55-71.

Glennie, E.A., 1958 : Nameless strams : proposed new terms. *Cave Research Group of Great Britain Newsletter* 72-7, 22-3.

Halliday, W.R., 1960 : Changing concepts of speleogenesis. *Bulletin of the National Speleological Society* 22, 23-9.

High, C. and Hanna, E.K. 1970 : A method for the direct measurement of erosion on rock surfaces. British Geomorphological Research Group Technical Bulletin 5.

Howard, A.D., 1964 : Process of limestone cavern development. *International Journal of Speleology* 1, pp. 47-60.

Jakucs, L., 1977 : *Morphogenetics of Karst Regions*. Hilger, Bristol.

James, J.M., 1977 : Carbon dioxide in the cave atmosphere. *Transactions of the British Cave Research Association* 4, pp. 417-29.

James, J.M., Francis, G. and Jennings, J.N., 1978 : Bungonia Caves and Gorge : a new view of their geology and geomorphology. *Helictite* 16, pp. 53-63.

Jennings, J.N., 1967 : Some karst areas of Australia. pp. 256-92 in *Landform Studies from Australia and New Guinea*. (eds.) J.N. Jennings & J.A. Mabbutt. Australian National University Press, Canberra.

Jennings, J.N., 1968 : Underground geography. *New Zealand Geographical Society Record* 45, pp. 1-4.

Jennings, J.N., 1977a : Caves around Canberra. *Proceedings of the Eleventh Biennial Conference Australian Speleological Federation Canberra 1978*, 79-95.

Jennings, J.N., 1977b : Limestone tablet experiments at Cooleman Plain, New South Wales, Australia, and their implications. *Abhandlungen* zür *Karst—und Hohlenkunde A—Speläologie* 15, pp. 26-38.

Jennings, J.N., James, J.M., Counsell, W.J., and Whaite, T.M. 1972 : Geomorphology of Bungonia Caves and Gorge. *Sydney Speleological Society Occasional Paper* 4, 113-43.

Jennings, J.N., and Sweeting, M.M., 1959 : Underground breach of a divide. *Australian Journal of Science* 21, pp. 261-2.

Jennings, J.N., and Sweeting, M.M., 1966 : Old Napier Downs Cave, West Kimberley, W.A. *Helictite* 4, pp. 23-32.

Jones, L.J., 1940 : The Gulgong Goldfield. *Mineral Resources of New South Wales* 38, pp. 1-139.

Lowry, D.C. and Jennings, J.N., 1974 : The Nullarbor karst Australia. *Zeitschrift für Geomorphologie* 18, pp. 35-81.

Malott, C.A., 1937 : Invasion theory of cavern development. *Proceedings* of the *Geological Society of America for 1937*, 323.

Maucci, W., 1960 : La speleogenesi nel Carso Triestino. *Bolletino Societa Adriatica Scienze Naturali* 51, pp. 233-54.

Miotke, F.D. and Palmer, A N., 1972 : *Genetic Relationship between Caves and Landforms in the Mammoth Cave National Park Area.* Bohler, Würzburg.

Moore, G.W., 1960 : Introduction to the origin of limestone caves *Bulletin of the National speleological Society 22*, pp. 3-4.

Newson, M.D., 1971 : A model of subterranean limestone erosion in the British Isles based on hydrology. *Transaction of the Institute of British Geographers* 54, pp. 55-70.

Palmer, A.N., 1975 : The origin of maze caves. *Bulletin of the National Speleological Society* 37, pp. 56-76.

Parker, F., 1970 : Notes on a large cave in the Keriaka Limestone, Bougainville Island, Solomon Islands. *Helictite* 8, pp. 63-67.

Pengelly, W., 1864 The introduction of cavern accumulations. *Report and Transactions of the Devonshire Association* 3, pp. 31-41.

Picknett, R.G., 1972 : The pH of calcite solutions with and without magnesium carbonate present, and the implications concerning rejuvenated aggressiveness. *Transactions of the Cave Research Group of Great Britain* 14, pp. 141-50.

Picknett, R.G., Bray, L.G. and Stenner, R.D., 1976 : The chemistry of cave waters. pp. 213-66 in *The Science of Speleology.* (Eds.) T.D. Ford & C.H.D. Cullingford. Academic Press, London.

Popov, I.V., Gvozdetskiy, N.A., Chikishev, A.G. and Kudelin, B.I., 1972 : Karst of the USSR. pp. 355-416 in *Karst Important Karst Regions of the Northern Hemisphere.* (Eds.), M. Herak and V.T. Stringfield. Elsevier, Amsterdam.

Quinlan, J.F., 1970 : Central Kentucky karst. *Méditérranée* 7, pp. 235-53.

Renault, P., 1967 : Contribution a l'étude des actions méchniques et sedimentologiques dans la spéléogenesé. *Annales de Spéloléogie* 22, pp. 5-21, 209-67; 23, pp. 259-307, 529-96.

Renault, P., 1970 : *La formation des cavernes.* Presses Universitaires de France, Paris.

Sweeting, M.M., 1950 : Erosion cycles and limestone caverns in the Ingleborough district of Yorkshire. *Geographical Journal* 115, pp. 63-78.

Swinnerton, A.C., 1932 : Origin of limestone caverns. *Bulletin of the Geological Society of America* 43, pp. 662-93.

Trimmel, H., 1968 : *Hohlenkunde.* Vieweg, Brunswick.

Tricart, J., 1974 : *Structural Geomorphology.* Longman, London.

Trudgill, S.T., 1979 : Chemical polish of limestone and interactions between calcium and organic matter in Peat drainage waters. *Transactions of the British Cave Research Association* 6, pp. 30-35.

Waltham, A.C., 1970 : Cave development in the limestone of the Ingleborough district. *Geographical Journal* 136, pp. 574-84.

Waltham, A.C., 1972 : Geological environment of cave development *International Geography 1972*, 2, pp. 1328-330

White, W.B., 1960 : Terminations of passages in Appalachian caves as evidence for a shallow phreatic origin. *Bulletin of the National Speleological Society* 22, pp. 43-53.

White, W.B., 1969 : Conceptual models for carbonate aquifers. *Ground Water* 7, pp. 15-21.

White, W.B., 1976 : Cave minerals and speleothems. pp. 267-327 in *The Science of Speleology*. (Eds.) T.D. Ford & C.H.D. Cullingford. Academic Press, London.

White, W.B. and Longyear, J., 1972 : Some limitations on speleo-genetic Speculation imposed by the hydraulics of groundwater flow in limestone. *Nittany Grotto Newsletter* 10, pp. 155-67.

Wilford, G.E., 1964 : The geology of Sarawak and Sabah caves. *Geological Survey Borneo Region, Malaysia Bulletin* 6.

Williams, P.W., 1978 : Interpretation of Australasian karsts. pp. 259-88 in *Landform Evolution in Australasia*. (Eds.) J.L. Davies & M.A.J. Williams. Australian National University Press, Canberra.

Williams, P.W. & Dowling, R.K., 1979 : Solution of marble in the karst of the Pikikiruna Range, northwest Nelson, New Zealand. *Earth Surface Processes* 4, pp. 15-36.

THIRTEEN

L. I. Maruashvili

FEATURES OF MOUNTAINOUS KARST OF THE SOUTHERN PART OF THE USSR
Georgia as an Example

GEORGIA is a country on the Caucasian Isthmus, located at the southern slope of the Great Caucasian Range*. From the north to south its territory is subdivided into mountainous areas of the Caucasus, Anticaucasus and the intermontane depression between them. From the west to east it is subdivided into the western and eastern Georgia (Colchis and Iveria), which are located in the basins of the Black and Caspian Seas correspondingly. Georgia is the country where according to ancient Greek myths, the Argonauts had sailed and Prometheus was chained up to the rock by Zeus's order.

GENERAL CHARACTERISTICS

Karst is spread mainly in the northern part of the west Georgia—on the slopes and foothills of the Great Caucasian Range, making up a belt of 300 km long and 25-30 km wide. Vertically it is spread from the sea bottom (Black Sea) up to 2757 m above sea level (The Peak of Speleologists, Gagra Range).

The first records on Georgian karst are dated back to the first-half of the XVIII century. In 1745 Georgian prince Vakhushti Bagrationi thoroughly described his homeland in "Geographic Description of Georgia". Among many other

* Georgia is one of the fifteen republics of the USSR. Area 69,500 sq. km. (1.7 times larger than Switzerland).

geographical peculiarities of Georgia he has mentioned a number of karst formations, such as caves and Vauclusian springs. In the XIX century Georgian karst was described by geologist G. Abikh, I. Pantjukov, L. Leonov. At the beginning of the XX century the karst of Georgia was studied by the famous French explorer Edward-Alfred Martel and Russian scientist A. Kruber (the author of book "Karst Province of the mountainous Crimea").

After the Great October Revolution in 1917 the study of Georgian karst was sped up considerably, since with Russian scientists (N.A. Gvozdetski) the Georgian ones were participated (Sh. Kipiani, Z. Tintilozov, L. Maruashvili). A great deal of scientific works, books, etc., were dedicated to karst, caves karst hydrology, etc., karst phenomena. Presently about 600 karst caves are studied on the territory of Georgia. The standard of Georgian karst study is inferior to that one of such countries as the Crimea, France, Yugoslavia, USA but is superior to many countries of the Asia, Africa and America.

The geological substratum and geomorphological conditions of karst development in Georgia are of great variety. First of all, the Alpine karst on the southern slope of the Great Caucasian Range and the lowland karst of the Northern Kolkhida are to be marked out. The Alpine karst is composed of thick limestone strata dated from the Upper Jurassic to Eocene, while the lowland one is composed of the Cretaceous-Paleogene limestones as well as of Neogene clastic rocks (conglomerates, sandstones). The above formations are folded and their composition ratio is rising generally in compliance with the aging of rocks.

The climatic conditions of karst processes and the temperature regime, in particular, are also diverse. The mean annual temperature of karst regions varies from 15-16°C at sea level up to —2°, —3°C on the ridges of karst massives. The annual precipitations sum is up to 2000-3000 mm at external (seaward) slopes of karst massives and 800-1000 mm at intermontane depressions, at the eastern part of the west Georgian karst belt. All in all, the karst of Georgia is characterized with humid climate and heavy snowfalls in winter.

The river drainage of the karst belt is strong enough and

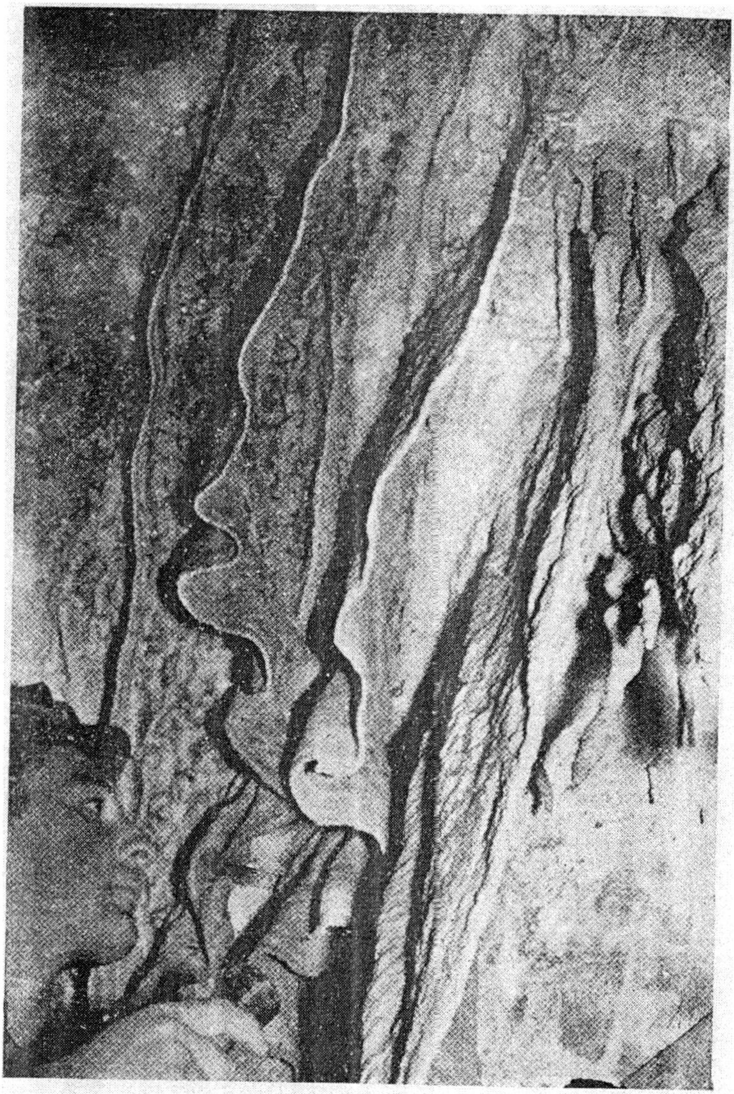

1. The goffers of calcite in the cave of Mtiskalta

2. The swallow holes on the Bzipi range

3. The macaroni type stalactites in the Abrskil cave

water abundant due to transit rivers flowing from the inner, non-karst regions of the Caucasian Range, through the karst belt. As a result, the belt is dissected by deep erosional gorges into numerous, separate massives (see below).

The vegetation cover of Georgian karst is rich. Mainly it is represented by Colchis flora with endemic species and relics of the Neogene subtropical flora. The greatest part of the karst belt (up to 1900-2100 m above sea level) is covered with forests, which in the low zone (up to 1200-1300 m) consist of leaf-sheddings species (oak, hornbeam, beech, chestnut), while in the upper zone of coniferous ones (spruce, fir). There are alpine meadows above the forest zone. In the lower part of the meadows, on the northward slopes in particular, thickets of Rhododendron Caucasicum with white rose-like flowers are widely spread.

The karst belt is inhabited by the peoples belonging to Iberia-Caucasian group. Its eastern part is populated by Georgians, the western by Abkhazians. The lower areas are densely populated and well cultivated, the higher ones are cultivated only in the East, while the rest are used mainly for timber exploitation and cattle breeding.

Georgian karst belongs to the so called "dressed" types, i.e., to the intermediate type between denuded or Mediterranean and mantled or the Central European*. It means that the surface of Georgian karst is mainly covered by soil and vegetation cover and is deprived of loose deposits (L. Maruashvili, Z. Tintilozov, 1971, "Geomorphology of Georgia", p. 468). In Georgia denuded and mantled karst make up separate sections. However, apart from the Alpine and Lowland karst, two lithologically stipulated types can be differentiated, such as—the limestone ("normal") and the clastic karst (or "clastokarst" according to G.A. Maksimovich, 1947).

Analytic Review

Going on to characteristics of the most interesting karst formations in Georgia, the author begins with the karst massives

* Some authors call this intermediate type—"turf-clad" or the Caucasian karst.

Under "karst massif" is meant compact mass of karstic upland which is not subjected to erosional, through cuts. Such a massif can be drained by several runoff systems having diverse directions. That is stipulated by the influence of river gorges limiting the massif and "intersecting" the subterranean runoff as well as by existence of several (usually two) independent karst strata divided by water-resisting layer. The total of karst massives in the western Georgia is over one hundred (L. Maruashvili, 1957).

It goes without saying that the largest karst massives belong to the Alpine karst, since the latter, due to its closeness to the Caucasian Range is much less subjected to fluvial erosion than the Lowland karst massives.

The largest Alpine karst massives are as follows : Askhi, Racha, Bzibi and Arabika. (Fig. 13.1). The area of each one is around several hundreds of square kilometres. The smaller massives are : Okhachkue, Migaria, Kvira, Khvamli, Gumishkha. Three ones, such as Arabika, Askhi and Bzibi are located over 2500 m above sea level. The Arabika and Bzibi massives bear well marked traces of ancient glacial activity (troughs, cirques, etc).

The largest Lowland karst massives are : Tskhaltubo, Okriba, Argveti, Tsebelda, Upper Imereti, etc. Their absolute height is not great in general, and just in some places reaches 1000 m above sea level.

The above ground karst formations are represented by corries, sink holes, funnels, kettleholes, poljes. For karst massives the canyons, deep scarps, waterfalls, Vauclusian springs, etc., formations are typical.

Magnificent *field of corries* is at the Arabika massif, near Gagra health resort, in the zone of Alpine meadows, near village Gelgeluk. Here bending, steep walled trenches divided by sharp-edged partitions occupy an area of six hectares, on the inclined by 20° stratum of the Urgonian limestone. In Zhoekvara through valley, at the same massif, the corries form something similar to decayed walls. The Bzibi massif is also rich with corries. In other karst regions the corries occupy comparatively small sections both, in the Alpine and forest zones.

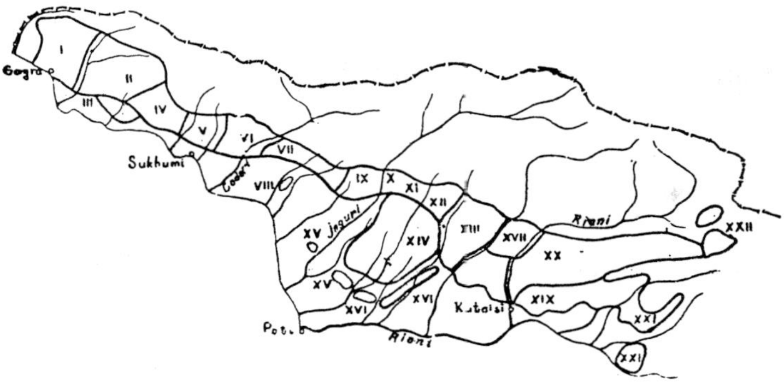

FIG. 13.1 : *Scheme of location of karst Massifs of Western Georgia*

I-Gagra
II-Bzibi
III-Gudauta
IV-Gumista-Psirtskha
V-Chaami
VI-Tsebelda
VII-Panavi Group
VIII-Dzali
IX-Okhachkue
X-Kvira
XI-Gaucha

XII-Mingaria
XIII-Askhi
XIV-Central Megrelia
XV-Urta
XVI-Unagira (group of Massifs)
XVII-Khwamli
XVIII-Tskhaltubo
XIX-Okribo-Argvete
XX-Radcha
XXI-Upper Imereti
XXII-Kudaro-Valkhokhi

Karst funnels and sinkholes are numerous at all massives. It is observed that their distribution is subjected to the certain law : at areas, which escaped the Pleistocene glaciation prevail "normal", conic funnels 10-20 m or 200-300 m in diameter, sometimes even more. However, those parts of Arabika and Bzibi massives which were affected by glaciation, prevail narrow types of sinkholes (with steep walls). This can be explained by absence of subglacial runoff and poor development of karst under glaciers. Since the last glaciation period was over about 10-15 millennia ago, it can be concluded that the development of large karst funnels requires many tens and hundreds of millennia.

There are *kettleholes and poljes* at some Alpine and Lowland massives. The largest one is Shaori polje, 137 km^2, at the Racha massif. Hitherto, the polje was considered as "cold trap" or "winter pole" of the whole Caucasus, due to temperature minimums up to —35 to —40°C. After construction of Shaori power station the climatic regime of the area has been changed. The Turchu polje and *Kvibia* kettleholes (Askhi massif), kettleholes of the *Okhachkue, Migaria,* etc. massives all belong to the Alpine karst. Among the poljes of the Lowland karst the Akhalsopeli polje is to be mentioned, which similar to Shaori polje is used as a reservoir for power station. All poljes and kettleholes in Georgia were formed by tectonic forces. Some of them have synclinal structure (Kvibia, Turchu, kettleholes at the Okhachkue and Migaria massives), the others were formed as a result of dolines partitioning by faults and thrusts (Shaori and Akhalsopeli poljes) Fig. 13.2

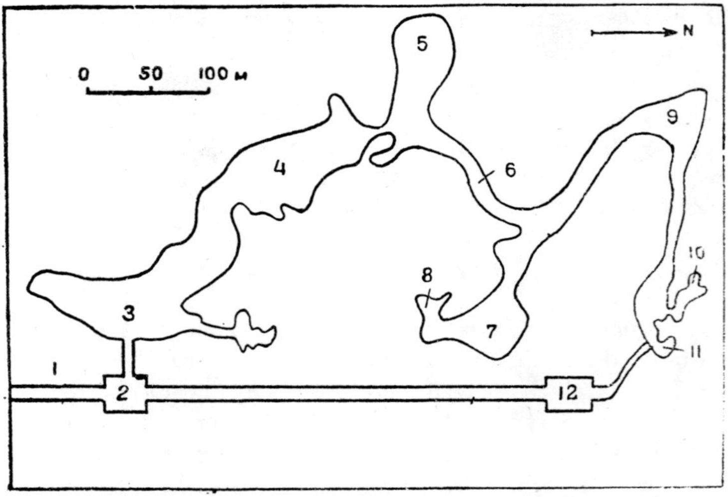

FIG. 13.2 : *The schematic plan of Akhali Atoni Cave*

1. The opening of transport tunnel and a platform.
2. A platform and the cave entrance.
3. The Hall of "Abkhazia".
4. The Hall of "Georgian Speleologists".
5. The Clay Hall.
6. "Canyon".
7. The Hall of "Moscow".
8. The Hall of "Sukhumi".
9. The Hall of "Iveria".
10. The Hall of "Tbilisi".
11. The Saloon of "Helictites".
12. The cave exit and a platform.

To complete characteristics of the above ground karst of Georgia, one is to mention its fluvial and erosional dissection as well as well development of such "satellites" of karst as relic dolines, canyons, scarps, waterfalls, etc. On average, the drainage of karst regions, is three times more sparse than in the non-karst regions, under similar climatic conditions. The difference in density of erosional dissection of karst and non-karst regions is comparatively less, since at limestone areas the relic dolines (erosional forms deprived of their streams due to karst drainage development) are quite common. The limestone canyons are common throughout Georgia. Among them the most remarkable are Akhatskhe, at river Bzibi, in north west part of the country and Okatse in the province of Imereti. The limestone scarps are also common, and the most deep ones (200-300 m) are located at the Alpine karst areas (the Khvamli, Askhi, Racha massives). The most effective waterfalls, sometimes over 100 m high, are at the Migaria and Askhi massives.

The whole karst belt of Georgia is rich with high flow streams (Vauclusian). The most remarkable one is to be mentioned below, in the paragraph about caves.

The underground world of Georgian karst, i.e., its "underground kingdom" is rich and diverse. The number of caves was already mentioned above. The caves being studied can be subdivided into several morphological types :

I. Subhorizontal caves :
   (a) Ascending
   (b) Descending in limestones
   (c) Trough
II. Abysses
III. Clastokarst caves :
   (a) Conglomerate
   (b) Sandstone

While making an attempt to distinguish the most interesting and important caves in Georgia, one may proceed from various points of view. A part of caves is notable for their dimensions (length, volume, depth). The rest for the richness of their natural decoration, hydrological regime, paleontologic and archaeologic treasures, etc. The complete characteristics of all caves of Georgia should require a volume.* That is why the author is to confine himself just to some of them.

Akhali Atoni Cave

(Novy Afon) is the largest by its volume (1,500,000 m^3) in the USSR (Fig. 13.2). It is located in the north-west part of Georgia, in Albkhazia, near towns Gagra, Gudauta and Sukhumi, at the karst massif Gumishkha-Psyrtskha. The cave consists of several giant halls. Some of them are richly decorated with calcite formation—stalagmites, stalagtites, columns, curtains, and many others. Before discovery of this huge cave only its sinkhole was known to the local population, who called it the "bottomless pit". In 1961 the members of expedition of Vakhushti Institute of Geography descended into the cave to the depth 180 m, and for the first time, broke an eternal silence of its underground palaces. The importance of this discovery was soon admitted by the society and government of the Republic. After thorough study (topographic, photo and cine shooting, etc.) of the cave, 1350 m long railway tunnel was cut into the depths of the massif, that was followed by the construction work. That made it possible to let the stream of sightseers into the cave, and to make their trip safe and sound. Now the cave is "the world champion" by the number of its visitors (up to one million a year). This is favoured by its location at the region of busy tourism and health resorts on a shore of the warm Black Sea.

As far as Akhali Atoni cave is distinguished by its volume, as much as Tovliani Abyss (Snezhnaya) is remarkable for its depth. Its sole, top entrance is at the crest of Bzibi Range (karst massif) at 1900 m above sea level, to the North from Gudauta. Its full depth is still unknown. The Soviet speleologists and mountaineers descended into Tovliani abyss to the depth 1180 m., thus setting the record of the USSR, that is the second by 200 m to the world record set in Pierre St. Marten, France. However, the prospects of Tovliani abyss are not

* The volume—"Kadastr of Karst Caves of Georgia" was published in 1966. Presently its second, revised edition is in the press.

exhausted yet, and the coming descends into the final huge well, where powerful stream is dashed down, will unvail its secrets.

Abrskil's Cave

At the first place by length among Georgian caves. It is located on the southern slope of Panavi Ridge, near the port of Ochamchire. The total length of its passages is up to 3 km. The cave consists of the main winding gallery where river Achkhitizgo flows and six halls. In the depth, the main gallery is divided into four passages. The cave is luxuriously decorated with calcite formations—stalagtites, helictites, columns, curtains, etc. According to the legend, Abrskil, the hero was chained to the wall with his stallion by the enraged God. In the near future the cave will be organized for mass tourism.

Cave Sathaplia

The vicinity of Kutaisi is very popular among tourists. It is well organized and decorated with calcite crystals. Near the cave the tracks of dinosaurs of the cretaceous era were found. The herds of ancient giant reptiles left deep tracks of their feet and tails on limestone slabs.

The following flooded caves are of interest :

Mchishta

A giant, still unexplored cave at the Bzibi Range on the Black Sea shore near Gudauta. It is a subterranean bed of river Mchishta, which has maximum discharge about 200 m^3/sec. The river is used for trout breeding. The river drains almost the whole Bzibi Range. It has been calculated theoretically, that the length of the underground system's main passage is at least 16 km., while its total length is supposed to be about some dozens of kilometres. The very narrow entrance and fast contrary river flow lay obstacles to cave exploration. A man will be able to set his foot inside the cave only after enlargement of its entrance, lowering of river level or its draining.

Tkibula-Dzevrula

A descending cave, in general, with succession of vertical and subhorizontal sections. Before construction of Tkibuli power station it was unpassable for the man due to deep river flowing through the cave. After construction of the dam, the water level has been decreased and speleologists from Kutaisi have used the chance, reaching depth 290 m. However, one is to be very careful while descending into the cave, since in rainy weather the cave is flooded, that caused a human death once.

Mothena Cave

Is located in Samegrelo, near town of Gegechkori. It is interested from the point of view of history and water supply. A clear, cold river is flowing through the cave, that is used for water supply of village Balda. Due to its water provision the cave was used by the medieval Georgian feudals as a fortress. There is a wall with loop-holes and sites for the rifle-men at the entrance. In XVI-XVIII centuries the cave was used for road defence along which the cattle herds were driven from Kolkhida lowland to the Alpine pastures at the Askhi massif.

Arsena Okrojanashvili's Cave

Is located at the scarp of Askhi massif, just above a beautiful waterfall. The cave is a final part of subterranean "diving" river Turchu-Toba. It has been explored up to 1350 m (the total length of the cave from river Turchu's sinkhole to the cave entrance is three kilometres). There is 25 m high waterfall inside the cave. The cave was named after young, zealous speleologist A. Okrojanashvili who perished during explorations at the Askhi massif. The memorial plaque in the honour of the brave "underground pilot" is installed in the upper reaches of river Toba.

There are many ice caves in Georgia. They are distributed at the altitude 1200-1300 m and over. The ice caves are called "Sakhinule" in Georgian. There are dozens of hundreds of them at the Arabika and Bzibi as well as Racha, Khvamli, etc. massives (Fig. 13.3). Ice is formed in the sinkholes and at the top parts of abysses due to snow drifts in winter time and its successsive melting and freezing, as well as due to freezing of subterranean water flowing from cracks and passages. The Arabika massif is particularly rich with sinkholes decorated with ice stalagmites, columns, stalactites, mushrooms, etc., ice forms. Some ice caves being explored are described below.

Boga Ice Cave

It is at the Khvamli massif, at 1700 m above sea level. Its entrance is at the bottom of a funnel. A visitor finds himself in sub-horizontal passage where walls are armoured with ice, stalagtites and curtains of ice are hanging from the ceiling and the floor is covered with ice sheets. There is mouth of a pit, at the end of the passage, its walls encrusted with ice. The ice in Boga cave survives the whole summer and only in September suddenly thaws away. At the same time a rapid water stream bursts out from caves Verdzistava, located by 1300 m below Boga cave, near village Tvishi, in the valley of river Rioni. This indicates at subterranean link between these two caves.

Skhvavi Ice Cave

It is at the Racha massif, at 1450 m above sea level. It begins with snow intake funnel that turns into descending passage. The caves total depth is 80 m. The cave is decorated with ice stalactites, curtains, columns (among them the junction of a thin and long stalagtite with a massive stalagmite 15-20 m high attracts attention). The fantastic world of ice is completed with underground glacier with a terminal moraine. Ice from Skhvavi cave is used for medical and practical purposes long since.

However, the caves accumulate not the ice only, but snow as well. From this point of view Vakhushti Bagrationi's and Martell's abysses at the Arabika massif, as well as the above mentioned Tovliani abyss at the Bzibi massif are of particular interest. Martell's abyss, at least 100 m deep, is a crater-like cylindrical well. In winter it is filled with snow that melts around the walls in warm seasons and forms pointed and steep snow cone up to 35 m high. A cone of less regular shape is present at the top part of Vakhushti Bagrationi's abyss. The steps are to be cut down on the slopes of the snow cone to descend. In Tovliani abyss the snow is present at the depth 180 m.

Many karst (limestone) caves in Georgia attract attention due to *paleontologic, paleoanthpologic, archaeologic, climatic-stratigraphical, etc. finds.* In this respect Tsutskhvati, Kudaro, Tsona, Sakajia, Jruchula, Sagvarjile, Gvarjila's Klde, Tskhrajvari, Apiancha, etc., caves are of great interest.

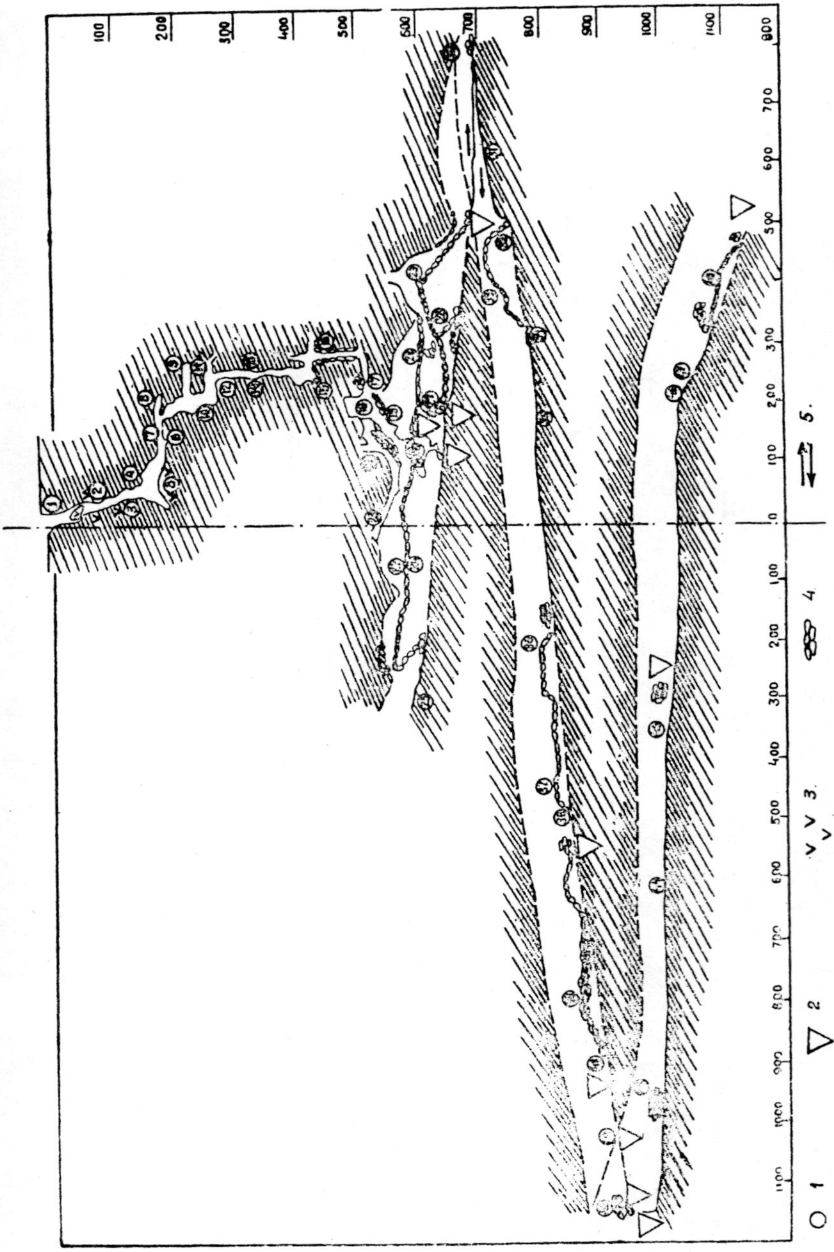

FIG. 13.3 : *Vertical section of Tooliani Precipice on the Bzibi Range*

CONVENTIONS

Absolute height of the entrance—1975 m, reached depth of the descent—1180 m.

1—halls, wells, galleries.
2—waterfalls.
3—snow and ice.
4—stone obstructions.
5—course of the underground river.

Meaning of figures in circles

1. Entrance well
2. Hall of "Gvozdetski"
3. Crooked well
4. Great Hall
5. Ice cone
6. Minor Hall
7. Flayer
8. Spotted deer's track
9. Labyrinth
10. Corallite well
11. Gallery
12. Prewell
13. Great well
14. Black scarp
15. Hall of "University"
16. Neutral obstruction
17. Waterfall streamlet
18. First obstruction
19. Second obstruction
20. Hall of "Good Hope"
21. Flower track
22. Suite of rooms
23. Track ascending from the suite
24. Sink
25. Upper river
26. Third obstruction
27. Fourth obstruction
28. Fifth obstruction
29. Hall of "Victory"
30. Stream of error
31. "Khum" on the deep river
32. Sixth obstruction
33. Dolmen Hall
34. Streamlet-small jet
35. Small river
36. Seventh obstruction
37. Forgotten seat
38. Pot hole
39. Canyon
40. Thundering hall
41. Zigzag of success
42. Hall of clay stalagmites
43. Clay obstruction
44. Akvaflayer
45. Lake waterfall
46. Handle
47. Roaring cascade
48. Hall of "Institute of Geography"

Tsutskhvati Multi-tier Cave System

It is located to the east from the town of Kutaisi, at low Okriba-Argveti Ridge, at 250-350 m above sea level (Fig. 13.4). It consists of 11-13 tiers, distributed within the vertical range of 70-80 m, but not precisely above each other. Length of the main, through passage is about 200 m, volume 75000 m^3. Above the third tier, the Pleistocene deposits are present at all tiers. The complex study of Tsutskhvati cave, carried out by Vakhushti Institute of Geography during four years, has established 39 types of fossil fauna*, the residua of the Mousterian, Upper Paleolithic, Eneolithic, Bronze, Antic and Medieval cultures, as well as spore and pollen complexes of the Middle and Upper Pleistocene and Holocene. However, the most remarkable finds at the complex were the special cult cave, a tooth of Neaderthal man and the most important—the geological section, that reflects the climatic history of Kolkhida during the last 100-120 millennia (see "The Study of Kolkhida Caves..." 1978). For the last 40,000 years the climatic range consisting of 5,800 year cycles is pictured. The each cycle included a short period of temperature drop and much longer period of comparative warmth.

The Complex of Kudaro Caves

It is located at the Alpine zone, on the southern slope of the Central Caucasus, to the east from Oni, at 1600-1700 m above sea level. The complex is composed of six tiers. In the majority of caves the pleistocene deposits with residua fossil fauna, cultures of the Lower and Middle Paleolith, Mizolith and later periods were excavated. The most ancient stratum belongs to the Middle Acheulean. Among pleontologic finds the most interesting are monkey (Macaca sp.), a great deal of salmons (Salmo sp.) and rodents that prove arid climatic conditions. Kudaro caves were inhabited in warm periods only, while during the periods of temperature falls the human beings have deserted them due to severe climatic conditions (V.P. Ljubin, 1959).

Tskhrajvari Cave

Is is at the crest of the Racha karst massif, at 1100 m above sea

* The presence of arid fauna species, such as Vormela peregužna and rodents, peculiar for forestless landscapes is of special interest.

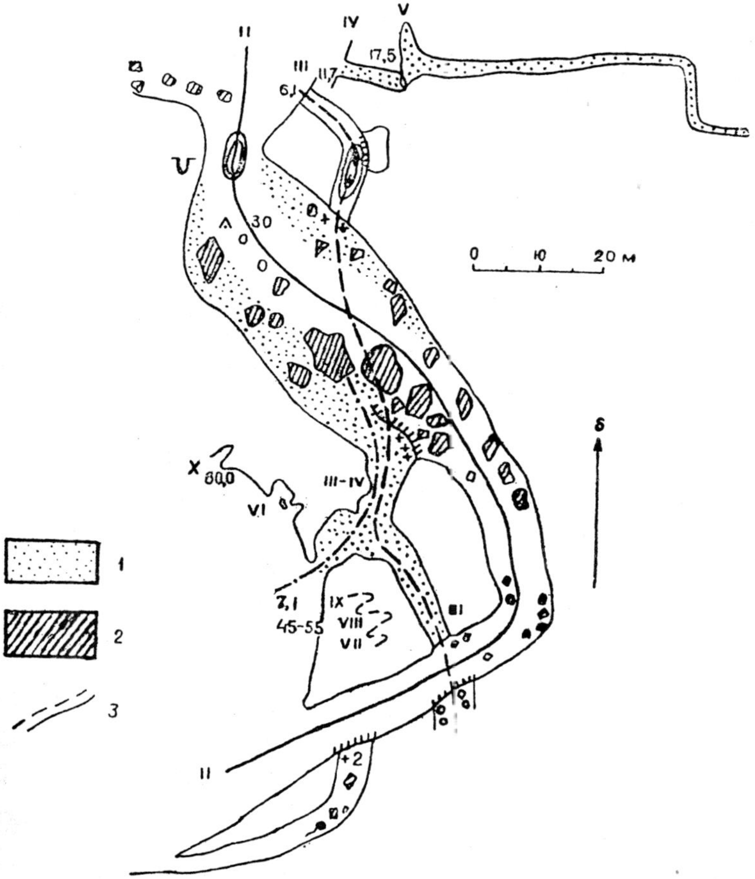

FIG. 13.4 : *Plan of the multi-storeyed cave of Tsutskhvati*

1—unconsolidated deposits

2—limestone blocks

3—recent and former courses of the cave river

Roman numerals—number of storeys

Arabic numerals—relative height of caves

a group of stalagmits "Conspirators".

level. In the course of centuries it was bear's trap. The animals entered the cave looking for water and failed to find the exit. A lot of skeletons of cave and brown caucasian bears (Ursus spelaeus and Ursus Arctos) and intact sculls were accumulated in the cave.

Sakazhia Cave

Near the town of Kutaisi—A place where first Paleolithic culture was found in Transcaucasia (1914). The loose filling of the cave is composed of three strata. The top and bottom ones include tools of the ancient man and residual fauna, while the middle one is blank. R. Shmidt and G. Nioradze have excavated the top stratum at different times and found the traces of the Upper Paleolith. After many years, a daughter of the latter, archaeologist woman Medea Nioradze, has removed the middle horizon and found the remnants of Mousterian time in the bottom stratum. Thus, two periods of the man inhabitance in the cave, divided by a period of his absence were proved.

Tekenteri Cave

It is the unique one. It is located at the Khvamli massif, at 1400 m above sea level. It has only one entrance—through the hollow of a large beech tree. The hollow has no bottom but the karst sinkhole 18 m deep, with a lake of cold, clean water. There are many metal buckets and pitchers floating on the surface or sank into the lake, that indicates, that cave water was used by herdsmen and woodcutters during many years. The bottom of sinkhole is connected with horizontal passage where water stream forms a waterfall, judging by a noise from the depths. The lake at the bottom of sinkhole is two metres deep.

All the above said concerns limestone caves only. However, up to 30 *clastic karst caves* are investigated in Georgia. They are scattered at the western and central parts of the karst belt at Western Georgia—at ethnographic provinces of Abkhazia and Odishi. The caves are developed in Meotian conglomerates (Lower Pliocene) and Karaganian sandstones (Middle Miocene).

Kalichona Cave

The longest clastic karst cave in Georgia is located to the east from Chkhorotskhu. It is 990 m long and has complex, almost circular outline, in plan. There are 14 hall-like chambers in it. The cave is mainly waterless, but at a place there is a stream disappearing shortly. The cave is developed in conglomerates.

Nazodelavo Cave

The total length is 600 m. It has a form of branching tree, in plan. There is a river with clean and cold water, which flows in erosional canyon, being cut down into flat bottom, with residual terraces. The canyon is 7-8 m deep in places.

Garakha Cave

It is 320 m long and is similar to the latter. However, it has some peculiarities, such as, the huge entrance ("vestibule") and spacious mouth of the main passage that resembles an underground terminal. The folds at the walls (anticlines and sinclines), less depth of erosional down cut into the primary bottom of the cave which is developed in the Pliocene conglomerates.

Savekuo Cave (conglomerates)

It is the exceptional one among clastic karst caves due to clearly developed tiers. There are two tiers in the cave: the young one—strongly twisted (meandering); and the ancient—almost rectilinear. The both tiers often cross each other.

Kortskheli Cave

It is 790 m long and is of great interest. As far as the author knows, it is the largest cave developed in the sandstones (miocene). The main passage is subhorizontal, rather like the highway tunnel, with a small stream on the bottom. Almost the whole roof is cut along by vertical crack—that is "guilty" of the cave development. Another reason of the cave origin is carboniferous cement that fastens together the sand grains. There is a lot of bats in cave. Cave is passable through. Its middle part is branched like a railway siding or funicular railway in Tbilisi.

HISTORY OF DEVELOPMENT

Georgian karst is relatively young formation as the whole relief of the country. The most ancient karst forms of relief, being dated paleontologically and archaeologically, like caves (Kudaro, Tsona, Tsutskhvati) in particular, would have been formed (reach dry-passage development stage) at the Early Pleistocene. According to correlation of diameters of post-glacial sinkholes at Arabika and those of non-glacial regions, the latter type of karst forms should be older than the Pleistocene. Much more younger are clastic karst caves, that have been formed after the Early Pleistocene or even after the Middle Pleistocene.

The scientists of many countries on the globe take a great interest in correlation of karst caves and their tiers with river and marine terraces. The most devoted supporters of this idea acknowledge the simultaneous development of terraces and caves at the same level above the flood or coastal plains. However, the existence of such a synchronous mechanism is disproved by the actual state of affairs and theoretical considerations.

The point is that both, the terrace and cave development are regulated by the same factors (tectonic movements and eustasy of effective reservoirs), and the each one responds to these factors in its own way. However, if terrace development is adapted to erosion basis, the karst cave development has its own karst basis. Depending on geological structure the karst basis is located either under the bottom of a flood plain and sea level or above them. The possibility of the first variant (i.e. deep location of karst basis) is provided by the ability of water, being under pressure higher than atmospheric, to percolate under the local erosion basis and even to overcome the sea water column pressure (this accounts for existence of sub-marine springs). The second variant (i.e. location of karst basis over the local denudation and erosion basis) is stipulated by the position of karst stratum contact line with underlying impervious rocks above the bottom of a valley or the bank of a reservoir. A striking example of the first variant is *Kelasuri cave* (in the vicinity of Sukhumi), where air filled sinkhole reaches depth 100 m lower the level of Kelasuri river and

4. The Arsen Okrojanashvili's cave

5. The karst landscape of the Okhachkue massif

changes into meandering subhorizontal passage with a stream on the bottom. An example of the second variant is the above mentioned *Mothena cave.* Its entrance (the source of a river) is on a steep slope, 250 m higher the level of river Abasha, at the contact line of the Upper Cretaceous limestones with underlying impervious stratum.

Prior to correlation of terraces with caves, one is to agree upon—how to determine their age? The point is solved quite simply for a terrace, since its age is recorded from the moment when its shelf (bench) is formed and a river starts to downcut it, thus making a slope of the terrace. As to caves, the point is more complicated, since there is no such a turning point which can be considered indisputably as the "zero" for estimation of their age. The zero can be considered when an intensive development of a cave volume is seased and a cave is getting filled with various deposits. This stage of cave development can be continued during millennia—while a cave is in the state of dynamic equilibrium, i.e., the accumulation of deposits is partially compensated by corrosion caused by condensed and infiltrated water and periodically due to wash out by a stream.

Naturally, a terrace development either at the flood plain or at the bank of a reservoir may coincide in time with a stage of cave development mentioned above. However, the altitudinal position of a terrace bench and the "synchronous" cave can be the same in exceptional cases only. The point is that a cave starts to develop when the preceding cave tier continues its active development (newly formed cave) and passes through the following stages and substages of evolution (according to the scheme by G.A. Maksimovich—L.I. Maruashvilis, 1963, 1968) :

I. Phreatic stage with substages :

(a) Crack
(b) Chink
(c) Channel

II. Vadose stage (water-passage) with substages :

(a) Vauclusian
(b) Stream-passage
(c) Periodical stream

III. Waterless stages with substages :

(a) Dry-passage
(b) Grotto-chamber

An "embryo cave" is lying under erosional surface, at a certain depth, and during the whole phreatic stage water in the cave is under pressure, since it has no free discharge. An opening of the bottom exit, reduces water pressure, lets the air inside, increases water flow and develops erosion. Naturally, cave opening takes place during the phase of deep surface erosion of a flood plain (or during the phase of sea level decrease). Then comes the Vauclusian stage of morphologic-genetic cycle. The following development of a cave (during the vadose stage) goes on as a process independent from morphological fate of the flood-plain (or the sea-shore). A terrace can be formed either prior or after reaching by a cave the full substage of vadose stage, and only in rare cases, quite incidentally, these two acts may coincide in time. Even more rarely (almost never) a terrace is developed at the same level with the synchronous cave.

Georgian karst caves are opened at sub-terrace, inter-terrace and over terrace slopes almost exclusively. The age and altitude of caves practically do not coincide with that ones of the terraces. Moreover, the number of levels in terrace series of the flood-plains and in multi-tier cave systems is not the same as a rule. The striking example of full disregard of cave development to terrace development is clearly expressed by multi-tier cave systems of Tsutskhvati and Kudaro. The number of tiers of the cave systems is considerably higher than the number of terrace levels in the area, while the heights of tiers (Tsutskhvati) are lower, than that of terraces on one hand, and higher (Kudaro caves) on the other.

In this manner, the karst caves of Georgia refute the idea simplifying the real state of affairs on synchronous development of caves and terraces at the same level and with the same rate. This refutation is valid at least for the karst of tectonically mobile regions. Development of tiers of the karst caves is the peculiar, rhythmic geomorphological process.

SUGGESTED READINGS

Geomorphology of Georgia, 1971 : Ed. by L.I. Maruashvili. Publishing House "Metsniereba", Tbilisi.

Kiknadze, T.Z., 1979 : *Geology, Hydrogeology and Activity of the Limestone Karst* (Abkhazia as an example). Publishing House "Metsniereba" Tbilisi.

Kipiani., 1974 : Sh. Karst of Georgia. (In Georgian language). Publishing House "Metsniereba".

Ljubin, V.P., 1959 : High mountain dwelling cave Kudaro I (The South Oseti). *Izvestija Vsesojuznogo Geograficheskogo Obschestiva*, No 2.

Ljubin, V.P., 1977 : *The Mousterian Cultures of the Caucasus*. Publishing House "Nauka", Leningrad.

Maksimovich, G.A., 1963 : *Fundamentals of karst study*. Perm, Vol. I, Vol. II-1969.

Maruashvili, L., 1957 : Introduction to Georgian translation of N. Kaster's book *Ten Years Underground*. Tbilisi, (In Georgian language).

Maruashvili, L. 1963: Light in the darkness (Caves of Georgia). Publishing House "Sabchote Sakartvelo", Tbilisi. (In Georgian language).

Maruashvili, L., 1968 : *Evolution of Karst caves*. 21st International Geographical Congress. "Abstracts of Papers". Edited by Siv Prasad Das Gupta and Teresa Romanovska-Lakshmanen . . . *National Committee for Geography*. Calcutta.

Maruashvili, L., 1969 : The morphological analysis *of Karst caves*. Collection "Ocherki Fizicheskoy Geograffi Kavkaża". Publishing House "Metsniereba", Tbilisi.

Maruashvili, L., 1972 : Tiers development in Karst Caves as the peculiar rhythmic geomorphological process. *Doklady AN SSSR* p. 202, No. 3.

Maruashvili, L.I., Tintilozov, Z.K., 1979 : *Clastic Karst Caves of Georgia*. Collection "Priorodnye resursy Gruzii i metody ikh issledovanija". Publishing House "Metsniereba, Tbilisi".

Soviet Georgia, 1972 : Progress Publishers, Moscow.

Study of Caves of Georgia, 1978 : Tsutskhvati multi-tier karst cave system. Publishing House "Metsniereba", Tbilisi,.

Tintilozov, Z.K., 1976 : Karst Caves of Georgia. Publishing House "Metsniereba", Tbilisi.

Vakhushti Bagrationi : Description of Georgian Kingdom. Pubiished by Geographical Society of Georgia. Tbilisi, 1941 (In Georgian Language).

SECTION THREE

GLACIAL AND PERI-GLACIAL PROCESSES AND LANDFORMS

INTRODUCTION

A considerable amount of research work have been done on the continental and mountain glaciation. The Pleistocene glaciation has received great attention by geologists and geomorphologists. The study of Caucasian isthmus from the point of view of paleo-glaciation began in the II part of the last century. At the end of 19th century the paleological studies of the Caucasus were developed under the influence of West European ideas based on the study of the Alps. The West European ideas were subjected to criticism since 1956 and first attempt was made to work out the objective scheme of Caucasian glaciation by Maruashvili. However, the Alpine tradition still exists in the works of some investigators of the Caucasus. The paper by Maruashvili and Khazardze has attempted to recognise ancient glaciation of the Caucasus and have tried to challenge the ideas of West European Alpine tradition.

The paper also attempts to give distribution of glaciers in different parts of the Caucasus. In fact the largest glaciers of the Caucasus are smaller than those of Pamir's, Hindukus, Himalayas and Alps. The authors have studied the Pleistocene glaciation at a great length. The glaciation of this region was entirely of mountain valley type except local areas, i.e., volcanic plateau of Keli where an entire ice-sheet was formed.

The area and length of largest pleistocene glaciers of the great Caucasus exceeded the present day glaciers of the Alps. In the last the authors have cited some problems of the Caucasian glaciation chronology. The Caucasian glaciation chronology is poorly studied—there is only some absolute dates mentioning of the Holocene. Hence the use of physical methods to solve the problem of local glaciation in time is rather limited. They have stated that under general characteristics and regional survey of the ancient glaciation at the Caucasus the age sequence of glacial formations was purposely ignored since it was impossible to determine it reliably for all basins and regions. The second problem of the Caucasian glaciation chronology is the number and absolute age of the last glaciation retreat stages as well as their relation to the analogous stages of the Alps and Russian plains.

Jaromir Demek in his paper on Periglacial geomorphology has reviewed the present problems and future prospects. The main development of Periglacial geomorphology took place only after the II World War. According to J. Tricart, W.M. Davis' geomorphological school regarded chronogenic forms as rarities. Prior to 1945 the attention of geomorphologists in most countries was restricted mainly to small chryogenic forms (patterned, ground, store and stream). Since 1930 only in the USSR, considerable attention has been paid to the study of permafrost. Since 1945 geomorphologists engaged in the study of processes and relationship in cold environments.

All over the world geomorphologists have concentrated their attention on the investigation of permafrost and incidental chryogenic phenomena. J. Demek has considered permafrost as a system and has presented a schematic model of Alas development as a result of permafrost degradation. He has concluded that in connection with the economic utilization of Arctic and sub-Arctic regions, better information has been obtained in recent decades concerning chryogenic forms and processes in the Periglacial zone. Simultaneously the practical significance of such knowledge in the utilization of these regions has become apparent. But the amount of information is still insufficient in comparision with knowledge of processes operating in other climatic zones.

The quantitative measurements of chryogenic processes using field equipment are still relatively undeveloped in counting many differences because of the unfavourable climatic conditions in the periglacial zone. According to him the future development of periglacial geomorphology will be more detailed knowledge of the process acting in the permafrost as in a dynamic natural system.

FOURTEEN

L. I. Maruashvili and *R. D. Khazaradze*

ANCIENT GLACIATION OF THE CAUCASUS (THE USSR)

THE STUDY of the Caucasian Isthmus, from the point of view of paleo-glaciation, began in the second part of the last century (Abich, Palgrave 1872). At the end of the XIX century paleological study of the Caucasus was developed under the influence of West European ideas based on the study of Alps. Since 1956 the West European ideas were subjected to criticism and the first attempts were made to work out the objective scheme of Caucasian glaciation (Maruashvili, 1956). However, the "Alpine tradition" still exists in the works of some investigators of the Caucasus. The Caucasus geologically is a complex region with diverse relief. Two mountain systems—the Great Caucasus (Kavkasioni) and the Caucasus Minor are spread from the East coasts of the Black and Asov seas to the west coast of the Caspian Sea. Among these two systems, in the same sublatitudinal direction (from N to S) lies the Transcaucasian midland, while to the south of the Caucasus Minor the Transcaucasian Upland is located. To the north of the Great Caucasus lies pre-Caucasian plain adjacent to the Russian plain. The main summits of the Great Caucasus by their absolute height exceed the Alps by 200-800 m. However, it does not mean that during the Pleistocene the conditions for glaciers development in the Caucasus were more favourable than in the Alps. On the contrary, the orographic conditions of the Great Caucasus (comparative simplicity of the mountainous belt's plain) and its climate are not, and never were favourable for the existence of such a powerful glacial cover, as in the Alps. The

Caucasus is located to the South from the Alps by 5° on the average, and consequently is warmer by several degrees. Moreover, the Eastern and partly the central regions of the Great Caucasus are characterized by dry, continental climate unfavourable for accumulation of snow.

Even less favourable conditions for ice accumulation are peculiar to the Caucasus Minor, which is located to the South from the Alps by 6-8°, is considerably lower and dryer in general.

The Transcaucasian volcanic Upland that is adjacent to the Caucasus Minor from the South is comparatively more favourable for glaciation. Its average height exceeds the Caucasus Minor as well as the Great Caucasus. However, the conditions for glaciers origin during the recent epoch are practically absent there, while during the Pleistocene they took place at the highest volcanic massives only.

THE GLACIERS OF THE CAUCASUS IN THE RECENT TIME

The recent glaciation is almost totally concentrated at the Great Caucasus and above all in its central part, between (Elbrus) Ialbuzi and Kazbegi mountains, on the northern slope of the Tergi (Terek) basin in particular. The glaciation occupies considerable area on the southern slope of the Great Caucasus in the basins of the Inguri and Rioni rivers and comparatively small area in the upper reaches of the Mtkvari (Kura) tributaries (Liakhvi and Aragvi). At the Wastern Caucasus (to the West from the upper stream of the Kubani and Nenskra rivers) there is a considerable number of glaciers, though the small ones prevail. Their distribution on the macroslopes of the Main ridge is just the same as in the Central Caucasus. And finally, the Eastern Caucasus though higher than the Western one (to the east of the meridional section of Tergi and Aragvi upstream) is poorer with glaciers, both in their size and number (See Table 14.1). The glaciers are almost absent on the southern slope. In the Caucasus Minor the glaciers exist only on the Kapudjikh mountain (the Zangezuri Range), while in the Transcaucasian highland—on the volcano Aragats (The Catalogue

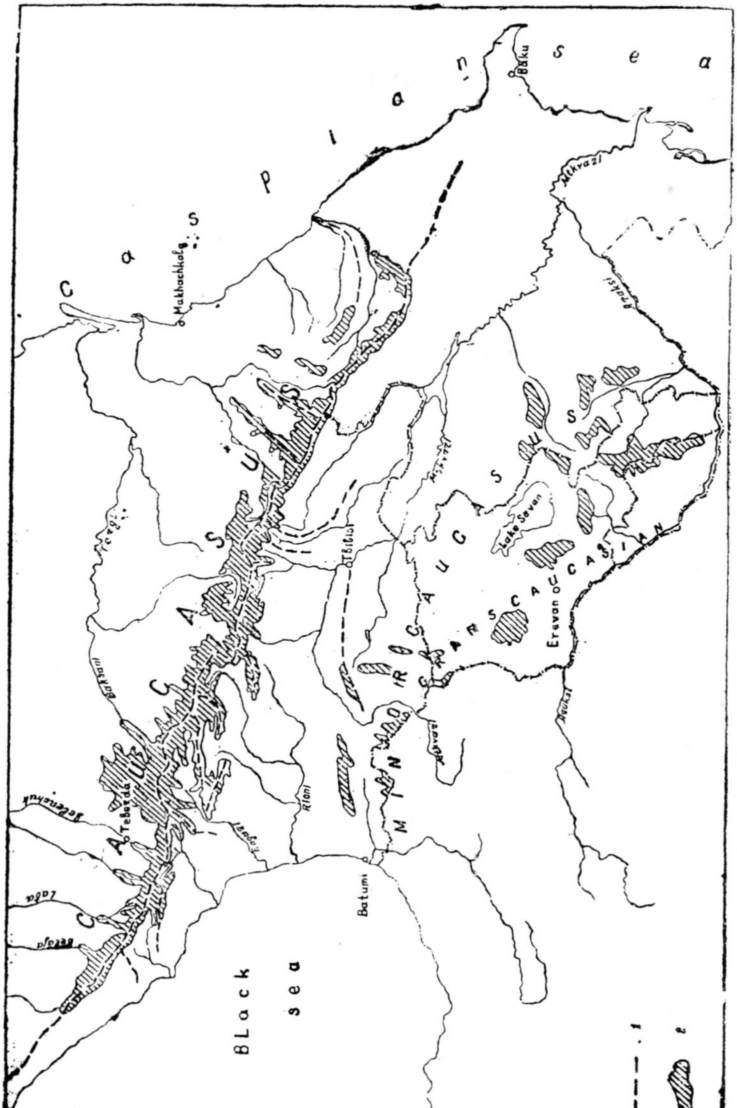

FIG. 14.1 : *Maximum pleistocene glaciation of the Caucasus*

1. Range Crests; 2. Glaciation areals

of Glaciers of the USSR 1967-1977, Podozerski 1911). (Fig. 14.1).

TABLE 14.1 : Distribution of the number and areas of the contemporary glaciers at the Caucasus

	The Great Caucasus					
	Number	Area km²	North slope		South slope	
			Number	Area	Number	Area
W. Caucasus	568	289.3	408	220.0	160.0	69.3
Centr. Caucasus	950	913.3	547	553.9	403	359.4
E. Caucasus	214	60.5	208	58.9	6	1.6
Total at Great Caucasus	1732	1263.1	1163	832.8	569	430.3
Caucasus Minor	Some	Insignificant				
Transcaucasian Upland	3+X	5				

The largest glaciers of the Caucasus are smaller than those of the Pamirs, Hindukush, and Himalayas, as well as of the Alps. Three largest glaciers are located on the northern slope of the Central Caucasus (Bezengi and Dikhsu in Balkaria, Karaugom in Oseti. All of them in the Tergi basin). The next three are on the southern slope of the Central Caucasus (Lekhziri, Tviber and Tsanner in Svaneti. All of them in the Enguri basin).

The largest glaciers at the western Caucasus are : Alibek and Southern Marukhi, while at the Eastern Caucasus-Kidegani, Belengi etc. (see Table 14.2).

TABLE 14.2 : Dimensions of the largest contemporary glaciers of the great Caucasus

Western Caucasus			Central Caucasus			Eastern Caucasus		
Name	L.	km²	Name	L.	km²	Name	L.	km²
Alibek	4,6	5.4	Bezengi	17,6	36.2	Kidegani	3,4	1.6
S. Marukh	3,9	2.2	Dykhsu	13,3	34.0	Belengi	3,2	2.9
			Karaugom	13,3	26.6	Murkar	2,7+ 0,7	1.6
			Lekhzir	11,8	33.7			
			Tviber	9,5	27.16			
			Tsanner	10,1	28.8			

Among the contemporary glaciers of the Caucasus there are polysynthetic ones, such as Tviberi in Svaneti. Up to 1930 it included five branches. (Presently one of them has separated and now exists independently). The Lekhziri glacier includes three main branches, merging at right angles.

The lower termini of the Caucasian glaciers are mainly located at the absolute height of 2000-3000 m. Only two glaciers Chalaati (1850 m)—on the southern slope, and Karaugomi (1830 m)—on the northern slope, are located below 2000 m above the sea level.

The peculiarity of the Caucasian glaciation is the existence of two radial glacial systems, being fed by firns of Ialbuzi and Kazbegi mountains. The upper reaches of these glaciers interflew into entire firn. The above peculiarity of the Great Caucasus is originated due to the young volcanizm (Pleistocene) and differs it from the majority of other muntainous regions.

Snow-line in the Caucasus is located at elevation 2700—3800 m. On the southern slopes of the western Caucasus (in Abkhazeti and near its boundaries, 2700-2900) it is very low. Rising to the east, the snow-line reaches 3700-3800 m in Dagestan. On the northern slope of the Great Caucasus the snow-line is generally higher than on the southern one, that is stipulated by distribution of precipitations. In the South of the Caucasian Isthmus, at the Caucasus Minor and the Transcaucasian Upland the snow-line rises to the south-east up to 3800 m on the Aragats massif and Zangezuri Range.

GENERAL CHARACTERISTICS OF THE PLEISTOCENE GLACIATION

Going on to the past glaciation of the Caucasus, it should be noted that the advance of the local glaciers down to the populated mountain valleys is reflected in the folklore of mountaineers. Svanetian poems recited by Nizharadze and put down by A. Krasnov tell us about the advance of Namkvami glacier from the upper reaches of Enguri through the community of Ushguli, up to the village of Davberi of Kala community, i.e. at a distance of 17 km (Krasnov 1891, pp. 363-365). This legend is based either upon the observation of the Svanetians of the signs of the ancient glaciation (troughs, moraines, blocks) located below the contemporary glaciation, or upon a recollection of a concrete event known by generations (one of the phases of glaciation, the "minor glacial epoch" in the XVII-XIX centuries, or the middle Holocene stage). The adherents of the application of the Alpine paleoglacial schemes to the Caucasus, with A.L. Reinghard at the head, admitted, and even now admit that during the glaciation stages of *the Quaternary period* the snow-line in the Caucasian mountains was lower by 1200-1300 m in comparison with the present one. The glaciers completely filled the river valleys of the Great Caucasus, passed to the foothills and covered some of the piedmont valleys. It is supposed that, the strongest glaciation took place in the Targi (the northern slope) and Kodori (the southern slope of the Great Caucasus) river basins as well as in the Enguri and Rioni basins. According to the above mentioned scientists, "The Great Tergi glacier" was even spread along the Osetian plain and reached the Kabardian highland, thus coming down to 500-600 m above the sea level. According to the same scientists, the Kodori glacier reached the boundary of mountainous belt and there (at 300-400 m above the sea level) joined a tributary—the Amtkeli glacier. These and some other ideas about the huge ancient glaciers of the Great Caucasus are corroborated by the presence of coarse-detrital material, resembling glacier moraines.

The latest investigations of many soviet scientists (L.A. Vardaniants, L.I. Maruashvili, A.N. Berezniakov, I.M. Velikovskaia, K.N. Pafenholts, R.D. Khazaradze etc.) proved, that sediments of different genesis, such as mudflow in the first place (as well as volcanic, deluvial, coluvial etc.) were taken for glacial moraines. The actual mudflow activity in the Caucasus is still rather intensive (especially in the east).

In the past it was even more intensive, due to the tectonic shoves, rock falls and catastrophic thaw of large glaciers. In some regions such as Kazbegi, it was even more intensive due to volcanizm. Hot lava streams thawed the huge snow and ice bodies. The mud-stone streams (lahars) gushed out along the valleys and carried away huge stone blocks. One of the largest boulders in the Caucasus-"Ermolov's block" (dimensions $39 \times 15 \times 13$ m), lying in the Tergi river bed, at the end of the Dariali gorge, was displaced by the mudflow in 1832, before the people's eyes. Blocks that littered the Kabardo Upland

and Oseti sloping plain were shifted at the beginning of the Pleistocene—during the development of the Valakhian orogenic phase, when the abrupt uplift cf the Great Caucasus took place with simultaneous downwarping of the present day Kabardo upland and the pre-Tergi river dipping at a high angle. At the same time the gigantic volcanoes were active in the Kazbegi region, that is proved by petrographical contents of the Rukhs-dzuar suite. In the Kodori river basin no volcanoes occurred and the boulders are smaller, than in Kazbegi. They were shifted by ordinary (not volcanoglacial) mudflow, that was caused, either by rapid uplift of the Chkalta and Bzibi ridges during the same Valakhian phaze, or by the intensive thaw of the glaciers located higher on their slopes. The original erratic boulders of glacial genesis at the Caucasus were moved lower 1000 m above the sea level, only in Abkhazeti and Svaneti. The largest of them are found in Khevsuretis, in the upper reaches of Khevsuretis-Aragvi, at the foot of a diabasic massif of Chaukhi, at the altitude 1800-2000 m. Some of them are larger than the erratic blocks in the Alps. Large glacier boulders, almost of the same size are found in Svaneti, (Perkhuliskva near village Khalde), in Racha ("Datusela" on the Lechkumi ridge at Chutkharo mountain), etc.

At the Caucasus ancient snow-line, determined according to the traces of glaciers, was located 700-1000 m lower than the contemporary one. In Abkhazeti it went to 2000-2100 m above the sea level, to the east it raised up in parallel with the present snow-line and reached 2900-3000 m in Tusheti. On the Ajara-Imereti Range of the Caucasus Minor, where glaciers don't exist nowadays, the snow-line was at 2200-2300 m, while in the depth of Anticaucasus* in Armenia and the south-west Azerbaijan, it raised up to 2900- 3100 m above the sea level.

The lower boundaries of the Pleistocene glaciation, i.e., hypsometric position of the lower termini of ancient glaciers was stipulated by various factors—mountains height, location of snow-line, topography of a glacier basin. The preservation

* Anticaucasus—the uplifted country of the South from Kolkhida, Kartli, and Azerbaijan lowlands. It includes the Caucasus Minor and Trans-caucasian Upland.

of traces of glacier's activity was considerably influenced by lithological factor. Therefore, the geographical distribution of the lower height boundaries of the ancient glaciation in the Caucasus, like in other countries, is less natural and more complicated than distribution of height of the snow-line. In Pleistocene, the best conditions for glaciers slide to the lower levels were in Abkhazeti and western Svaneti. It was favoured by the considerable height of mountains (though not the maximum in the Caucasus), excessive moistening (the modulus of the river runoff is the maximum for the Caucasus and equals to 100 m/sec. from 1 sq. km), as well and low level of the river beds near the main Range (valleys of Chkhata, Nenskra, etc.). There, the glaciers termini sliding down the mountains Dombai-Ulgen, Dalar and others reached the lowest heights in the Caucasus 700-800 m above the sea level. Lower boundaries of the glaciers of the Northern Caucasus, Eastern Svaneti and Racha lay at comparatively high levels (1000-1300 m and over). In the eastern Caucasus the glaciers generally never slided lower than 2000 m. However, at separate massives, such as the Chaukhi and Didgverdi they reached the level 1500-1800 m. At the Caucasus Minor and the Transcaucasian volcanic upland the glaciers ended above 2000 m. The glacier, that slided down the mountain of Sametskhvario (Ajara-Imereti Range) to the east, into the basin of Shavi-Tskhali river (left tributary of Chkherimela) and reached 1800 m above the sea level, was an exception. Submontane glaciation, similar to the contemporary glaciation of the Alaska or that of Quaternary Alpine one, never existed in the Caucasus. The glaciers ended in the depth of mountain belt. Glaciation was entirely of mountain-valley type, except local areas (e.g. volcanic plateau of Keli) where an entire ice sheet was formed. The area and length of the largest Pleistocene glaciers of the great Caucasus exceeded the present day glaciation of the Alps, but still was less than that of the Pamirs and Hindukush. The length of the largest Pleistocene glaciers of the Caucasus rarely exceeded 40 km. The greatest glaciers of that epoch, were located in the same place where now the largest glaciers of the Caucasus are—in Balkareti, in the valleys of two Chereks (on the northern slope of the Central Caucasus) as well as in Teberda valley (Western Caucasus), where contemporary glaciation is poor. On the opposite, southern slope, in Svaneti the length of Pleistocene glaciers reached 30—35 km and only Nenskra glacier was over 40 km long. Glaciers in other regions of the Great Caucasus were considerably smaller than these "champions". In the Western Caucasus, the length of the Bzibi glacier was 17 km, South-Ptyshi—14 km, Chkhalta —20 km, Sakeni—32 km. In the Eastern Caucasus only Abudelauri glacier was up to 12 km (Maruashvili, 1969).

Valley glaciers of the cold epoch had larger dimension and more complicated character than the contemporary ones. Their polysynthetic character was more obvious, in general. However, the former idea on huge, complex, branched glacial systems often proves incorrect. For example in Tergi, Engury and Kodori basins, the largest glacial systems were supposed to exist, but later due to thorough investigations the paleogeographic picture was changed. (See below the development of ideas on the ancient glaciation of these basins).

Contemporary glaciers of the Caucasus almost never cross saddles of mountain ranges. The powerful (up to 300-400 m) Pleistocene glaciers, in some places crossed the ridges confining their basins (transfluent glaciers). The Ancient Mulakhi glacier crossed over the Ugviri pass, Chkhalta glacier-Amtkeli pass, Sabertsi—Igliis gele, Abudelauri-Blos-gele pass, etc. Petrographic composition of the erratic material confirms these facts.

Contemporary glaciers of the Caucasus do not overpass the side valleys and never crossed the main valleys anywhere. In the Pleistocene, such phenomena occurred in many regions of the Great Caucasus. That was clearly expressed in the Upper Svaneti, where five glaciers from side valleys (Mulakkhi, Dolra, Baki, Nakra and Nenskra) passed into the Enguri valley from the right, while from the left came only one (Lailachala) glacier. The Enguri valley, below the Ipari community, was free of ice. The similar situation occurred in the other river basins—Rioni, Chkhalta, Liakhvi, Tergi, Big Laba etc.

At present, it is observed that in some regions of the Caucasus, the maximum development of glaciers and the minimum absolute height of their lower termini do not coincide with hypsometrically well developed basins. For example, in Svaneti, the largest of contemporary glaciers, sliding lower than others, are located not in the area of mountains Shkhara and Dzangha (5000-5200 m above the sea level), but around Ushba, Tikhtengen, Kulaktau etc., (4000-4700 m). In the Pleistocene this discrepancy was more evident. In Svaneti the largest Pleistocene glacier—Nenskra—was fed from the mountains up to 4000 m high. In the Eastern Caucasus the Didgverdi glacier starting from Didgverdi massif (the absolute height only 3334 m), was larger than the glaciers of the mountain Shavikldе (3576 m), Khochaldag (3484 m), etc. The reasons of such a discrepancy are the following : (a) topographic conditions of snow accumulation in combination with climate, and (b) low position of maximum precipitation level during glacial epochs (at present it is located at 2500-3000 m, on average).

Poor development of glacial relief forms in the areas of gigantic extinct volcanoes of the Great Caucasus-Ialbuzi and Kazbegi are to be noted. Development of troughs, moraines, and other traces of glaciation around these massifs does not correspond to the power and large extension of their firns. Troughs are poorly expressed along the valleys of Baksani and Tergi rivers. This fact suggests an idea that the Ialbuzi and Kazbegi cones were uplifted not before, but after the maximum glaciation epoch. However, poor development of trough valleys in these regions, can be partly explained by filling of many troughs with lava streams, and that còntemporary rivers flow along the young canyons, being cut down along the contact line of lavas with the rocks of their substratum.

Regional Survey of Ancient Glaciation

The Great Caucasus

The geographical range of the Pleistocene glaciation of the Great Caucasus begins in the west, at the meridian of the health resort Sochi—in the region of Fisht and Oshten mountains (39°31′ the east longitude) and spreads to the east for 760 km, up to the mountain of Babadag, at the meridian of town Kjurdamir in Azerbaijan (47°31′)

From the region of Fisht-Oshten, up to the upper reaches of the river Bzibi (125 km), where contemporary glaciation is almost absent, the largest Pleistocene glaciers were those of the rivers Mzimta, Bzibi and Avadkhara on the southern slope, and White and Great Laba on the northern one. Their length ranged between 12-33 km. Low altitude of these mountains (up to 3000-3500 m) didn't hinder the development of glaciers, since it was compensated by humid climate and low location of the maximum precipitation belt in the Glacial epoch. The ancient glaciers slided down also from the side ridges of Gagra and Bzibi on the southern slope, and from the forerange on the northern slope where at present insignificant remnants of eternal ice and snow are preserved.

To the east from the upper reaches of river Bzibi (41°00′ E.L.), up to the junction of the Main and Kodori Ranges (42°00′ E.L., distance 75 km), large Pleistocene glaciers slided down Teberda river on the northern slope, and Sakeni on the southern one, their length ranged between 10-45 km. Presently the Main Range is armoured with an unbroken cover of glaciers. The maximum height of mountains in this area is up to 4046 m (Dombai-Ulgen), climate is humid. The side ranges (Chkhalta and Kodori on the southern slope, Lateral glacier on the northern slope) also underwent glaciation. The contemporary glaciation is represented there by small separated glaciers. The problem of the Ancient Glaciation of the river Kodori basin is debatable. A part of the scientists (Reinghard, Feodorov, Tsereteli and others) admit that the Amtkeli glacier probably slided down to the altitude of 300-400 m above the sea level. Other (Maruashvili, Khazaradze, Bereznjakov, Velikovskaja) neglect the above, and consider, that the glacier ended one km higher. The rudaceous material in the area of Teberda, considered by the first group of scientists as moraines, is actually the outwash of glaciers mudflow.

To the east of the Kodori Range up to the Mamisoni Pass (43°30′ E.L., distance 146 km), spreads that part of the Central Caicasus, where the Main Range reaches the maximum height (up to 4500-5200 m) and bears on its slopes the complete cover of large glaciers. There the Forerange is crowned with the summits of Ialbuzi (5633 m), Dikhtau (5203 m), Koshtantau (5201m), Dzangitau (5049 m), Donghuzorun (4452 m), etc. Some ancient glaciers located on the both slopes were over 40 km long. In the above paragraph "General characteristics of the Pleistocene glaciation" the peculiar distribution of ancient glaciers in Upper Svaneti (Enguri basin) has been mentioned. Namely, glacier protrusion from the side valley into the main valley free of ice, as well as ice sliding across the Ugviri pass etc. These facts have been confirmed by detailed petrographic study of moraines and their sources (Khazaradze 1971, 1974).

To the east from the Mamisoni pass, at a distance of 156 km, up to Shavi-klde (45°30′ E.L.) the Main Range is up to 4000 m above the sea level and its contemporary glaciers are scattered on separate, highest mountains (Khalatsa, Chaukhi etc). On the Forerange (Khokhi, Pirikitis etc. ranges) the contemporary glaciers cover a considerable area. The Ancient Glaciation of this part of the Great Caucasus was not entire. It was broken at low areas of the Main Range and river valleys crossing the Forerange (rivers Ardon, Tergi, Assa, Argun). The largest Pleistocene glaciers of the Main Range were connected with the Chaukhi, Didgverdi, Khalatsa Zilgakhokh mountains and were 8-12 km long (Abudelauri, Alazani, Tergi and other glaciers). Among the side, southern ranges, only the Germikha Range underwent a considerable glaciation. The glaciers Britata, Knogi, Batsgadi etc., 5-7 km long slided down from Germikha Range to the valleys of Large and Small Liakhvi, (at present the Germukha Range is totally deprived of ice cover). The larger glaciers lay on the slopes of the Forerange. They slided down the Dzimaraikhokh, Kazbegi, Dakuekhi, Shan and other mountains, along the river valleys of Gizeldon, Suatisidon, Genaldon, Maitsis-tskali. Tergi basin has been the subject of discussion among the scientists, as it was mentioned above. Some authors (Reinghard, Krivolutski, Bereznjakov and others) contended, that "The Great Tergi Glacier" 80 km long existed in the past, reached Tergi and Sunzhen ranges in the Precaucasus being branched into numerous tributaries in the upper part. The study of lithological composition of deposits of Tergi valley, Osetian sloping plain and the above mentioned ranges made the opposite group of scientists (Vardaniants, Maruashvili, Velikovskaja, etc.) to conclude, that such a powerful glaciation in Tergi basin was unreal. Following L.A. Vardaniants the scientists consider the volcanoglacial mudflows (lahars) as the main factor of transportation of huge blocks of Kazbegi and Kabarjina massives, along the Tergi valley.

To the east from Shavi-klde and Babadag mountains

(255 km) the Main Range, except Bazardjuzi mountain (4466 m), is free of contemporary glaciation. (Now the glaciers exist only on the northern side ranges and especially on the Bogozi range.) The Pleistocene glaciation covered the greatest part of the Main Range and most of the northern side ranges. Valley glaciers slided down the rivers Kusarchai, Samur, all four Koisus and some of their tributaries.

The Caucasus Minor

Glaciers located in the extreme west of the caicasus minor were located at the mountain Khino, in the western part of the Ajara-Imereti Range to the north-east of Batumi (42°00′ E.L.). From this point a chain of small (cirque, corrie, hanging) glaciers were spread to the east at 70-75 km, up to the Sametskhvario and Meghvriki (43°12′ E.L.) mountains, in the eastern part of a range (to the north from Akhaltsikhe). The glaciers were mostly of cirque and corrie type. They predominated on the northern slope (in the basins of Natanebi, Gubazouli, Supsa, Khonistsqali and Shavi-tsqali rivers). There were glaciers on the southern slope of Ajara-Imereti range as well., e.g., at the Persati Plateau (to the west of the health resort Abastumani) that is a relic of the Pliocene lava sheet. The glaciers covered also the eastern part of the Shavshveti Range.

To the east of Borjomi gorge, of the river Mtkvari, on the Trialeti Range only several small glaciers existed, while on the mountains of the Middle Khrami group and the Lokhi Range there were none. Further, to the south-east, in the part of the Caucasus Minor, which is a borderline between Armenia and Azerbaijan (on the Shakkhdag, Mrovdag and other ranges) a separate group of cirque and small valley glaciers appeared, owing to the high mountains. At the higher Zangezuri Range there were large glaciers (e.g. Vokhchi), Talysh (the Soviet part of the South Caspian mountain system), bears no traces of glacial influence.

The Transcaucasian Highland

The ancient glaciation of this volcanic region in the west, began to the south from Goderdzi Pass, at the Frusheti highland. The ancient glacial cirques are cut into the dome-like mountain massives of Tlili, Vandag and Shuakaravi. While at the upper reaches of river Uraveli a trough over 5 km long was spared.

To the east of the upper-stream of the Mtkvari river, in Javakheti, the ancient glaciation developed on the Sumsari and Javakheti volcanic ranges, the absolute height 3000-3200 m. Glaciation of the first one was more considerable, where side by side with large cirque glaciers there was a valley glacier 5 km long, that slided down the Samsar mountain to the north-west. Further to the south, in Armenia, a gigantic volcano of Aragats, (Alagez) 4095 m, was covered with considerable ice sheet. The glaciers branched off down from the crater, similar to the contemporary glacier systems of Ialbuzi and Kazbegi. The length of separate glaciers was up to 10-15 km.

Gegami (Akhmanganian), Vardeni (South Gogcha) and Karabakh volcanic highlands of Armenia and Azerbaijan were also affected by the ancient glaciation.

Problems of the Caucasian Glaciation Chronology

Glaciation chronology is one of the most difficult problems of the ancient glaciation study. It is comparatively well studied in regions with ice cover, such as the North Europe and North America, where the radiocarbonic dating of the last glaciation and its stages is already made. The analogues data are obtained in the Alps (Serebjany, 1978). The Caucasian glaciation chronology is poorly studied—there are only some absolute dates mainly on the Holocene (the post-glacial epoch). Hence, the use of physical methods to solve the problem of local glaciation distribution in time is rather limited.

The first problem is—the number of the Pleistocene glacial and interglacial periods in the Caucasus. Above, under general characteristics and regional survey of the ancient glaciation at the Caucasus, the age sequence of glacial formations was purposely ignored, since it was impossible to determine it reliably and systematically for all basins and regions.

Apart from the last glaciation, being correlated with that of the Würm in the Alps, in some valleys of the Great Caucasus one can observe the traces of more ancient glaciation, lower than the former. The recurrent glaciation is also observed in the basins of Kodori (mouths of Gvandra and Sakeni rivers),

Enguri (Mulkhura, Dolra, Menskra valleys), Tskenistskhali (Gobishuri valley), Aragvi (Abudelauri valley), etc. However, one can't state definitely that these glaciations were separated by interglacial epoch i.e. the period when glaciers reduced more than their present-day dimensions or totally disappeared. Interglacial deposits reflecting the presence of warm period between two glaciations are not established.

From this point of view, the Khevsuretis-Aragvi basin in the Eastern Caucasus is of the utmost interest, (to the east of Tbilisi meridian). Abudelauri river, the tributary of Khevsuretis-Aragvi, starts from the rocky massif of Chaukhi (3940 m) formed by diabases and porphyrites. Due to glacial climate these, very strong rocks were disintegrated into large blocks, producing reliable, marking material for determination of glaciers distribution of the above massif. The Abudelauri glacier was the largest among the glaciers of Chaukhi. Its vast trough is dotted with erratic blocks reaching enormous dimensions in the vecinity of Roskka village. The history of Abudelauri glacier is often subdivided into 2 epochs or megastages. During the interval between them, a considerable erosional downcut took place, hence the Kmosti and Ghelisvake positions of the glacier's terminus.

At Kmosti position, the Abudelauri glacier ended at the altitude of 1850 m, between villages Roskha and Kmosti, near the church, built on a crest of a terminal moraine. Above this place, Abudelauri river flows along the trough bottom. Between the Kmosti moraine and the terminus of contemporary Abudelauri glacier, 8-9 ancient recessional moraines are marked.

The glacier at Ghelisvake position was lower and passed into the valley of Khevsuretis-Aragvi. Diabase blocks, left by the glacier are scattered at the absolute height 1650-1700 m (near villages Ghelisvake and Guli). The surface, these blocks lie on, is cut through by the Khevsuretis-Aragvi erosional gorge, to the depth of 200-250 m. It is indicative of the tectonic uplift of the area to the certain height, after existence of the Ghelisvake glacier. The river-bed of Khevsuretis-Aragvi is filled with huge water-polished boulders, originated from the Chaukhi Massif, that apparently rolled down from the terrace-like residual surface. Consequently, the glacier terminus near village Ghelisvake located initially at the altitude about 1400 m.

The snow-line at the altitude 2600-2750 m, and its depression by 750-800 m, corresponds to the Kmosti position of the ancient Abudelauri glacier's terminus, while that one at the altitude 2400-2450 m, and its depression by 900-950 m, to the Ghelisvake position (taking into consideration the tectonic uplift).

The morphology of Abudelauri trough, above the Kmosti moraine, is deprived of erosional downcut traces at the interval between the Kmosti and Ghelisvake stages, (glacials?). The trough has wide non-segmented bed, has neither arms (shoulders) nor any other signs of double glaciation with the separating interglacial. After Ghelisvake stage the glacier reduced, though its terminus has not moved far away from the site, where later, Kmosti moraine was formed. Above this moraine the trough was filled with ice and thus avoided an erosional downcut.

The considerable value of erosional downcut (200-250 m) into the valley floor, between Ghelisvake and Kmosti moraines, testifies the old age of Ghelisvake stage. Consequently, the continuous glaciation epoch in this region (in Abudelauri valley above Kmosti moraine in particular was of long duration. At the Akhuni Range, located near Khevsuretis-Aragvi basin, that has a series of "step-like" erosion surfaces at the crestal zone, the surface at 300 m of the relative height is dated by the lower Pleistocene (Maruashvili, 1975). The surface which had been covered by the glacier of the Ghelisvake stage, is to be attributed also to this phase of plain pedimentation. Consequently, this stage is dated by the Middle Pleistocene as well.

The absence of clear-cut interglacial stage during the period covering Middle and Upper Pleistocene is caused either by the tectonic uplift of the Great Caucasus, at the end of the Middle Pleistocene, or according to some authors (Gromov, 1948; Maruashvili, 1977), due to absence of the Riss-Würm interglacial since there was only one—the Riss-Würm glaciation, that lasted over 1,40,000 years. The above problem requires further investigations.

Traces of more ancient, the Early Pleistocene and Late Pliocene glaciations are present just at some sites of the Caucasus (Milanivski 1966, Budagov, 1963), that are located at high altitudes. According to these sites it is very difficult to have

1. Chalaati glacier

2. Peak Ushba

3. Sakeni erratic block in Apkhazeti

5. Discharging of Trtishi glacier into Triberi glacier

4. View of one part of central caucasus from a slope of Mt. Elbrus

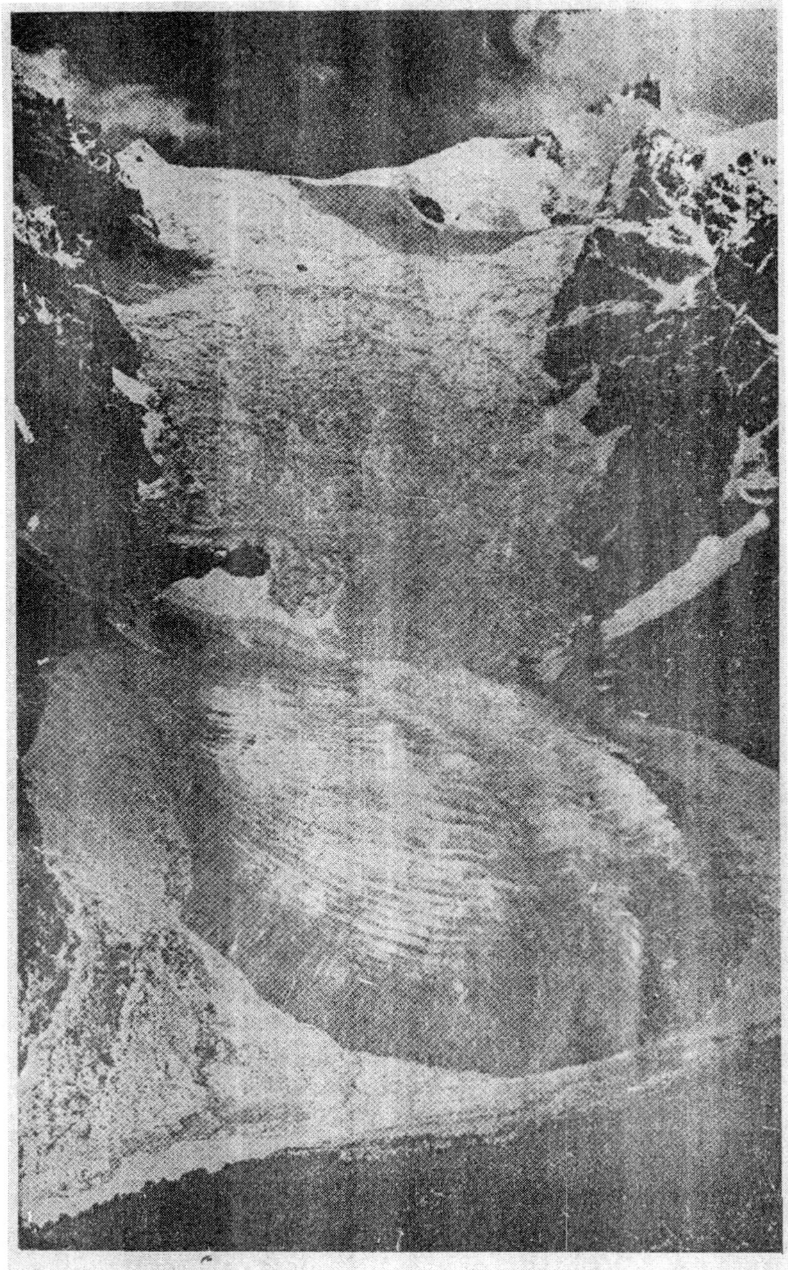

6. Adishi glacier

an idea on the character of the Early Pleistocene and Late Pliocene glaciations, as a whole. However, if these glaciations have really existed in the Caucasus, their areas must have been situated within the boundaries of the Riss-Würm glaciation i.e. to be of less dimensions. The idea on the Early Pleistocene cover glaciation of the Great Caucasus, suggested by some authors resting their arguments on the groundless idea of the Late Pliocene peneplanation of the mountain system, is beneath criticism.

The second problem of the Caucasian glacian chronology is the number and absolute age of the last glaciation retreat stages as well as their relation to the analogues stages of the Alps, Russian Plain etc. It is obvious that side by side with glaciers fluctuations caused by the global climatic changes, there were fluctuations of glaciers during the Holocene connected with the local factors—first of all with the geomorphological one. The establishment of the global glacial stages and inter-stages is of utmost importance for determination of the post-glacial changes of the earth's climate. The problem seems even more complicated due to existence of the local fluctuations and different preservation of recessional moraines. In the areas of denudation resistant mountain rocks (diabase, granite, andezite, limestone, etc.), moraines preserve their morphologic expressiveness for a larger time, while in the regions of sedimentary strata development (shales in particular), they are quickly flattened and completely obliterated. At present the Caucasian recessional moraines are dated with the radiocarbon method in some places only (Serebrjanny, 1978).

The last stage of glaciers advance in the Caucasus is comparatively well-known. This stage of the XVII-XIX centuries, is called sometimes as "Minor glacial epoch" or the Fernau stage. It is explained by well preserved corresponding moraines and the existence of Russian topographic maps of the second half of the XIX century. During the last 90-100 years, large valley glaciers Bezengi, Tsanner, Karaugom, etc., reduced by 1—1.5 km, while some small glaciers disappeared completely. During the last 50 years, Kvitlod and Laboda glaciers separated from Tviber and Zopkhito glaciers and became independent. Practicability of glacier passes has also changed (e.g. Midagrabin Pass between Tergi and Gezeldoni upper reaches, once very convenient for horseman, became hardly passable for foot-passengers due to reduction of the Midagrabin glacier). However, the maximum advance of the Caucasian glaciers to the stage of the XVII-XIX centuries, is still scantily studied, as in the XVIII century and the first-half of the XIX century, the large scale topographic survey was not practised in the Caucasus.

Some authors stick to the idea that Caucasian glaciation were caused or intensified by the influence of South European covering glaciation and hence were delayed, in comparison with the latter (Tsereteli, 1966). Such an outlook can hardly be admitted due to the large distance between the Caucasus and the side of European glacial shield (above 800 km during the Riss, and 1700 km during Würm). It is well-known, that the cooling influence of large ice sheets on periglacial zone spreads up to 150-200 km only. The author believes that mountain glaciers of the Caucasus and the Alps reached their maximum development earlier than the covering glaciation of the high latitudes, the growth of which was delayed by many centuries, in connection with the autofluctuation process. The Caucasian glaciations were influenced by the global changes of atmospheric circulation and regional tectonic movements.

SUGGESTED READINGS

Abich, H. Etudes sur les glaciers actuels et anciens du Caucase.

Bereznjakov, A. I., 1965 : Petrographic Characteristic of Problematic Coarse deposits in the Vicinity of village Khevi. *Materialy Kavkazskoy expeditsii* (*po programe Mezhdunarodnogo Geofizicheskogo goda*), Vol. 5, Kharkov, pp. 112-113.

Budagov, B. A., 1963 : The ancient glaciation of the southern slope of the great caucasus (between rivers Mazymchai and Kishchai). *Trudy Instituta Geografii AN Azerbaijanskoi SSR*, Vol. II, Baku, pp. 115-132.

Dinnik, N. J., 1890 : Contemporary and ancient glaciers of the caucasus. *Zapiski Kavkazskogo otdela Russkogo Geograficheskogo obschestva*, 14, Vyp. I, pp. 282-416.

Dumitrashko, N. V., Lilienberg, D. A. Antonov, B. A., Baljan, S. P., 1961 : Ancient glaciations of the caucasus aud their correlation with glaciations of the Russian plain. *Trudy Komisii po izucheniju chetvertichnogo perioda*, Vol. XIX, pp. 170-180.

Favre, Ern. Recherches geologiques dans partie centrale de la chaine du Caucase. *Genéve-Bãle-Lyon*. pp. (1-4%) t (I-VIII) + (I-117).

Geology of the armenian, SSR. Vol. I., 1962 : *The Geomorphology*. Frevan, Published by the Academy of Sciences of the Armenian SSR, pp. 5-584.

Geomorphology of Azerbaijan. 1959 : Baku, Published by the Academy of Sciences of the Azerbaijan SSR. pp. 5-371.

Geomorphology of Georgia. 1971 : Tbilisi, Publishing House 'Metsniereba' Tbilisi, pp. 5-609.

Gromov, V. I., 1948 : Paleontological and archaeological substantiation of continental deposits Stratigraphy in Quaternary perion at the territory of the USSR (Mammals, Paleolith). Moscow, Published by the Academy of Sciences of the USSR. pp. 1-521.

Khazaradze, R. D., 1971 : Methods of Determination of the ancient glaciers direction in the Upper Svaneti based on the study of Morainic material lithology. *Soobschemija Akademii Nauk Gruzinskoi SSR*, Vol. 61, No 3, pp. 621-624. (In Georgian Language).

Khazaradze, R. D. : The boundaries of the quaternary glaciation in the Upper Svaneti. *Soobschenija Akademii Nauk Gruzinskoi SSR*. Vol. 76, No 2, pp. 377-380.

Khazaradze, R. D., 1979 : The problem of existence of the ancient glacial deposits in the vicinity of village Khevi (the right bank of river Amtkeli, the West Georgia). *Soobschenija Akdemii Nauk Gruzinskoi SSR*, Vol. 94, No 1, pp. 109-112.

Kovalev, P. V., 1965 : The problems of quaternary glaciation at the caucasus. *Materialy Kavkazskoy ekspeditsii* (*po programme Mezhdunarodnogo Geofizicheskogo goda*). Vol. 6, Kharkov. pp. 3-162.

Krasnov, A.N., 1891 : The upland flora of Svaneti and peculiarities of its grouping, depending on the present day living conditions and the influence of glacial period. *Izvestija Russkogo Geografichekogo Obschestva,* Vyp. 5, pp. 357-383.

Maruashvili, L.I., 1938 : On the ancient glaciation of the caucasus minor. *Priroda*, No 7-8, pp. 21-30.

Maruashvili, L.I., 1956 : *Revision advisability of the existing ideas on paleogeographical conditions of the glaciation period at the caucasus*. Published by the Academy of Sciences of the Georgian SSR, Tbilisi, pp. 3-125.

Maruashvili, L.I., 1975 : The valley planation of uplifted mountain systems. The Caucasus as an Example. *Geomorphologija*, No 4, pp. 12-21.

Merzbacher, G., 1901 : *Aus den kochregionen des kaukasus*. Vols. I-II, Leipzig.

Milanovski, E.E., 1966 : The basic problems of the ancient glaciation history of the central caucasus. In the book *Problemy geologii paleogeografii antropogena*. Published by Moscow University, pp. 5-49.

Oiunin, V.N., 1953 : For the history of glaciation of the south-east mountainous part of the Kabardino-Balkarsk autonomous republic. *Trudy*

Iustituta Geografii An SSR, Vyp. 58, *Materialy po geomorfologii i paleogeografii*, Vyp. 10. Published by the Academy of Sciences of the USSR, Moscow, pp. 90-178.

Palgrave, W.G., 1872 : Vestiges of the glacial period in the north-eastern Anatolia. *Nature*, t.V.

Paffengol'tz, K.N., 1958 : New data on stratigraphy of lavas of the kazbegi region and Kheli volcanic plateau (The Central Caucasus) and the ancient glaciation of the area. *Sovetskaja geologija*, No 12, pp. 115-139.

Podozerski, K.I., 1911 : Glaciers of the caucasian range. *Zapiski Kavkazskogo otdela imperatorskogo Russkogo Geograficheskogo obschestva*, Vol. 29, Vyp. I, Tiflis, pp. 1-200.

Reingard, A.L., 1933 : The glacial period at the caucasus and its relation with glaciation of the Alps and Altai. *Trudy II Mezhdunarodnoi konferentsii Assotsiatsii po izucheniju chetvertichnogo periods Evropy*, Vyp. II,

Safronov, I.N., 1969 : *The geomorphology of the northern caucasus.* Published by the Rostov University, pp. 3-218.

Serebrjannyi, L. R., 1978 : *Dynamics of the Cover glaciation and the glacial eustacy in the late quaternary period.* Publishing House "Nauka", Moscow, pp. 3-268.

Sharma, V. K., 1977 : Preliminary investigation of pleistocene valley glaciation in Part of Kangra district, Himachal Pradesh. *J. Geol. Soc. India*, 18, No 1, pp. 23-30.

Scherbakova, E. M., 1973 : *The ancient Glaciation of the great caucasus.* Published by Moscow University, pp. 3-271.

The Catalogue of Glaciers of the USSR. Vols. 8, 9, Leningrad, 1967-1977.

Tsereteli. D. V., 1966 : *Pleistocene deposits in Georgia.* Publishing House "Metsniereba", Tbilisi, pp. 1-583.

Velikovskaja, E. M., Kozhevnikov, A. V., 1959 : On the origin of moraine like strata in the valleys of Terek, Gizeldon and Urukh. "Vestnik Moskovskogo Universiteta", *Ser. Geol.*, No 4, pp. 125-134.

Vardanjantz, L. A., 1948 : *The post-pliocene history of the Gaucasian—Black Sea—Caspian Sea region.* Erevan, published by the Academy of Sciences of the Armenian SSR, pp. 4-184.

FIFTEEN

Jaromir Demek

PERIGLACIAL GEOMORPHOLOGY
Present Problems and Future Prospects

CRYOGENIC phenomena and forms have been known for a relatively long time. As early as 1640, the Yakutian dukes reported that in Yakutia 'ground does not thaw completely even in summer', and knew about the characteristics of permafrost. In 1828, F. Shergin, a worker of the Russian-American Company, began to sink a well on the bank of a branch of the Lena River in the town of Yakutsk; he finished his work 9 years later in frozen deposits at a depth of 116.4 m. In 1843-46 A.F. Middendorf measured the temperature of the frozen deposits in this well and scientifically demonstrated the presence of permafrost.

More than 100 years ago, geologists and geographers mapping in Southern England and Wales noticed superficial non-stratified and coarse deposits which could not have originated by processes active at that time. This unsorted heterogeneous debris has been referred to as 'head', 'coombe rock', 'warp', and other terms. It was obvious that these deposits had developed in a much colder climate than that of the present time. This knowledge became the basis of a new branch of geomorphology dealing with the forms and processes of a cold environment.

However, the main development of periglacial geomorphology took place only after the Second World War. According to J. Tricart (1970, p. xv), W.M. Davis's geomorphological school regarded cryogenic forms as rarities. Prior to 1945 the attention of geomorphologists in most countries was restricted mainly to small cryogenic forms (patterned ground, stone streams). Only in the USSR since the 1930s has considerable attention been paid to the study of permafrost. Since 1945, geomorphologists engaged in the study of processes and relationships in cold environments all over the world have concentrated their attention on the investigation of permafrost and incidental cryogenic phenomena. Extensive economic utilization of the natural resources of polar regions, as well as the resulting intensification of investigations and the increasing number of expeditions in these often relatively inaccessible regions, contributed greatly to the development of periglacial geomorphology.

The problems of periglacial geomorphology may be divided into two basic groups :

1. problems of phenomena linked to processes and forms conditioned by permafrost, and
2. problems connected with cold climates affecting the operation of geomorphological processes.

TERMINOLOGY

It is first of all necessary to consider a terminological question. The term 'periglacial' was introduced by the Polish explorer Walery Lozinski in 1909 and is now firmly entrenched. In the last 3 years it has appeared in the names of significant monographs on periglacial geomorphology, namely those by A.L. Washburn (*Periglacial processes and environments.* 1973), C. Embleton and C.A.M. King (*Periglacial geomorphology.* 1975), and most recently by by A. Jahn, (*Problems of the periglacial zone.* 1975). But since the term 'periglacial' still tends to evoke the idea that we are dealing with a region, a zone, at the margin of a glacier, it is quite often wrongly thought that the periglacial zone is the result of the climatic effects of the glacier. The term 'periglacial zone' also creates the impression that the climate of this zone is a more temperate variant of a glacial climate and that in the case of climatic, cooling, a glacier or ice sheet will develop in the present periglacial zone.

But most typical periglacial zones with characteristic landforms and the greatest thicknesses of permafrost occur in the interiors of continents or in high mountains where either there was no glaciation at all or the glaciers were insignificant. A typical periglacial climate differs substantially from a glacial climate mainly by the aridity and the negative heat balance of the ground surface. Unfortunately, the periglacial zone has not been well-defined climatically, being delimited so far rather on the basis of the occurrence of cryogenic phenomena. The development of cryogenic phenomena is connected with the freezing of soil and the phase transformation of water. Ground freezing can be perennial (ground temperature below freezing point for more than 2 years), seasonal (several months) and short-period (from several hours up to days).

The core of the periglacial zone comprises the region of perennial ground freezing with permafrost.

SPATIAL DIFFERENTIATION OF THE PERIGLACIAL ZONE

In the spatial differentiation of the periglacial zone, the difference between the Arctic and Antarctic periglacial zone on the one hand, and the periglacial zone in mountain ranges on the other, is of the greatest significance. In the Arctic and Antarctic periglacial zone, there is also a significant difference between maritime and continental climatic conditions. In the maritime zone (for example, Iceland, Spitzbergen) the temperature of the permafrost approaches 0°C, higher humidity causing cryogenic processes to be active over much of the year. The active layer is of considerable thickness and therefore solifluction is rather intensive. In the continental zone, however, the permafrost exhibits much lower temperatures. The activity of cryogenic processes is concentrated in a short period in spring only when the area becomes dry, and cryogenic processes become less active. In autumn, the temperature decreases rapidly, the active layer freezes, and the effects of geomorphological processes are small. A typical feature of this continental periglacial zone is the occurrence of extensive cryogenic planation surfaces (cryopediments, cryoplanation terraces).

PERMAFROST AS A SYSTEM

Over about 24 per cent of the surface of continents and in some marginal parts of the oceans, the rocks exhibit a negative thermal balance and a temperature below 0°C for more than 2 years. If these rocks contain water then this water usually freezes and is transformed into the solid state—ice. The freezing of water in rocks substantially changes their permeability and their physical, mechanical, thermal and other properties. Loose sediments become consolidated. Owing to the freezing of rocks and the transformation of water into the solid state, a complicated system of solid (mineral or organo-mineral skeleton ice), liquid and gaseous phases develops. This system is called permafrost. The origin of permafrost and the related changes of rock properties manifest themselves naturally during the operation of geomorphological processes and the development of specific cryogenic forms.

Most permafrost is linked to present conditions, mainly climate, relief, vegetation, hydrological conditions and the activities of man. It is accordingly a dynamic system adapting itself to the changing conditions of the environment. Unfortunately, relatively little is known about the dynamics of permafrost in spite of the fact that research has made considerable progress, especially in the U.S.S.R., the U.S.A. and Canada.

In any consideration of permafrost dynamics, and also, therefore, of the development of landforms, the temperature, thickness and origin of the permafrost are of greatest importance. In low temperatures permafrost is stable and the development of landforms is slow. Around 0°C, permafrost dynamics are, on the contrary, much more active and the development of specific cryogenic forms takes place. Permafrost thickness also plays an important part. The greatest permafrost thicknesses known are to be found in the U.S.S.R. : in the Udokan Ridge, where a thickness of 1300 m has been recorded, and in the basin of the Marcha River (a tributary of the Viljuj River) where a thickness of 1500 m has been established (P.I. Melnikov and N.I. Tolstikhin, 1974, p. 96).

The origin of permafrost is of considerable geomorphological significance, mainly because it affects the distribution of ground-ice in frozen rocks, and especially in sediments. In epigenetic permafrost, ground-ice develops mainly in the upper parts. On the other hand, in syngenetic permafrost, ground-ice is fairly regularly distributed through the whole permafrost thickness. The significance of this for periglacial geomorphology follows from the fact that ground-ice forms from 50 to 80 per cent by volume of the fine-grained sediments of polar regions. The dynamics of permafrost, that is, its aggradation and degradation, are of unusual significance for the development of landforms in the periglacial zone.

Increase in the negative thermal soil balance results in a decrease of permafrost temperature Because of this, permafrost aggradation also begins to result in a decrease of the thickness of the active layer and this in turn affects the dynamics of the geomorphological processes. But the aggradation of permafrost also results in the freezing of taliks and in this way some typical forms, such as pingos, originate. An acceleration of the growth of ice wedges and the development of landforms, such as low-centred polygons, also takes place. In some places in polar regions, ice-wedge polygons occupy as much as 60 per cent of the surface. In the deposits at the foot of the Kular Ridge, ice wedges up to depths of 6 m amount to 71 per cent by volume, and in the lowland of Yanolndigirskaya-nizmennost as much as 76 per cent by volume, of all deposits (T. Czudek and Demek, 1973, p. 10). The ice wedges in exposures on the Indigirka River bank near the community of Sypnyy Yar are as much as 40 to 50 m deep.

However, the growth of ice wedges can result simultaneously in local degradation of the permafrost, and the pushing out of soils on the margins and the development of ramparts. Low-centred polygons with a concave core develop. The growth of the ramparts breaks the turf cover, thus exposing the soils to the direct effects of insolation. In places where polygon boundaries cross, and in the concave cores, water accumulates. The temperature of the water in the depressions is always higher than that of permafrost. Owing to the thermal and chemical effects of the water, ice wedges begin to thaw, even in very cold climates where permafrost temperatures range between —9° and —11°C.

Forest fires in the taiga may also be a cause of local permafrost degradation. In eastern Siberia, disturbance of the vegetation cover changes the thermal conditions of the ground to depths of 10 to 15 m and can result in the onset of thermokarst processes which considerably affect the development of the relief.

Regional permafrost degradation leads to marked changes in the relief of polar lowlands where the permafrost is often characterised by a high ground-ice content. Permafrost degradation by thermokarst processes results in local subsidence of the soil, owing to the thawing of ground-ice, and the development of depressions (Fig. 15.1). When flying over the Central Yakutian Lowland between the Lena and Aldan Rivers, numerous circular or oval-shaped thermokarst depressions called *alases* occupying 40 to 50 per cent of the surface of the lowland can be seen. Most are closed, often containing a lake, but some merge into elongated dry depressions with flat grassy floors. The light-green grass on the floor of the *alas* contrasts sharply with the surrounding dark green of the larch taiga. Many thermokarst depressions in the Central Yakutian Lowland developed as a result of regional permafrost degradation during the Holocene climatic optimum, 4000-5000 years ago.

Geomorphologists have so far paid relatively little attention to thermokarst phenomena in spite of the fact that they are of considerable geomorphological significance. An important result of thermokarst processes is the degradation of the surface of lowlands in the periglacial zone. This process is designated 'thermo-planation', which can take place either laterally or vertically.

Lateral permafrost degradation and thermo-planation connected with this phenomenon are more significant in areas of dissected relief. These processes are usually linked to the thermo-erosion of streams or to thermo-abrasion. Characteristic forms developing by lateral thermo-planation are arcuate depressions on slopes, termed 'thermocirques'. The sides of the thermocirques usually exhibit parallel retreat. The thermocirque floor becomes larger, and adjacent thermocirques thus often coalesce with each other. In this way a terrace develops on the slope, increasing in size by backwearing of the thermocirque walls. Gradually a lower level of the polar lowland develops.

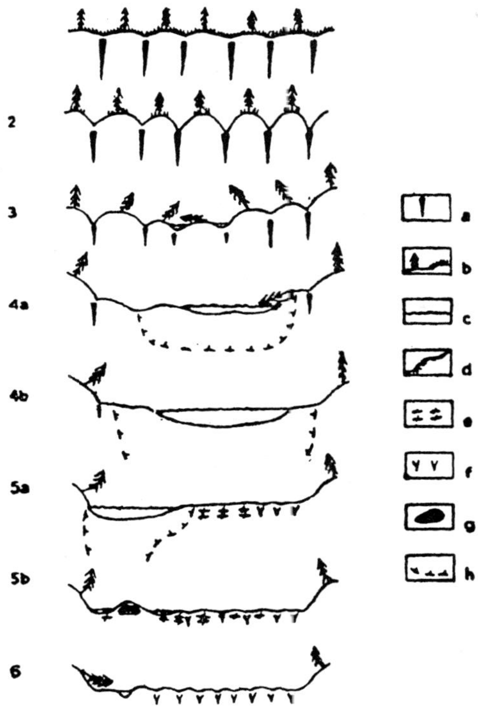

FIG. 15.1 : *Schematic representation of alas development as a result of permafrost degradation from above (T. Czudek and J. Demek, 1973).*

Key : 1. original surface with syngenetic ice wedges; 2. beginning of permafrost degradation and origin of baydjarakhs; 3. duyoda; 4 a and b, alas, 5 a and b, khonu with pingo; 6. thermokarst valley.

Vertical permafrost degradation occurs mainly in flat undissected terrain, often on watersheds. The extent of relief change owing to permafrost downwearing depends on the amount of ground-ice in the deposits, and on its type. Considerable changes will take place if there is a high ground-ice content in the deposits, especially if the ice-wedge network is well-developed. The resulting landforms are designated by

Yakutian names. In the first stage, at the beginning of down-wearing, the ice wedges begin to thaw and high-centred polygons develop. The vegetation usually remains undisturbed. In the second phase, the depression above the ice wedges becomes deeper, the humus layer breaks and slips down into the depressions. The polygon cores change into conical elevations called *baydzharakhs*. The vegetation cover is disturbed and deposits are eroded. In the third stage, the *baydzharakhs* are gradually disturbed and a depression called a *duyoda* develops, in whose centre water accumulates. If water remains in the depression for the greater part of the year, thermokarst processes begin to accelerate. On the slopes of the *duyodas*, *baydzharakhs* develop, and sliding and solifluction take place owing to oversaturation of the ground with water from the melting ground-ice.

In the fourth stage, a distinct depression called an *alas* develops, with steep slopes and a flat floor. In the Yakutian language this term means an oval or circular depression with steep slopes and a flat floor covered with green grass and a lake. When the water in the lake attains a certain depth, freezing of the water to the bottom in winter ceases and a lake talik develops. The ground-ice thaws out slowly and the deposits settle, so that the *alas* becomes deeper. On the shores of the lake, thermo-abrasion takes place. Due to wind action, the shores of lakes and *alases* shift.

In the fifth phase the lake progressively disappears and renewed permafrost aggradation takes place. Owing to freezing of the talik, a pingo usually develops. The depression with secondary ice wedges and a pingo is called a *khonu*.

The coalescence of *alases* and *khonu* results in thermokarst valleys. In ground-plan, the thermokarst valleys consist of wider sections (*alases, khonu*) and narrower sections marking the position of former watershed ridges. The thermokarst valleys have numerous meanders, blind projections, and may run counter to the general dip of the surface in places.

Permafrost degradation thus causes substantial changes in the relief of the polar lowlands and the thermokarst processes lower the surface of the terrain to create a lower level.

PERMAFROST STRUCTURE AND CRYOGENIC TEXTURES

Depending upon its mode of origin and development, permafrost usually contains various kinds of ground-ice. 'Texture ice', involving ice crystals from some tens of millimetres up to a few centimetres in size, forms a typical cryogenic texture in the case of rock freezing. This texture is defined by its form, dimensions and the distribution of ground-ice crystals in the frozen rock. The kind of texture depends on the genesis of the permafrost and the conditions under which it developed.

The main differences in cryogenic textures are related to the different origin of syngenetic or epigenetic permafrost (Fig. 15.2). In the case of syngenetic permafrost, the cryogenic textures develop as the permafrost thickens and grows upwards and as, at the same time, sediments accumulate under subaerial or sub-aqueous conditions (Fig. 15.3). Under subaerial conditions the lower part of the active layer of colluvial, eolian proluvial and fluvial deposits changes gradually with accumulation into permafrost. Under subaqueous conditions, the deposits of river beds, ox-bow lakes, lacustrine and marine deposits, freeze. Freezing often takes place not only in the vertical but also in the horizontal direction. Texture-ice in the case of this type of permafrost develops from the water above the upper permafrost line which freely migrates towards the freezing-point. In clays, loams and silts, the water of adhesion migrates. The growing ice crystals push off rock particles and form lenses and bands of texture ice. In sands, gravels and coarser debris, water penetrates as far as the freezing-front only by gravity or precipitation from vapour. The ice fills the free spaces among the mineral particles. This is why, in syngenetic permafrost, rather complicated textures of segregation type prevail. The principal types of texture are shown in Figure 15.2.

In the case of the development of epigenetic permafrost, vertical freezing dominates (Fig. 15.4). Most cryogenic textures therefore develop under conditions of water migrating under pressure towards the freezing-front. Ice develops from water under the permafrost. The water migrating under pressure occurs not only in water-bearing rocks but even in rocks normally considered impermeable in hydrogeology. Epigenetic

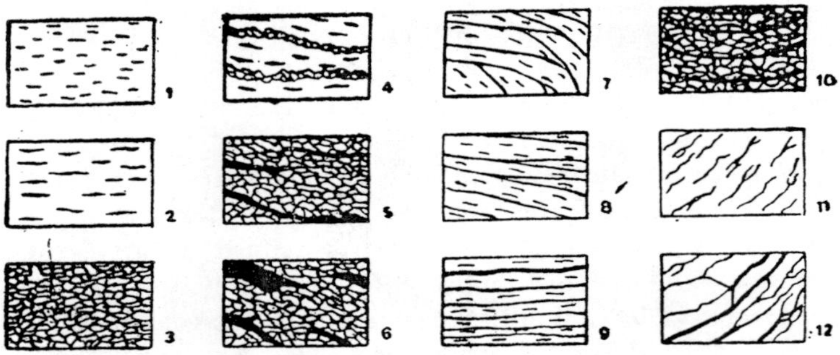

FIG. 15.2 : *Main cryogenic textures caused by syngenetic freezing of unconsolidated deposits* (*P. I. Melnikov and N. I. Tolstikhin, 1974*)

Key : *cryogenic textures of subaerially freezing deposits. Simple textures* : (1) texture of fine ice lenses (typical of solifluction deposits). (2) lenticular texture (typical of dry slopes deposits). (3) reticulate texture (typical of solifluction terraces and stream). *Complicated textures* : (4) texture of undulating bands (typical of humid slope deposits). (5) reticulate, texture, tilted (typical of rather humid slope deposits). (6) texture of interrupted bands and nets (typical of creep deposits. (7) strongly undulating lenticular texture (typical of mobile solifluction terraces and streams). (8) fanlike lenticular texture (typical of alluvial cones). (9) horizontal stratified texture typical of floodplains). (10) stratified texture with concave curvature (typical of higher levels of floodplain with ice- wedges).
Cryogenic texture of subaquaeously freezing deposits : (11) Texture of oblique ice lenses (typical of ox-bow deposits). (12) obliquely stratified texture (typical of ox-bows).

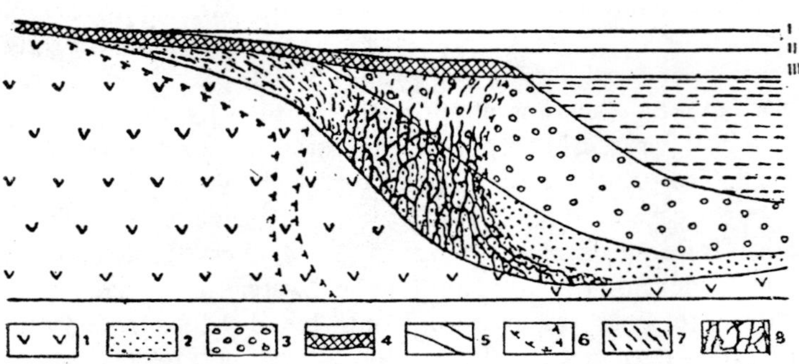

FIG. 15.3 : *Syngenetic freezing of lake-floor deposits* (*P.I. Melnikov and N. I. Tolstikhin, 1974*)

Key : (1) bedrock. (2) deposits of first generation on lake floor. (3) deposits of second generation on lake floor. (4) active layer. (5) lake floor in various stages of filling with deposits. (6) upper permafrost table in various stages of lake filling with deposits. (7) lake-floor deposits with a dense pattern of relatively fine ice lenses. (8) deposits on lake floor with a widely spaced pattern of relatively broad ice bands. I-III : water level in the lake in various stages of development.

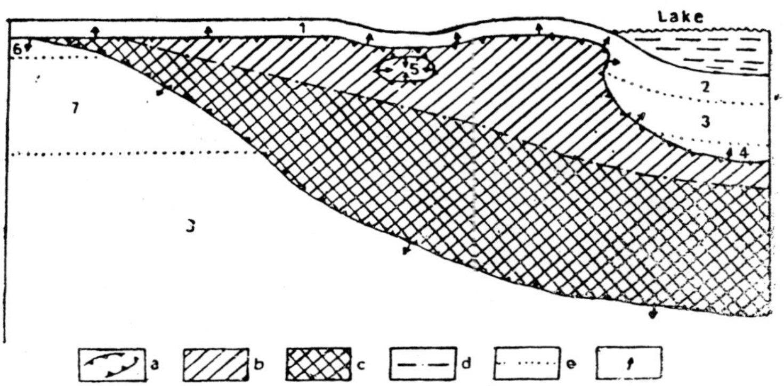

FIG. 15.4 : *Diagram to show freezing of loose deposits (P.I. Melnikov and N. I. Tolstikhin, 1974).*

Key : *Syngenetic type.* (a) upper and lower permafrost tables, (b) region of syngenetically freezing deposits, (c) region of epigenetically freezing deposits, (d) borders between regions b. and c., (e) borders of horizons and lenses, (f) direction of migration of permafrost tables. (1) Horizon of subaerially freezing deposits with superimposed cryogenic textures of segregation type. (2) Horizon of subaqueously freezing deposits with superimposed cryogenic textures of segregation type. (3) Horizon of subaqueously freezing deposits with syngenetic cryogenic textures of segregation type. (4) Horizon of passive ice formation with lithogenic cryotextures. (5) Lense with cryogenic of injection type.

Epigenetic type : (6) Horizon of ground-ice formation under conditions of free migration of water towards the border of freezing with cryogenic textures of segregation type. (7) Horizon of ice formation under conditions of water migration under pressure towards the freezing border with cryogenic textures of injection type. (8) Horizon of passive ice formation with lithogenic cryotextures.

freezing of rocks with ground ice development leads to an increase in their volume. Freezing rocks therefore exert intense differential pressures on the underlying unfrozen rocks. Since the compression and consolidation of underlying rocks is restricted, uplift and compression of the permafrost take place.

In the case of thin permafrost, deformation can easily take place and the pressure on the substratum is small. The conditions of water migration therefore resemble those of syngenetic permafrost. But with increasing permafrost thickness, the pressure on unfrozen rocks increases. Clays, loams and silts become consolidated under this pressure. Some of the groundwater is freed and is forced out into the less consolidated part. This water then gradually freezes in fissures, larger pores or in less compressed rocks. This is why in epigenetic permafrost, both segregation and injection textures can be found, but pressure migration of water predominates. In epigenetic permafrost, three texture types therefore occur in a vertical direction (from above) :

(a) a horizon of ground-ice originating under conditions of free-water migration towards the freezing-front and showing textures of segregation type,
(b) a horizon of ground-ice developing under conditions of water migration under pressure towards the freezing front, showing cryogenic textures of injection type, and
(c) a horizon of passive ground-ice with textures depending on the rock properties (Melnikov and Tolstikhin, 1974, p. 151).

The cryogenic textures of epigenetic permafrost thus differ substantially from those of syngenetic permafrost. Cryogenic textures can therefore be utilized for an analysis of permafrost facies and to reconstruct the geomorphological conditions of permafrost development. In geomorphology, most attention has been paid to phenomena showing disturbance of the original rock structures, such as ice wedges, frost wedges and cryoturbation. Little attention has so far been paid to analysis of permafrost facies on the basis of cryogenic textures.

POST-CRYOGENIC TEXTURES AND THEIR PALAEOGEOGRAPHICAL SIGNIFICANCE

During permafrost degradation, ground-ice thawing and changes of rock properties take place. On steeper slopes, solifluction and mudflows develop owing to oversaturation of soils with water from melting ground ice. These movements naturally result in the disturbance of cryogenic textures. On gentle slopes or in the case of a lower degree of saturation of soils with water, rock consolidation takes place without any disturbance of the textures. In place of ice crystals, fine cracks are preserved, forming a so-called post-cryogenic texture. The walls of the cracks are often covered with a fine film of precipitated ferrous oxide or clay. Study of post-cryogenic textures in deposits allows the presence of permafrost in the Pleistocene periglacial zone, and its type, to be established.

An analysis of airphotos of various parts of the Pleistocene periglacial zone and detailed geomorphological investigations, for instance in Sweden and the U.S.S.R., shows that, in the case of permafrost degradation at the end of the Pleistocene, numerous landforms were preserved in the former periglacial zone, such as pseudomorphs of ice wedges and thermokarst phenomena (H. Svensson, 1964; A.A. Velichko, 1969, 1972, 1973). A study of exposures has shown that even post-cryogenic textures have been preserved, enabling us to analyse the palaeographical conditions of the origin of these surface forms under Permafrost conditions.

In Central Europe, post-cryogenic textures have been found mainly in slope deposits and loess. In loess especially, these investigations raise several questions. The thickness of loess deposits separated by buried soils in some locations in Central Europe reaches several tens of metres. The loess in the present-day periglacial zone in Siberia and Alaska (T.L. Péwé, 1968) exhibits a considerable ground-ice content. Ice wedges in the loess deposits of the polar lowlands in the north-eastern U.S.S.R. attain vertical thickness between 40 and 50 m (S.V. Tomirdiaro *et al.*, 1974, p. 47). In Central Europe, pseudomorphs of ice wedges of considerable size have also been found in loess. Simultaneously, however, the post-cryogenic textures point to a significant content of texture ice. During permafrost thawing, considerable changes in volume, consolidation of sediments and relief changes must have taken place (Demek and J. Kulla, 1969) which have not yet been sufficiently evaluated by geomorphologists and Quaternary geologists.

TALIKS AND THEIR GEOMORPHOLOGICAL SIGNIFICANCE

Taliks occur even in northern regions where temperatures in the permafrost are very low. Their number and size naturally increase towards the southern permafrost border. Taliks most often occur below rivers and lakes. Open taliks exist under large rivers such as the Ob, Yenisey, Lena, Indigirka and Kolyma. Beneath smaller rivers and lakes, closed taliks can be found. Sometimes an interesting phenomenon can be observed: in the valley-head below the stream there may be no talik; farther downstream a closed talik may appear; while in the lower reaches, an open talik appears. Taliks can also be found where strong groundwater spring, mainly salt-water, emerge.

Taliks are of considerable importance as regards the geomorphological activity of streams in the periglacial zone. The thermal effects of the water accelerate erosion, evoking the phenomenon termed thermo-erosion, that is, the combined thermal and mechanical activity of running water. Thermo-erosion is the cause of the wide valleys of Siberian rivers.

Taliks also determine the origin of other landforms such as icings and various types of ground-ice mounds. Little attention has so far been paid to the geomorphological significance of taliks in the development of periglacial geomorphology.

EROSION AND PLANATION IN THE PERIGLACIAL ZONE

In his geomorphological division of continents J. Büdel (1963, 1969a) used the designation 'zone of intensive linear erosion and pronounced valley formation'. This specification has appeared even in some significant textbooks and in the *Encyclopaedia of Geomorphology*, edited by R.W. Fairbridge (1968). Büdel claimed that in the periglacial zone linear erosion is the dominant activity, while lateral planation takes place only to a small extent or even not at all. This idea is wrong.

Intensive linear erosion undoubtedly takes place in the periglacial zone. And Büdel's (1969b) idea of the mechanism of linear erosion in the periglacial zone taking place along ice crusts (*Eisrindeneffekt*) is valid provided it is restricted to the uppermost valley sections without a talik. Nevertheless, linear erosion in the periglacial zone is favoured by the common mechanical and thermal effects of running water (so-called 'thermo-erosion'). Especially in regions with a well-developed ice-wedge system, linear erosion is affected and controlled by these ice wedges. As ice wedges on slopes thaw out, thermo-erosion gullies develop, not only on syngenetic ice wedges in loose deposits but also on epigenetic ice wedges in solid rocks. On the steep banks of the Lena River incised in the limestones and mudstones of the Vysokoe Prilenskoe Plateau, gullies as much as 30 m in depth have developed at regular intervals, undoubtedly as a result of thermo-erosion on a system of epigenetic ice wedges. The extent of lateral thermo-erosion is also considerable, sapping the valley sides and leading to the development of broad trough-shaped valleys.

Another phenomenon connected with thermo-erosion is the development of one of the most characteristic forms of the periglacial zone—dells. Dells are small dry valleys shallowly cut in large numbers along the slopes of the periglacial zone, running straight down the direction of slope. They have flat floors, gently inclined sides and reach lengths of several hundreds of metres, widths of several tens of metres and depths of several metres. Often they began to develop as a result of local permafrost degradation and the thawing of ice wedges. Their further modelling is controlled by solifluction, sheet wash and other cryogenic processes. With respect to the greater humidity along their floor lines, the dells act as the main denudation and transport lines on periglacial slopes. Owing to their frequency on slopes, dells are an important aspect of subaerial modelling.

As well as linear erosion in the periglacial zone, extensive planation takes place. This has been stressed in the last 25 years in the papers of numerous geomorphologists. In some regions, mainly with extremely continental periglacial climates, linear erosion is less active than planation. Sheet denudation and pediplanation owing to cryogenic processes such as frost creep, frost heaving, nivation, solifluction, suffosion and sheet wash, attain considerable importance in the periglacial zone.

Planation takes place here both at the foot of slopes, and on summits and watershed ridges. At the foot of slopes, piedmont planation surfaces designated 'cryopediments' develop owing to cryogenic processes in the periglacial zone. Extensive pediments on the margins of mountain ranges in eastern Siberia are often linked to Tertiary pediments. In valleys, valley cryopediments occur, these being continuations of Quaternary terraces. The valley cryopediments in the flysch Carpathians in Czechoslovakia are usually linked to Upper Pleistocene river terraces, sometimes even to the floodplain. The cryopediments have developed in various rocks and cut across rocks of varying resistance (H.M. French, 1973; Czudek and Demek, 1973; P. Macar, 1969).

The principal process leading to the development of cryopediments is the retreat of steep slopes caused by frost weathering and the sapping of slopes by nivation. In the modelling of the gentle foot-slope, many different processes play a part, especially solifluction, suffosion and sheet wash. Of great significance are the dells along whose axes there is greater moisture content in the floor deposits, and consequently the active layer above the permafrost is thicker. The gentle foot-slope is protected by permafrost from further denudation, the slope serving only as a surface of transport of the material that has descended the scarp.

The cryopediments extend from the main valleys into lateral valleys, even joining by pediment passes over watersheds. As the cryopediments coalesce, a regional planation surface develops called a cryopediplain.

Planation surfaces in the periglacial zone also develop on summits and watershed ridges. The summits and watershed ridges of the periglacial zone often exhibit a distinct stepped topography. The planation process usually begins by the formation of nivation hollows. As they widen and merge, and because of the retreat of rock steps termed 'frost-riven cliffs' and 'frost-riven scarps', cryoplanation terraces develop. On the summits, several steps of successive cryoplanation terraces are often to be seen. As they merge, a cryogenic planation surface termed a 'cryoplain' develops.

APPLIED PERIGLACIAL GEOMORPHOLOGY

For several decades, periglacial geomorphology has been considered a science dealing with phenomena and processes without any special practical significance. As late as the Second World War, and especially in the last two decades, the practical aspects of periglacial geomorphology have come to the forefront. This is connected above all with the economic utilisation of polar regions, mainly the Arctic. When man utilises the natural resources of extensive regions of Canada and Siberia, he encounters the characteristic natural complexes linked to permafrost along with the widespread occurrence of ground-ice and typical natural processes. The economic utilization of the periglacial zone requires careful planning of construction, the use of mineral resources, and the economic utilisation of the whole region. Permafrost and the associated natural complexes are very sensitive systems. Small mistakes which might have only negligible effects on construction in other climatic zones may result in catastrophic effects in the periglacial zone where permafrost occurs. The disturbance of the permafrost equilibrium by man leads to an irreparable disturbance of the surface of the terrain owing to thermokarst thermo-erosion or thermo-abrasion processes. Under such conditions, the prediction of the development of the surface during construction work is of great significance.

Geomorphologists are able nowadays, with the aid of geomorphological mapping based mainly on air and space photographs, to establish the extent of permafrost, delimit taliks, establish the course of cryogenic processes and make predictions about possible changes in natural complexes resulting from the development of certain projects. Most constructional projects (for instance, gas and oil pipelines) are designed with conservation of the permafrost in mind. From the evidence of landforms, the disturbance of the thermal balance of the permafrost can be monitored, the beginning of its degradation can be established, and protective measures can be recommended. In the case of major structures (railways, roads, long oil and gas pipelines), the basic principles governing the periglacial zone must be studied and small-scale geomorphological mapping of extensive areas is necessary. As well as general geomorphological maps, special geomorphological maps for particular purposes must be compiled.

Geomorphology also plays an important part in prospecting for mineral deposits in the periglacial zone. To establish the occurrences of gold, tin and other deposits, geomorphological analysis aimed at distinguishing regions with different tectonic regimes is important, that is, regions in tectonic equilibrium or regions of minor uplift in which cryopediments develop and gold placers are lacking, or subsiding regions where talus deposits and extensive gold placers may develop, or regions of strong uplift with small placers in river beds.

In mining from placers in slope or fluvial sediments, economically favourable modes of permafrost thawing for systematic mining must be sought. Geomorphologists can give valuable recommendations on the basis of their knowledge of thermokarst phenomena. When mining on the margins of taliks, artificial freezing is often necessary to prevent water from flowing from the talik into the deposit.

Geomorphologists are facing special problems resulting from the development of modern forms of agriculture in the periglacial zone. In the continental zone especially, agriculture often requires irrigation which leads to permafrost degradation. Owing to the increase in the thickness of the active layer when fields are irrigated, suffosion takes place, and suffosion wells, sinks and sub-surface hollows develop.

Recently, the construction of large water reservoirs and dams has been started in the periglacial zone. The accumulation of water causes degradation of the permafrost and disturbance of the bottom and shores built of frozen rocks. Prediction of the rate of permafrost thawing on the bottom and the shores, and the related extent of seepage and loss of water from the reservoir are of great practical significance. Permafrost thawing on the shores can result in dangerous slides and mudflows. Filling of major reservoirs is usually accompanied by earth tremors not only in the solid rocks but also in the frozen deposits because frozen unconsolidated rock possess a stability modulus closely resembling that of unfrozen solid rock. Special problems occur in the construction of dams in frozen calcareous rocks with karst phenomena. Ground-ice in karst spaces may prevent their being filled by injections of concrete and later, in the case of permafrost degradation, leakage of water can take place.

The protection of the environment in the periglacial zone is therefore of immediate economic significance, and geomorphologists have great possibilities of helping in the rational utilisation of the natural resources of polar regions, by the study of cryogenic forms and processes.

CONCLUSION

In connection with the economic utilisation of arctic and subarctic areas, better information has been obtained in recent decades concerning cryogenic processes and forms in the periglacial zone. Simultaneously, the practical significance of such knowledge in the utilisation of these regions has become apparent. But the amount of information is still insufficient in comparison with knowledge of processes operating in other climatic zones. Of principal significance for the future development of periglacial geomorphology will be a more detailed knowledge of the processes acting in the permafrost as in a dynamic natural system. Quantitative measurements of cryogenic processes using field equipment are still relatively undeveloped, encountering many difficulties because of the unfavourable climatic conditions in the periglacial zone. Qualitative and quantitative measurements of the processes will, however, become still more important because further progress will require the development of mathematical models of frozen rocks and the processes active in them. The increase in the amount of information will also require other ways of data processing, with the use of statistical analysis and computers. In this connection the immediate practical significance of periglacial geomorphology will further increase.

SUGGESTED READINGS

Büdel, J., 1963 : 'Klimagenetische Geomorphologie', *Geogr, Rdsch,* **15,** pp. 269-85.

Büdel, J., 1969a : 'Das System der klimagenetischen Geomorphologie', *Erdkunde* 23, pp. 175-83.

Büdel; J., 1969b : 'Der Eisrinden-Eflekt als Motor der Tiefenerosion in der excessiven Talbildungszone', *Würzb geogr. Arb.* 25, pp. 1-41.

Czudek, T. and Demek, J., 1970 : 'Thermokarst in Siberia and its influence on the development of lowland relief', *Quatern. Res.* 1 (1), pp. 103-20.

Czudek, T. and Demek, J., 1971 : 'Pleistocene cryoplanation in the Ceská Vysocina Highlands, Czechoslovakia', *Trans. Inst. Br. Geogr.* 52, pp. 95-112.

Czudek, T. and Demek, J., 1973 : 'The valley cryopediments in eastern Siberia', *Biul. peryglac.* 22, pp. 117-30.

Czudek, T. and Demek, J., 1973 : 'Die Reliefentwicklung während der Dauerfrostbodendegradation', *Rozpr. csl. Akad. Ved. rada* MPV 83(2), pp. 1-69.

Davies, J.L., 1969 : *Landforms of cold climates* (M.I.T. Press, Cambridge, Massachusetts and London), p. 200.

Dedkov, A., 1965 : 'Das Problem der Oberflächenverebnungen', *Petermanns geogr. Mitt.* 109, p. 258-64.

Demek, J., 1968 : 'Cryoplanation terraces in Yakutia', *Biul. peryglac.* 17, pp. 91-116.

Demek, J., 1969 : 'Cryoplanation terraces, their geographical distribution, genesis and development', *Rozpr. csl. Akad. Ved. rada* MPV 79(4), pp. 1-80.

Demek, J., 1972 : 'Die Pedimentation im subnivalen Bereich', *Göttinger geogr. Abh.* 60, pp. 145-54.

Demek. J., and Kukla J. (eds.) 1969 : *Periglazialzone, Löss und Paläolithikum der Tschechoslowakei* (Czechoslovak Academy of Sciences, Institute of Geography, Brno), p. 155.

Dostovalov, B.N. and Kudrjavcev, V. A., 1967 : *Obshchee merzlotovedenie* (Izdatel stvo Moskovskogo universiteta, Moscow), p. 403.

Embleton, C. and King, C. A. M. 1975 : *Periglacial geomorphology* (Arnold, London), p. 203.

Fairbridge, R. W. (ed.) 1968 : *The encyclopedia of geomorphology* (Reinhold, New York-Amsterdam-London), p. 1295.

Ferrians, O. J. Kachadoorian, R. and Greene, G.W. 1969 : 'Permafrost and related engineering problems in Alaska', *U.S geol. Surv. Prof. Pap.* 678, p. 37.

French, H.M., 1973 : 'Cryopediments in the chalk of Southern England', *Biul. peryglac.* 23, pp. 149-56.

Jahn, A., 1965 : *Problems of the glacial zone* (PWN, Warszawa), p. 223.

Karrasch, H., 1972 : 'Flachenbildung unter periglazialen Klimabedingungen." *Göttinger geogr. Abh.* 60, pp. 155-68.

Macar, P., 1969 : 'Actions periglaciaires et évolution des pentes en Belgique', *Biul. peryglac.* 19, pp. 137-52.

Melnikov, P.I. and Tolstikhin, N.I., 1974 : *Obshchee merzlotovedenie* (Nauka, Novosibirsk), p. 291.

Melnikov, P., 1973 : 'Permafrost, North American contribution', *2nd int. Permafrost Conf. Yakutsk*, 1973 : National Academy of Sciences, Washington D C., p. 782.

Péwé, T.L., 1968 ; 'Loess deposits of Alaska', *23rd int. geol. Congr. Czechoslovakia* (1968), Proceedings of Section 8 (Academia, Prague) pp. 297-309.

Péwé, T.L., (ed.) 1969 : *The Periglacial environment* (McGil-Queen's Univ. Press, Montreal), p. 487.

Schunke, E., 1975 : 'Die Periglazialerscheinungen Islands in Abhängigkeit von Klima und Substrat', *Abh. Akad. Wiss. Göttingen* (Mathematisch-physikalische Klasse), Dritte Folge 30, pp. 1-273.

Shvecov, P.F. and Dostovalov, B.N. 1959: *Osnovy geokriologii (merzlotovedenie)* (Izdatelstvo Akad nauk S.S.S.R., Moscow) p. 459.

Svensson H., 1964 : 'Aerial Photographs for tracing and investigation fossil tundra ground in Scandinavia', *Biul. peryglac.* 14, pp. 321-25.

Tomirdiaro, S.V. *et. al*, 1974 : 'Fiziko-geograficheskaja obstanavka i osobennosti formirovanija lessovo-ledovogo pokrova na ravninakh severovostoka Azii'. (Physico-geographical environment and peculiarities in formation of loessglacial cap, north-east Asiatic Plains), *Geologiya geof.* 7, pp. 47-61 (Novosibirsk).

Tricart, J., 1970 : *Geomorphology of cold environments* (Macmillan, London), p. 320.

Velitchko, A.A., 1969 : 'Milieu géologique et géomorphologique de la zone périglaciaire de la plaine est-européenne', *Biul. peryglac.* 18, pp. 183-93.

Velitchko, A.A., 1972 : 'La morphologie cryogéne relicte : caractères fondamentaux et cartographie', *Z. Geomorph. Suppl.* 13, pp. 59-72.

Velitchko, A.A., 1973 : 'Osnovnye osobennosti reliktovou kriogennoi morfoskulptury i obschchiie principy ee kartirovanya' (The main peculiarities of relict cryogenic morphosculpture and general principles of its mapping) in *Paleokrilogya v chetvertichnoi stratigrafü i paleografü* (Nauka, Moscow), pp. 121-34.

Washburn, A.L., 1973 : *Periglacial processes and environments*, (Arnold, London), p. 320.

SECTION FOUR

COASTAL LANDFORMS

INTRODUCTION

Coastal Landforms and Processes

Modern work on the coastal processes and landforms has increasingly sought to understand the processes that create the forms out of the materials available. Since the mid-60s the momentum of research into coastal processes has increased, and most of this work is beginning to bear fruit. Like many contemporary advances the origin of this research can be termed to the Second World War, when it became important to determine beach slope for planned amphibious landings. Today, many parties participate in research in this area—mathematicians, engineers, geologists, biologists and even a few geomorphologists. It is usually sandy beaches that have attributed the attention of the field workers.

In following two papers in this volume elaborate systematically the coastal processes and forms. C.A.M. King in her paper has reviewed the coastal landforms, their classification, morphology and processes have been discussed. According to her the landforms of the coast are the result of the processes interacting with the material of the coast. The coastal zone is becoming increasingly important, as its use becomes more intensive for a wide range of purposes. Beaches become more vulnerable as sediment supply decreases, and coastal protection becomes more necessary. It is best achieved by beach replenishment, which preserves the natural state of the system. A wide, high beach is by far the best means of coastal defence.

E.C.F. Bird in his paper on coastal landforms in the fluvial tropics has pointed out that in the field of coastal geomorphology, stress has often been given to the fact that marine processes are essentially a zonal, the effects of waves, tides and currents showing little variation between hot and cold or wet and dry environments. Nevertheless, it is acknowledged that there are distinctive features on arctic coasts, subject to the effects of freezing and thaw and of the accumulation and melting of snow, and on arid coasts, where lack of runoff and sparsity of vegetation result in the development of characteristic landforms. He has further distinguished the coastal feature of Australian humid tropical sectors with that of cooler and drier sectors. The author has delimited the humid tropical coasts on the basis of climate and presented a systematic account of humid tropical coastal features.

The sub-marine terrain forms one of the significant aspect of coastal studies. The paper by Jag Mohan Agrawal deals with evaluation of sub-marine terrain. The author has proposed a new concept of classification of sub-marine terrain mainly on 'patterns' and 'Facts. Such studies are highly useful for sub-marine navigation, surface navigation, fisheries, mineral exploration and oil exploration.

With the increasing population and diminishing resources on land, man is diverting his attention to ocean and scientist will be demanding charts for spatial distribution, quantification and comparative analytic studies. The scientists of different field will though require specific charts in their field, the basic needs of everybody is sea floor relief charts. The paper by Satya Prakash and Jag Mohan Agrawal deals with cartographic practices used in representation of sub-marine terrain. The paper lays emphasis on the need for development of effective graphics of the sea by the imaginative and innovative cartographers.

SIXTEEN

C. A. M. King

COASTAL LANDFORMS

INTRODUCTION

COASTAL LANDFORMS are a function of the action of processes on the materials of the coastal zone, thereby producing the many and varied morphological forms that occur at the junction of land and sea. In order to analyse the forms, they will first be classified in various ways that illustrate both the forms and the processes that have shaped them. This section leads on to a brief review of the morphology of the coastal zone, leading in turn to a consideration of geometrical and rhythmic forms that have been the subject of much recent research. Next, the sediment that is of great importance in so many coastal features and in coastal defence will be briefly examined. Finally, the action of the processes that shape the sediment and the coastal rocks into the great variety of forms found in this environment will be summarised.

Modern work on the coast has increasingly sought to understand the processes that create the forms out of the materials available. This work has involved detailed field-work with elaborate equipment, as well as laboratory studies. Theoretical analyses of the processes, expressed in the form of differential equations, and based on a sound understanding of their dynamic action, have also been attempted. This work has become more urgent and significant as the development of the coastal zone for a wide range of uses has accelerated, affecting a great variety of coasts.

Classification

The great variety of coastal features can be classified in several different ways and at different scales. Some of the classifications are based on the structural character of the coast, and others on the morphology. Yet others concentrate on the processes operating to shape the coast. Examples of these different types of classification will illustrate their varying approaches. Some are based on the character of the land and others on that of the sea. Both have their value, while in others these aspects may be combined.

The structural, world-wide classification of Inman and Nordstrom (1971) provides a valuable method of relating coastal character to the new global tectonic theory. It is in some ways a development of the very early classification of Suess (1888), which divided the coasts into Pacific and Atlantic types, depending on the relationship between the structure and the coast. The two were parallel in the Pacific type, but at an angle in the Atlantic type.

In the more recent classification coasts are subdivided according to their relationship with the plates and their margins. The classification consists of the following groups :

Inman and Nordstrom (1971)

1. *Collision coasts* : — where two plates converge
   - (a) Continental collision coasts along the zone of convergence. — where a continental margin occurs.
   - (b) Island arc collision coasts. — where no continental margin is involved.
2. *Trailing edge coasts* : — where a plate-imbedded coast faces a spreading zone.
   - (a) Neo-trailing edge coasts. — where a new zone of spreading is taking place.
   - (b) Afro-trailing edge coasts. — where the opposite continental coast is also trailing
   - (c) Amero-trailing edge coasts. — where the opposite continental coast is a collision coast.

3. *Marginal sea coasts* : — where a plate-imbedded coast faces an island arc.

The collision coasts have structural lines parallel to the coast, as in the Suess' Pacific type. They mostly occur around the Pacific. The island arc type occurs mainly in the western Pacific, apart from the two island arcs that protrude into the Atlantic, the Caribbean and the Scotia arcs. The continental collision coastal type is exemplified by the western coasts of North and South America. Both types of collision coast are tectonically active, and vertical movements are common, in association with earthquake activity. Shelves are usually narrow and coastal slopes inland are steep.

The trailing edge coasts are imbedded within a plate and do not occur at their margins. They are, therefore, much less active tectonically. The neo-trailing edge coasts may have steep hinterlands because they represent active zones of spreading. The Red Sea coasts provide an example.

The essential difference between the Afro- and Amero-trailing edge coasts is that the Amero-trailing edge type has a collision coast on the opposite side of the continent. The high ground associated with the collision coast produces active erosion and a heavy sediment load in the streams, such as the Amazon and the Mississippi. The Afro-trailing edge type, on the other hand, has trailing edge coasts on both sides, and the country is more stable as a result. Sediment loads are much lower, as in the Zaire and Niger rivers.

The Indian coast is classed as Amero-trailing edge type, because of the plate collision along the Himalayas, which provides massive sediment loads to the Indus and Ganges rivers. Coasts of the Amero-trailing edge type tend to have large amounts of beach material, with wide shelves and relatively low hinterlands.

The tectonic character of the coast can be related to the major morphological elements, such as the shelf width, hinterland relief and amount of beach sediment. Thus global tectonics can be associated with large scale coastal features, in relation to the character of the land.

It is also necessary to consider the coast from the oceanic point of view. The dynamic coastal classification of J.L. Davies

(1972, 1980) provides a useful analysis of coastal features in relation to the dominant wave pattern. He points out that wave generation depends on the major climatic wind systems.

The classification includes a storm wave coastal environment, which is situated in high latitudes, in the belt of strong westerly winds. This belt is more variable in the northern hemisphere. The mean latitude of strongest winds is situated in 46°N in winter and 62°N in the northern summer. In the southern hemisphere the latitudinal range is much smaller, the mean latitude being 54°S in winter and 56°S in the southern summer. The continuous belt of sea around the Southern Ocean allows the southern westerly winds to generate larger and longer waves than in the north. Only a short stretch of coast in the extreme south of South America, however, is directly affected by them. The northern hemisphere coastal zone of storm wave dominance is considerably more extensive. It covers much of the northern Pacific and Atlantic coasts of North America and Eurasia.

The storm wave coasts are dominated by the action of short, high seas, which are generally destructive in their effect. This destructive action is especially strong when they are accompanied by strong onshore winds, as is often the case.

The coasts of very high latitudes, around Antarctica and the Arctic Ocean are generally coasts of low energy. Offshore ice in these situations prevents the approach of effective waves. Other low energy coasts are those of enclosed seas, such as the Baltic and Mediterranean. The fetches in these enclosed seas are too short for really effective waves to be generated. Waves tend to be short and to come from widely varying directions in these areas.

The coasts of the lower latitudes are mainly affected by long, low swells that have travelled far from the stormy generating areas of higher latitudes. The swells tend to be higher on the west-facing coasts. Thus west coast swell shores are distinguished in Davies' classification from east coast swell shores. On the east coasts waves are also long and generally constructive, but may be less high, owing to longer travel distances.

In the lower latitudes the steady Trade Winds are important in the generation of the dominant waves in some areas, especially on east coasts. The west coast of India combines west coast swell waves and waves generated by the Monsoon winds. The east coast of India is also affected by Monsoon generated waves and east coast swell waves.

These long swells often have periods of about 14 seconds, which is equivalent to a length of about 300 m. They are generally constructive in their action on sandy coasts, and sand is the most common beach material in the lower latitudes. In higher latitudes shingle plays a greater part, in conjunction with the action of the storm waves.

Davies' classification is based on the wave climate. It is a world-wide classification which recognises the very significant role of waves in shaping the coast. The action of tidal processes should not, however, be forgotten. Davies also draws attention to the world-wide variation of tidal range and type. He distinguishes coasts with a macrotidal range from those with a mesotidal range and those with a microtidal range. The dividing ranges are 4 m and 2 m, respectively.

Morphology

The older classification of Johnson (1919) and Shepard (1973) drew attention to the great variation in coastal morphology. Johnson divided coasts into four categories, coasts of submergence, coasts of emergence, neutral and compound coasts. Nearly all coasts, however, fall into the last category, as sea level has changed so rapidly recently. Evidence of both submergence and emergence is often found on the same coast. This reduces the value of the classification.

Shepard divided coasts into those shaped mainly by terrestrial agencies and those shaped mainly by marine processes. Many coasts, however, show evidence of both forces. His classification did, however, draw attention to the wide range of coastal features. That of Valentin (1952) is also useful from this point of view. Valentin's is based on the dichotomy between coasts that have advanced and those that have retreated. The difference may be due to the relation between accretion and erosion, on one hand, and submergence and emergence, on the other.

The wide range of coastal morphology will be described simply in relation to the main character of the coast. This can in turn be related to the classifications already discussed. Thus the coasts will be broadly divided into hard rock coasts, soft rock coasts, and low marine coasts. The major characteristics of these three types will be briefly described.

Hard Rock Coasts

Hard rock coasts are characterised by cliffs, rock platforms, stacks, arches and similar features. They occur widely in the storm wave belt, where destructive waves are dominant, and on the collision coasts where the hinterland is high and water offshore is deep. They may also occur on other types of coasts, where beach sediment is locally in short supply and hard rocks reach the coast. This situation is more likely to occur on the Afro-trailing edge coasts, and in the low energy environments, with their variable wave pattern.

Relatively little work has been done on the processes forming cliffs and wave-cut platforms, compared with depositional forms. The cliff profile is indicative of the relative activity of marine and subaerial processes. Cliff profiles indicate that they are often very old features. Many were probably formed in early Pleistocene or even Pliocene times. They have had a complex history of alternating marine erosion, in periods of high sea level, and subaerial modification, in times of low sea level. The result may be a bevelled cliff, in which the upper part is at a lower angle than the wave-steepened lower section.

The cutting of the platform at the base of the cliff is achieved by a combination of processes which include abrasion, water layer weathering and solution, as well as various biological processes. Abrasion is more likely to occur in exposed areas, where there is a small amount of large beach material. Water layer weathering is more common in less exposed areas that are alternately wet and dry with tidal changes. This process and solution both depend on the rock type and structure.

Soft Rock Coasts

When the rocks are soft and unconsolidated, such as coasts of glacial deposits, the form of the cliff is much more dependent on subaerial processes than marine ones. The action of the sea is mainly confined to the removal of the material that moves down the cliff by slumping or land-sliding.

These cliffs provide much beach material, although often most of it is too fine to remain on the foreshore to protect the cliff from further loss. They often suffer rapid erosion, as exemplified by the Holderness cliffs on the east coast of England. The rate of erosion of the Holderness coast is about 1 to 2m/year, (Valent, 1954).

Soft rocks tend to produce smoother coastal outlines than hard rocks. The forms of deposition often associated with the two types of coast also tend to be different. The hard rock coasts often have a crenulate outline, or are indented with alternating headlands and bays, while the soft rock coasts have considerably smoother outlines.

The greater the degree of indentation, the more varied and complex the features of deposition usually are. Thus on an indented or crenulate hard rock coast a wide variety of spits, bars and forelands are often found. These features reflect the varying angles of wave approach relative to the shore. This variation results in a complex pattern of longshore movement of material, and hence of depositional forms and incidence of erosion. Erosion tends to concentrate on the headlands, where wave refraction causes convergence of wave energy, while in the bays energy is dissipated as a result of refraction, and accumulation can occur where longshore currents carry material to the zones of minimum energy in the bays.

On the smoother soft rock coasts a depositional form is a spit that forms across the mouth of an estuary. Spurn Head to the south of Holderness has extended partially across the Humber exemplifies such spits. It only receives about 3 per cent of the material eroded from Holderness updrift of it. Spurn Head passes through a 250-year cycle of elongation, breaching and decay, as elucidated by de Boer (1964).

Low Marine Coasts

Where the hinterland is low-lying, marine processes tend to dominate both in the backshore and in the foreshore zones. One of the major features of the low coast is the offshore barrier island complex. Barrier islands extend along about 13 per cent of the world's coastline, according to Gierloff-Emden (1961). They are extensive along the southeast coast of the United States of America, along parts of the African and

Australian coasts and elsewhere. They normally occur where the tidal range is not very large and where there is a plentiful supply of sand. These conditions occur especially in the swell wave environments of relatively low latitudes. Barrier islands tend to be associated with wide shelves, they are thus to be expected on the Amero-trailing edge coastal type, and to a lesser extent, on the Afro-trailing edge type.

According to Johnson's (1919) classification, barriers are typical of emerged coasts, but in many instances this is not true. In fact barrier islands probably form by a number of different processes, (Bird, 1961; Gilbert, 1890; Hoyt, 1967; Pierce, 1970; Swift, 1975; Van Straaten, 1965). Some of the barriers, probably including most of the eastern United States of America barriers, formed when sea level was much lower. They have gradually been moved landwards by overwash as sea level has risen in the Flandrian transgression.

Overwash is still an important process, especially under hurricane conditions of abnormally high water levels. Barriers on the coast of southeast Australia also probably formed by this process, (Thom, 1978). Another method of barrier formation is by the elongation of spits from headlands, and their eventual extension across an estuary or inlet, (Gilbert, 1890). The barrier island complex consists of several distinct elements, which together make it a complex coastal system.

Behind the barrier island a lagoon is often enclosed, within which salt marsh growth, aided by suitable halophytic vegetation, is a dominant process. The barrier itself is usually raised in elevation by wind action, and dune formation is an important process. Again suitable vegetation plays an important part in trapping sand and consolidating the barrier surface. Under natural conditions sand is washed over the barrier through the dunes during storms. This does not destroy the system, although it may slowly move landward by this process.

Many barriers are broken at intervals by tidal inlets. These are kept open where the tidal range is sufficient to generate a large tidal prism, which will depend partly on the lagoon depth and size, Ebb and flood tidal deltas often form on the seaward and landward side of the inlet, respectively. Sand can usually be transmitted across the inlet by the action of tidal currents. Thus the barrier island system can be a very long one, and it contains many features.

Geometrical and Rhythmic Forms

Geometrical and rhythmic forms occur on a wide range of scales in the coastal zone. The larger scale features include those described as a half-heart or zeta curve in shape, (Silvester, 1962). They are more precisely defined by the logarithmic spiral, (Yasso, 1965). All these shapes vary in curvature, being more open at one end and curving more tightly at the other.

The forms can occur either convex or concave to the sea. The convex form occurs on spits that are shaped by waves coming at such an angle that they refract around the tip of the spit. The logarithmic spiral form is imposed on the spit outline in response to the direction of refracted wave approach.

The concave features are usually the outline of beaches within fairly long bays. The form is again related to wave refraction and is most common in those areas where swell approaches from one predominant direction. There are good examples on the Pacific coast of North America, where long swells reach the coast of California from the northwest. Similar forms also occur on the coast of South Africa and Australia, which are also situated in the swell wave environment.

The long, swell waves suffer considerable refraction before reaching the foreshore. The amount of refraction is greatest in the lee of a headland, so that it is here that the logarithmic curve is tightest. On the more exposed part of the bay the curvature is flatter, and the waves are less refracted. Until the spiral curvature has developed material will move along the beach in response to the oblique approach of the waves. Eventually, however, the curvature is developed that fits the direction of approach of the refracted swells. In this state there will be relatively little longshore movement of material, provided that the wave approach direction is consistent.

One of the advantages in recognising these forms is that they give an indication of the direction of wave approach, and hence the direction of sediment movement along the coast. The pattern suggests that the movement of material alongshore over the world's coasts can be determined as suggested by

Silvester, (1962). The results agree closely with the pattern of wave approach and type as depicted by Davies (1980).

In general there appears to be an equatorward movement from temperate latitudes in response to the swell waves and Trade winds. This pattern may help to account for the widespread barrier islands in low latitudes.

Rhythmic forms vary in size from small beach cusps on the foreshore, to larger protuberances, which are most common on sandy beaches. These cuspate features are concave to the sea, and occur in the swash zone on the upper foreshore. They differ from the crescentic features that are convex to the sea, and develop in the surf zone as submarine bars. The two types of rhythmic features are, nevertheless, related to each other, and to the pattern of water circulation in the near shore zone.

Both sets of features are usually ephemeral in their presence, depending on the level of wave power acting on the beach, (Short, 1979). They are related to the pattern of cell circulation that develops under some conditions on the beach. The shoreward transfer of water by mass transport sometimes develops a horizontally arranged pattern of circulation. Water moves slowly predominantly shoreward over part of the beach, but the return flow is concentrated into rip currents. These flow fast through the surf zone in narrow channels. Currents move laterally along the foreshore to feed the rips. A cellular pattern of water is thus set up, and it is reflected in the morphology of the rhythmic features on the beach.

Where water is arriving on the foreshore by mass transport there is a set-up of level, which is compensated by a wave set-down in the surf zone. A variation of water level along the shore results. In turn this variation can generate edge waves that travel along the shore. The edge waves are standing waves formed by resonance between the incident waves and those reflected from the shore. The edge waves usually operate at the same period as the incident waves or at harmonics of this period.

The result of the interference of the various waves is to increase wave height at the set of antinodes that are in phase. Wave height decreases at the intervening, out of phase, antinodes. The horns of the cusps will occcur where the breakers are higher, and the bays where they are lower. The pattern of water circulation is, therefore, reflected in the rhythmic beach morphology.

The significance of edge waves in accounting for the rhythmic features has been pointed out by Bowen and Inman (1969), Guza and Inman (1975) and Huntley and Bowen (1975). The latter have observed the presence of edge waves on the beach near Slapton on the south coast of England. They can occur both in short bays, and on more open coasts, when the wave conditions are favourable.

Recent work in Australia by Short (1979) and Wright *et al.*, (1979) has made progress towards the development of a model relating variations in wave power to the pattern of rhythmic shoreline features. In this way process can be related to small scale beach morphology. They recognise two major types of wave systems which relate to the nature of the beach morphology.

On the beaches of southeast Australia, when the wave power is low, long, low swells reach the beach. These waves are constructive and build up a steep berm, with a smooth and deep offshore profile. Under these conditions the waves are reflective, and a considerable amount of energy is reflected offshore out to sea. The beach crest is smooth and cusps are usually absent.

At the other extreme the waves have maximum power. A wide surf zone develops that absorbs all the wave energy and gives a dissipative system. These powerful, high waves tend to be destructive. Material is carried offshore to the break-point to form an offshore submarine bar. A low concave beach profile develops to landward. All flow is normal to the shore, and there is no longshore variation in profile, which is in equilibrium. The bar is linear at this stage of maximum wave power.

As wave power decreases edge waves may develop, and these produce cusps and crescentic bars. With continued decrease in wave power the cusps are modified. Eventually, at the minimum power situation, a continuous berm is built up. With increasing wave power the sequence of stages is reversed, until again a straight submarine bar lies off a concave straight beach.

These sequences and patterns have been related by short to the major wave environments that Davies suggested. Thus in the storm wave environment, the stages incorporating the straight submarine bar and major rip currents are the most common, owing to the predominance of high energy storm waves.

West coast swell environments also tend to have high wave power, and stages associated with maximum wave power occur. In the calmer season, however, the berm stage and intervening ones with cusps are fairly common. In the east coast swell environment large cusps and crescentic bars are common, owing to the generally moderate wave power. All stages can, however, be experienced, as wave power varies.

Monsoon coasts, such as those of India, have moderate summer wave power with the onshore monsoon. Rhythmic forms occur at this season. The low waves of the offshore monsoon, on the other hand, creates the berm pattern. Protected, low energy coasts usually have the berm or cusp profiles, which form under reflective conditions.

Beach Sediment

A beach can receive sediment from alongshore, from cliff erosion, from inland via rivers and glaciers or from offshore. Sediment can be lost alongshore or offshore, on onshore into dunes or estuaries. Sediment actually on the beach can be modified by attrition, or chemical and biological activity.

Apart from suitable soft rock cliffs, the supply of material from cliff erosion is probably rarely an important source of sediment. In a few areas, such as the Holderness coast, already mentioned, where the sea is cutting into unconsolidated glacial deposits, much sediment is derived from marine erosion. Another example is the coast of India south of Cochin, where old coastal barriers are being eroded by the sea and recycled (Manohar, 1961). Material derived from hard rock cliffs is usually very coarse, and requires further attrition before it can be moved far. On most coasts at present probably the main supply of new sediment is from inshore via the rivers.

At any one point most of the sediment reaches the coast from alongshore. The distance that sediment can move alongshore is very important, and can be used to define the coastal system. The system can be extensive, on relatively straight open coasts. The long barrier island coasts provided an example. On indented coasts, however, the systems are usually small. In some instances it has been shown that each small bay has its own distinctive material, thus implying minimal longshore movement.

Coastal erosion and accretion can usually be accounted for by the net amount of material reaching the area from alongshore. Where more moves out of the area than comes in, erosion will ensue. The reverse results in accretion. Thus the longshore movement of material is of great significance.

The supply of material from offshore has probably varied with time and in relation to changes in sea level. During the last 15,000 years sea level has been rising rapidly in the Flandrian transgression. The total rise amounts to about 100 m. It reached its present level about 3000 to 4000 years ago in the North Atlantic, but probably about 6000 years ago in Australia. The rise of sea level is significant from the point of view of sediment supply to the coast.

While sea level was rising fast, it is likely that much sand was swept shorewards to replenish and build up the beaches. It has been estimated that the barriers of the coast of Holland in the North Sea received half their sediment from alongshore, from the rivers Rhine and Meuse, and half from offshore, from the abundant supply left by the ice sheets in the North Sea, (Van Straaten, 1965). Similarly the barrier islands of southeast United States of America and Australia derived their sediment from the continental shelf as sea level rose, sweeping material landwards over the gently sloping shelf.

Once sea level had stabilised, however, the shelf would be steepened to wave base, and the supply of sediment is then much reduced. In some areas the beaches now have little fresh source of beach material. It is thus vital to preserve the existing supply of material, which will slowly be reduced by attrition. This situation applies to some of the shingle beaches of Britain, for example Chesil beach on the south coast. Chesil beach shows excellent lateral sorting, as the waves have adjusted the material to their energy and direction of approach, (Carr, 1969). The coarser material tends to accumulate in the zones of greater energy. The lack of new material allows such an adjustment to take place. The supply of new beach material from offshore has been reduced in many areas in the last few thousand years, as sea level has stabilised. Beaches are, therefore, more than ever reliant on inland sources.

Various opinions have been put forward regarding the likely change in beach sediment supply. On the one hand it has been argued by Curray (1964) that beaches will receive less sediment in the short term from inland. The postglacial submergence has left many deep embayments and estuaries, in which sediment will be deposited, and it will not reach the open coast. At a later period, however, when the estuaries are filled, sediment from inland will once more be able to reach the open coast and depositional regression will take place.

It has been argued by Russell (1967) that the supply of sediment from inland will be reduced through river control, dam building and similar devices inland. The offshore supply will also be restricted as the offshore slope steepens. Beaches are, therefore, likely to be starved of material. On the coast of California the supply of material to the coast has been reduced through the building of dams along the rivers reaching the sea in this area.

On the other hand human activity inland, in leading to accelerated erosion, has increased the load of many streams. This excess material may reach the sea, and will add to the nourishment of beaches. An example of this type of relation between inland and coastal changes has been described on the coast of India at the Godavari River mouth, (Mahadevan and Prasada Rao, 1958). A spit started to develop at the mouth of the Godavari River between 1864 and 1878, soon after deforestation inland started in 1874. The spit grew to a length of 16 km. It resulted in erosion further down drift, north of the river mouth, by disturbing the natural movement of material along the coast.

Sediment on beaches varies greatly in size. Some small beaches consist entirely of large boulders, while at the other extreme fine mud can accumulate in some sheltered situations. Mud is found in the runnels of a ridge and runnel macrotidal beach, for example on the east coast of England, and in sheltered salt marsh environments. In between these extremes by far the most abundant beach material is sand, but locally shingle is plentiful.

Shingle is generally more plentiful in higher latitudes, because much of it has been derived from glacial deposits. Shingle may occur where periglacial processes bring coarse material to the coast, or where heavy seasonal precipitation brings coarse material to the sea at one season.

The size of the sediment on a beach is significant, as its reaction to the waves depends partly on its size. Thus destructive waves can throw shingle well above the reach of normal waves to form high, long-lasting shingle ridges. Chesil beach has been built up to 13 m above high tide where the shingle is about 6 cm diameter at Portland, and 7 m at Abbotsbury, where the shingle is about 1.25 cm diameter.

Most sand-sized beach material is inorganic in origin, consisting of quartz or more rarely of basaltic grains. Locally, however, where terrestrial material cannot reach the coast, beach sediment consists of organic material, in the form of shell, coral or foraminiferal sand. Such material is usually less resistant than quartz sand and less dense.

The analysis of sediment character can provide useful information concerning both process and morphology. The relationship between material size and swash slope gradient is well known and significant, (Bascom, 1959 and Wiegel, 1964). Lateral sorting is also of interest. The coarser material is normally found in the zone of greatest energy. On a logarithmic spiral beach the material will be coarser at the open end of the curve, where the waves are higher and the beach is both steeper and higher. Thus a stream entering such a bay will tend to be deflected to the sheltered part of the beach, where the curvature is tighter and the beach crest lower, with a gentler gradient and finer material. Lateral sorting can only take place effectively where longshore movement is restricted, and the beach outline is in equilibrium with the processes.

The analysis of the sorting and skewness of the sediment can also provide valuable information concerning. process. Where the Φ notation is used, a negative skewness is characteristic of areas of strong currents and maximum wave energy. A positive skewness is often found in sheltered sites and in wind-blown coastal dune sand, where a tail of fine grains can settle.

Processes

The major processes that affect coastal landforms have already been mentioned; they include waves, wind, tide and biotic factors. In many areas the waves are the most important process, but in some areas the tide also plays a predominant part, while in others the biotic element is very significant. Locally the wind can play an important part, especially where dunes are present.

Waves

The distinction between storm waves and swell waves in Davies' coastal classification draws attention to the fundamental importance of wave character in determining beach response. The short, high, irregular waves of the storm wave environment are steep, and they exert a destructive influence on sand beaches, although they can build high shingle ridges. The long, low, regular swells, on the other hand, are mainly constructive. They build up the extensive beaches and barrier islands of lower latitudes. The waves are reduced to low steepness values during their long travel from the areas of generation in higher latitudes, by the westerly storm winds. They produce the high berms, steeper swash slopes and smooth offshore profiles of the reflective systems.

The storm waves tend to produce a flatter swash slope profile with a submarine bar in the nearshore zone, characteristic of the dissipative system.

Important modifications affect waves as they enter shallow water. The long swell waves, in particular, feel the bottom in relatively deep water. A wave starts to be influenced by the water depth when it is about half the wavelength. The long swells are often 300 m in length or more. Where the continental shelf is wide and shallow, the waves will be considerably refracted before they reach the shore. The refracted waves give rise to the logarithmic spiral forms that have already been mentioned. They are also responsible for the alignment of beaches affected by swell, especially when the waves come from one predominant direction.

The shorter waves of the storm wave and low energy environments, on the other hand, will suffer less refraction. The direction of movement of beach material alongshore will be more variable, both in time and space. More complex features typical of these areas will form as a result. The pattern of the beach ridges on these complex spits and forelands often give important clues concerning the pattern of wave attack during their formation. In their adjustment to the varying wave patterns, spits often become broad at their distal end and narrow at their proximal end in response to longshore movement of material.

The longshore transport of material by the waves is one of their most important functions. It is due to the generation of longshore currents in the surf and swash zones, which then move the sediment alongshore.

Longshore movement of material can be studied either theoretically, or by field experiment, using fluorescent tracers for example, or by laboratory studies in model tanks. The latter method has shown that with steep waves most of the longshore movement takes place in the surf zone, while with flat waves most moves alongshore in the swash zone by beach drifting, (Saville, 1950). This general pattern has been confirmed by full scale experiments carried out mainly with fluorescent tracers on the beach, (Ingle, 1966).

These observations are not easy to make. They can only be carried out over a limited extent of time and space, so that the results may not be representative and must be interpreted with caution. They have, however, provided useful data that allow some estimate of the amount of longshore sediment transport to be determined. This is very important, as on it depends the state of the beaches in terms of erosion and accretion in the longer term. Theoretical studies have provided differential equations that can provide an estimate of longshore sediment movement, based on observed wave characteristics.

The direction of wave approach is one of the most important variables. The wave power depends on the wave height and length, which together determine its energy. The beach slope determines the distance over which transport takes place, because transport is restricted mainly to the surf zone. Equations giving values for the volume of material carried alongshore have been developed. The longshore energy flux, P_{ls}, is given by

$P_{ls} = \frac{\rho g}{16} . H^2_b . C. \sin 2\alpha_b$ in the surf zone, P is the density, g is gravity, H_b is the breaking height of the waves, C is the wave phase velocity, and α_b is the angle of breaking wave approach. The volume of longshore transport, Q, in cubic yard per year, is given by $Q=(7.5\times10^3)\ P_{ls}$, where P_{ls} is given in ft. lbs./sec/ linear foot of beach, according to C.E.R.C. (1977).

Wind

The wind is fundamental in generating the waves in the first instance, and it also plays a part in generating the longshore current when it blows along the shore. Strong onshore winds have been shown, both in laboratory experiments and field observations, to enhance the destructive effect of steep waves. This effect may extend offshore beyond the break point of the waves, where, under calm conditions, the movement is normally shoreward. A strong onshore wind will also facilitate the movement of sand inland to the backshore zone to construct sand dunes. The process is most effective where a wide sandy foreshore is exposed to wind action at low tide, (Belly, 1964).

An offshore wind will, on the other hand, half to enhance the constructive effect of long, low swells, and reduce the destructive effect of storm waves. By blowing water offshore on the surface, a landward under-current is generated with an offshore wind. The reverse applies with an onshore wind, (King, 1953).

The Tide

The tide affects coastal landforms both through variations in water level, which are determined by its range, and through currents. The tidal streams tend to be stronger where the range is great. Ridge and runnel beaches are characteristic of high tidal ranges where there is an abundance of sandy sediment. They are found on some of the coasts of the Irish and North Seas. The ridges are built by the waves in their attempt to establish their equilibrium gradient on a foreshore that is too flat overall. The ridges often form at the mean positions of neap and spring tide levels, where they are parallel to the shore.

The tidal streams are also responsible for the development of a distinctive nearshore morphology in places, where they flow rectilinearly, and where there is a large amount of sand-sized material. Both large size features and characteristic bedforms are associated with tidal streams. These features can give useful data concerning the nature of the streams, and of the net direction of sediment transport in the nearshore zone.

The larger features are in the form of elongated banks and channels. The channels are associated with ebb or flood tidal streams. In the ebb channels the ebb flow continues for longer than the flood, and the net sediment transport is in the direction of the ebb stream. The channel tends to become shallower seaward or in the ebb flow direction. The reverse applies to the flood stream channels.

These banks and channels are particularly well developed in estuaries, including the Thames, Humber and the Wash in eastern England.

The bedforms include flood and ebb shields, which develop on the wider banks that dry out at low water. They form semi-circular ridges convex to the tidal stream that forms them, and they protect an area inside the semicircle from strong tidal action. There are also megaripples and sand waves and ribbons.

The sand waves are usually asymmetrical and sediment moves up the gentle slope towards the steep face. A survey of the sand waves, thus, provides valuable evidence of sediment transport direction. The crests of the waves are aligned normal to the direction of the reversing tidal streams. Sand waves occur throughout the continental shelf where there is enough sand, (Cartwright and Stride, 1958; Stride, 1963).

The position and movement of the nearshore channels and banks affect sediment supply to the foreshore in some areas, such as south Lincolnshire. They are also associated with nesses of accretion on the generally eroding coast of East Anglia to the south, (Robinson, 1966). One of these, Banacreness, moves north contrary to the normal southerly movement of material along this coast. The tidal streams are also responsible for the considerable area of accretion around Gibraltar Point in south Lincolnshire on the northwest corner of the Wash.

Biotic Processes

Plants and animals play an active part in some coastal landforms. The two landforms that are strongly influenced by plants are coastal marshes and sand dunes. Both are exemplified at Gibraltar Point in Lincolnshire, where beach ridges become stabilised on the upper foreshore and colonised by sand-loving plants to become foredunes. The muddy runnels to landward of the ridges can develop into strip saltings with the aid of salt marsh plants. These include *Spartina* which is very efficient silt trapper.

Salt marshes consist of very fine sediment, and can only accumulate in very sheltered areas. Thus marshes are found in estuaries, in the shelter of spits or in the runnels landward of beach ridges, and also in the lagoons behind barrier islands.

The type of marsh plants that flourish in these areas depends in part on the character of the sediment, whether it is fine sand, silt or clay, and on the climate. In temperate areas the main plants are salt marsh halophytes, such as the grass, *Spartina*, or annuals, such as *Salicornia* and *Suaeda*. These plants form a dense covering that helps to trap the sediment and aids in the upward growth of the marsh as tidal creeks develop. In tropical areas mangroves are common and also aid in trapping fine sediment in sheltered areas.

Coastal dunes form another environment where biotic processes are significant. Sand-loving grasses help to build up the coastal dunes. Marram grass, *Ammophila arenaria*, is the best known of these, but there are others that can grow even closer to high water level. Marram can grow effectively only where it is being continuously covered by fresh blown sand. Again there are species that are important in tropical waters, and others that grow in colder conditions. They help to build the foredunes, which can develop into parabolic or U-dunes. There are also forms including large transgressive sheets and climbing dunes.

Amongst the animals that influence coastal landforms corals are important. They are essential in the construction of substantial coastal features in the warmer seas. Corals are particularly prolific in the western Pacific Ocean, where they are mainly responsible for the many atolls, which develop from barrier and fringing reefs, (Stoddart, 1969). Algae are also important in the growth of some calcareous platforms in warm seas. Marine invertebrates can also exert a destructive role by boring through rocks and helping to erode shore platforms.

CONCLUSION

Within the restricted space available some of the important features of the coastal landform system have been mentioned. Other have been only very briefly aluded to, including the biotic elements of the system. The landforms of the coast are the result of the processes interacting with the material at the coast. The coastal zone is becoming increasingly important, as its use becomes more intensive for a wide range of purposes. Beaches become more vulnerable as sediment supply decreases, and coastal protection becomes more necessary. It is best achieved by beach replenishment, which preserves the natural state of the system. A wide, high beach is by far the best means of coastal defence.

SUGGESTED READINGS

Bascom, W.N., 1951 : The relationship between sand size and beach-face slope. *Transactions American Geophysical Union.* 32, (6), 866-74.

Bird, E.C.F., 1961 : The coastal barriers of East Gippsland, Australia. *Geographical Journal* 127, 460-80.

Bird, E.C.F., 1968 : (2nd Edition, 1976) *Coasts.* Australian National University Press, Canberra.

Belly, P.Y., 1964 : Sand movement by wind. *Coastal Eng. Res. Center.* Tech., Memo. 1.

Boer, G. de, 1964 : Spurn head : its history and evolution. *Transactions Institute of British Geographers* 34, 71-89.

Bowen, A.J. and Inman, D.L., 1969 : Rip currents 2. Laboratory and field observations, *Journal of Geophysical Research.* 74, 5479-90.

Carr, A.P., 1969 : Size grading along a pebble beach; Chesil beach, England. *Journal Sedimentary Petrology* 39, 297-311.

Cartwright, D.E. and Stride, A.H., 1958 : Large sand waves near the edge of the continental shelf. *Nature* 181, 41.

Coastal Engineering Research Centre, U.S. Army, 1977 : *Shore Protection Manual* 3rd Edition Washington, D.C.

Curray, J.R., 1964 : Transgression and regressions. In Miller, R.L. (Ed.) *Papers in marine geology : Shepard commemorative volume.* New York, MacMillan, 175-203.

Davies, J.L., 1958 : Wave refraction and the evolution of shoreline curves. *Geographical studies* 5, 1-14.

Davies, J.L., 1960 : Beach alignment in southern Australia. *Australian Geographer* 8, 42-44.

Davies, J.L., 1980 : *Geographical variation in coastal development.* 2nd Edition. Longman, London.

Draper, L., 1967: Wave activity at the sea bed around north-western Europe. *Marine Geology* 5, 133-40.

Folk, R.L., 1966 : A review of grain size parameters, *Sedimentology* 6, 73-93.

Gierloff-Emden, H.G., 1961 : Nehrungen and Lagunen. *Petermanns Geographische Mitteilungen* 105, 81-92, and 161-76.

Gilbert, G.K., 1890 : *Lake Bonneville* U.S. Geological Survey monograph.

Guza, R.T. and Inman, D.L., 1975 : Edge waves and beach cusps. *Journal Geophysical Research* 80, 2997-3012.

Hails, J.R. and Carr, A.P., (Eds.) 1975 : *Nearshore sediment dynamics and sedimentation.*

Hoyt, J.H., 1967 : Barrier island formation. *Bulletin Geological Society America* 78, 1125-36.

Ingle, J.C., 1966 : *The movement of beach sand.* Developments in Sedimentology. 5, Elsevier, Amsterdam.

Inman, D.L.; Komar P.D. and Bowen, A.J., 1969 : Longshore transport of sand. *Proceedings 11th Conference on Coastal Engineering,* London.

Inman, D.L. and Nordstrom, C.E., 1971 : On the tectonic and morphologic classification of coasts. *Journal of Geology.* 79, 1-21.

Johnson, D.W., 1919 : *Shore processes and shoreline development.* New York. (Facsimile edition, Hafner, New York; 1965).

King, C.A.M.., 1953 : The relationship between wave incidence, wind direction and beach changes at Marsden Bay, Co. Durham. *Transactions Institute British Geographers* 19, 13-23.

King, C.A.M., 1963 : Some problems concerning marine planation and the formation of erosion surfaces. *Transactions Institute British Geographers* 33, 29-43.

King, C.A.M., 1972 : *Beaches and coasts.* 2nd Edition. Arnold, London.

Komar, P.D,, 1971 : The mechanics of sand transport on beaches. *Journal of Geophysical Research* 76, (3).

Komar, P.D. and Inman, D.L., 1970 : Longshore sand transport on beaches. *Journal of Geophysical Research* 75 (30), 5914-5927.

Komar, P.D., 1976 : *Beach processes and sedimentation.* Prentice-Hall, Englewood Cliffs.

Longuet-Higgins, M.S., 1970 : Longshore currents generated by obliquely incident sea waves, 1, and 2. *Journal Geophysical Research* 75, (33), 6788-6801 and 6790-6801.

Manohar, M., 1961 : Sediment movement at Indian ports. *Proceedings 7th Conference on Coastal Engineering,* 342-74.

Mahadevan, C. and Prasada Rao, R., 1958 : Causes of the growth of a sand spit north of the Godavari confluence. *Andhra University Memoirs Oceanography*, 62, 69-74.

Orme, A.R., 1962 : Abandoned and composite sea cliffs in Britain and Ireland. *Irish Geography* 4, 279-91.

Pierce, J.W., 1970 : Tidal inlets and washover fans. *Journal Geology* 78, 230-34.

Robinson, A.H.W., 1960 : Ebb-flood channel systems in sandy bays and estuaries. *Geography* 45, 183-99.

Robinson, A.H.W., 1966 : Residual currents in relation to shoreline evolution of the East Anglian coast. *Marine Geology*. 4, 57-84.

Russell, R.J., 1967 : *River Plains and Sea Coasts*. Berkeley and Los Angeles University of California Press.

Saville, T., 1950 : Model study of sand transport along an infinitely long straight beach. *Transactions American Geophysical Union* 31, 555-65.

Shepard, F.P., 1973 : *Submarine Geology*. 3rd Edition Harper and Row, New York.

Short, A.D., 1979 : Wave power and beach stages : a global model. *Proceedings 16th Conference Coastal Engineering* 1145-62.

Silvester, R., 1960 : Stabilization of sedimentary coastlines. *Nature* 188, 467-9.

Silvester, R,, 1962 : Sediment movement around the coastlines of the world. *Proceedings Conference on Civil Engineering Problems Overseas*, London, Paper 14.

Sonu, C.J., 1972 : Field observations of nearshore and meandering currents. *Journal Geophysical Research* 77, 3232-47.

Stoddart, D.R., 1969 : Ecology and morphology of recent coral reefs. *Biological Review* 44, 433-98.

Stride, A.H., 1963 : Current-swept sea floors near the southern half of Great Britain. *Quarterly Journal Geological Society*. London 119, 175-99.

Suess, E., 1888 : *The face of the earth*, II (English translation by H.B.C. Sollas, 1906) Oxford University Press.

Swift, D.J.P., 1970 : Quaternary shelves and the return to grade. *Marine Geology* 8, 5-30.

Swift, D.J.P., 1975 : Barrier island genesis : evidence from the central Atlantic shelf, eastern U.S.A. *Sedimentary Geology*, 14, 1-43.

Thom, B.G., 1978 : Coastal sand deposition in southeast Australia during the Holocene. In *Landform Evolution in Australia*, 197-214, Davies, J.L. and Williams, M.A.J. (Eds.).

Valentin, H., 1952 : Die Küsten der Erde. Supplement to *Petermanns Geographschie Mitteilungen* No. 246. Erganzungsheft.

Valentin; H., 1954 : Der Landverlust in Holderness, Ostengland von 1852 bis 1952. *Die Erde* 3, 296-315.

Van Straaten, L.M.J.U., 1965 : Coastal barrier deposits in south and north Holland. *Medel. Geol. Sticht*. NS 17, 41-75.

Wiegel, R.L., 1964 : *Oceanographical Engineering*, Prentice-Hall, New York.

Wright, L.W., 1970 : Variation in the level of the cliff-shore platform junction along the south coast of Great Britain. *Marine Geology* 9, 347-53.

Wright, L.D., Chappel; J. Thom, B.G., Bradshaw, M.P. and Cowell, P., 1979 : Morphodynamics of reflective and dissipative beach and inshore systems : south eastern Australia. *Marine Geology* 32, 105-40.

Yasso, W.E., 1965 : Plan geometry of headland—bay beaches. *Journal Geology* 73, 702-14.

Zenkovich, V.P., 1967 : *Processes of coastal development*. Oliver and Boyd, Edinburgh.

SEVENTEEN

E. C. F. Bird

COASTAL LANDFORMS IN THE HUMID TROPICS

INTRODUCTION

ASSESSMENTS of the extent to which climate has influenced the geomorphological processes that affect the land surface, and has therefore been a factor in the evolution of landforms, have varied through the literature of geomorphology (e.g., Büdel, 1963; Stoddart, 1969; Tricart and Cailleux 1972; Derbyshire 1973). In the field of coastal geomorphology, stress has often been given to the fact that marine processes are essentially azonal, the effects of waves, tides and currents showing little variation between hot and cold or wet and dry environments. Nevertheless, it is acknowledged that there are distinctive features on arctic coasts, subject to the effects of freeze-and-thaw and of the accumulation and melting of snow, and on arid coasts, where lack of runoff and sparsity of vegetation result in the development of characteristic landforms (Anfrére, 1934). On the Australian coast, the humid tropical sector is scenically contrasted with cooler and drier sectors, partly because of geomorphological features (Bird and Hopley, 1969).

The world's shorelines have had comparatively brief histories, having been established only since the Holocene marine transgression brought the sea up to approximately its present level about 6000 years ago. Their features of weathering, erosion and deposition are thus related to processes that have operated in climatic environments that have not changed greatly within this period. Certainly, many coasts include landform features inherited from past environments, when climatic conditions were different, and when the sea stood above or below its present level relative to the land, but the dynamic shore zone can be related to processes that have operated under climatic conditions similar to those of the present. On this basis, it is worth reviewing the features characteristic of coastal landforms in the humid tropics, examining how far they are distinctively related to their perennially warm and wet environments.

Delimitation of Humid Tropical Coasts

Various attempts have been made to define humid tropical climates and indicate their extent (e.g. Fosberg *et al.*, 1961; Jackson, 1977). As there are lateral transitions to sub-humid and semi-arid environments, and to cooler, subtropical environments, any delimitation based on selected climatic parameters places linear boundaries through transitional zones. Some have attempted to overcome this problem by using natural vegetation (tropical rain forest) as an indication of humid tropical environments (e.g., Axelrod, 1958), but detailed studies of the plant communities show variations in species composition and luxuriance, some of which reflect climatic transitions while others relate to edaphic, biotic, or anthropogenic factors (Richards, 1952). Vegetation can be a misleading indication of climatic conditions in coastal regions, where tropical rain forest for example extends from the wettest areas into sectors with a distinct dry season, particularly on coastal plains, where high water tables maintain a sufficient supply of moisture to sustain forest growth through the dry season. Mangroves are characteristic features of humid tropical coasts, but they also extend far into sectors where the climate is arid or temperate : in Australia, for instance, mangroves extend as far south as Corner Inlet (latitude 38°55′ S), more than 2000 kilometres from the humid tropical sector in northeast Queensland.

Delimitations based on the boundaries of vegetation features can thus be misleading, and for the purpose of environmental description and analysis it is preferable to use a climatic classification. Bailey (1959) devised a scheme for classifying coastal climates on the basis of statistical analyses of temperature and rainfall data, and derived a map showing the extent of each of 14 categories, including 'rainy tropical' and 'sub-humid tropical'. Comparison with maps showing the distribution of each of the 15 major categories of Köppen's (1936) classification that extend to the coast indicates an essentially similar breakdown, especially within the tropics, where Bailey's 'rainy tropical' sector has much the same lateral extent as Köppen's *Af* and *Am* climates, and where his 'sub-humid' approximates Köppen's *Aw* category. In view of this similarity, and the simplicity and ease of application of the Köppen classification (Wilcock, 1958), it will be used here to locate the boundaries of humid tropical sectors around the world's coastline.

Tropical coasts may be defined as those with climatic stations having a mean temperature in the coolest month of at least 18°C. This includes the sectors with Köppen's *A* category (which is thus defined), and some of the sectors that fall within his *B* category (the dry climates); notably those parts of the arid (*BW*) and semi-arid (*BS*) areas where the mean temperature of the coolest month exceeds 18°C. All such places, since their mean annual temperature must exceed 18°C, fall in Köppen's *BWh* or *BSh*. The extent of tropical coasts, thus defined, is generally within the inter-tropical zone defined geographically by the Tropics of Cancer and Capricorn (23½°N and S); but the isotherm for 18°C in the coolest month extends beyond these limits to 25°S in Brazil, 26°N in southern Florida, and 26°S in Mozambique.

Köppen proposed four main subdivisions for *A* climates: *Af*, with no distinct dry season; *Am*, with a short dry season compensated by a large annual rainfall; *Aw*, with a low-sun dry season not thus compensated; and *As*, with a high-sun dry season not thus compensated. The *As* category is of little relevance here, and will be disregarded. *Af* climates, with mean rainfall in the driest month of at least 60 millimetres, are certainly humid tropical; *Am* climates, with a compensated dry season, may be included with them but *Aw* and *As* climates are essentially sub-humid, rather than humid tropical. Delimitation of humid tropical coasts is thus a matter of excluding the arid and semi-arid *BWh* and *BSh* sectors, and also the sub-humid *Aw* sectors, and retaining those with *Af* and *Am* climates.

An analysis of coastal climates can be made from data published in the British Air Ministry *Tables of Temperature, Relative Humidity and Precipitation for the World* (1958), and volumes 8 to 13 of *World Survey of Climatology* (1969), edited by H.E. Landsberg. The extent of the humid tropical sector, as here defined, is shown on Figure 17.1 and summarised in Table 17.1 (Bird and McArthur, 1978). Further details of the humid tropical sectors follow.

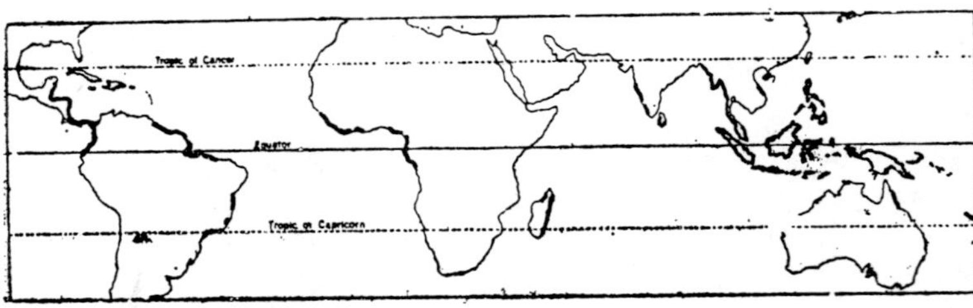

FIG. 17.1 : *Humid tropical coasts*

Moving southwards down the Pacific coast of Americas, tropical climates being in Mexico, the northwestern coasts of which have *Aw* climates with dry winter seasons. South from Mazatlan the winter becomes less dry, and humid tropical conditions may occur locally between Guatemala and Costa Rica, but the available information does not confirm them until Panama and Columbia. In Equador the winter dry season again becomes pronounced, and there is a transition towards the arid coasts of Peru. Guayaquil has an *Aw* classification with an annual rainfall of 988 millimetres, 933 of which fall between December and April.

On the Atlantic coast of South America, Florianopolis in southern Brazil is humid, but warm temperate (*Cf*) rather than tropical : the humid tropical sector begins just south of Santos. Along the Brazilian coast, *Af* and *Aw* sectors alternate, the dry season being more marked at Rio de Janeiro and Otoni, and from Recife round to Sao Luis, than in the truly humid tropical areas around Victoria and Salvador. A longer sector of humid tropical coast extends from Belem in northern Brazil to beyond Georgetown in Guyana, on to Trinidad, and there is a sharp transition from humid to arid conditions in eastern

TABLE 17.1 : **Humid Tropical Sectors**

1. The Pacific coast from the Gulf of Panama to northern Equador;
2. The Atlantic coast of Brazil near Santos, south of Rio de Janeiro, and Salvador, to the north, and from São Luis to the Orinoco Delta in north-eastern Venezuela;
3. The Caribbean coast from Turbo in Columbia to eastern Yucatan;
4. A small sector around Miami in southeastern Florida;
5. Parts of Jamaica, the Dominican Republic, Puerto Rico and the Lesser Antilles;
6. The west African coast between The Gambia and western Ghana, and from western Nigeria to Gabon;
7. The east and part of the northwest coast of Madagascar;
8. The west coast of India between Bombay and Kerala, and the south-western shores of Sri Lanka;
9. The Bangladesh coast south from Chittagong, and the whole of the coast of Burma and Malaya;
10. The coast of Vietnam north from Tourane, south western Hainan, and the southern tip of Taiwan;
11. Much of the coast of Indonesia New Guinea and the Philippines, and the inter-tropical Pacific Islands;
12. Northeastern Australia between Ingham and Mossman in Queensland.

Venezuela. Alternations of arid and *Aw* coastal climates occur along the Caribbean coast westward into Columbia, where there is another sharp transition from desert climate (*BS*) at Barranquilla to humid tropical (*Am*) at Turbo. The western Caribbean coast then shows humid tropical conditions through Panama northward to the east coast of Yucatan (Belize, *Am*), beyond which the Mexican coast has a more pronounced dry season (*Aw* climates). Garnier (in Fosberg *et al.*, 1961) mapped humid tropical climate on the shores of the Bay of Campeche, and Miami, in southernmost Florida, has a sufficiently warm and wet climate to attain a humid tropical (*Am*) classification.

Within the Caribbean, the island coasts of Cuba, Jamaica, the Bahamas and Haiti generally have relatively low rainfall and marked dry seasons (*Aw*), Punta Maisi in eastern Cuba falling within the semi-arid (*BS*) classification. Wetter conditions occur in northeastern Jamaica and on much of the coastline of the Dominican Republic and Puerto Rico, and the Lesser Antilles are generally humid tropical with local drier areas, for instance in Barbados, where Bridgetown (*Aw*) has a late winter dry season.

On the West African coast, arid conditions in Senegal give place to subhumid climates as the summer rainfall season develops and intensifies, and by Bokè on the Guinea coast an *Am* classification is attained. Humid tropical conditions extend through the Sierra Leone and Liberian coasts and on through the Ivory Coast (except for an *Aw* sector around Sassandra) to western Ghana. Between Takoradi in Ghana and Lagos in Nigeria the coast has a drier (mainly *Aw*) climate, but east of Lagos humid tropical conditions return, and with the exception of Santa Isabel in Cameroon (which has a short, but marked winter dry season, and is classified *Aw*), these are maintained to beyond Libreville in Gabon. Southward through Congo and Zaire, annual rainfall diminishes and the winter dry season lengthens : Mayumba and Loanga are *Aw* stations, Cabinda is semi-arid (*BS*), and south of the mouth of the Congo River the coast becomes arid (*BW*).

On the east coast of Africa the warm temperate climate of Natal province gives place to tropical conditions in Mozambique, Tanzania and Kenya, but the winter dry season remains sufficiently marked to give an *Aw* classification throughout, except for Porto Amelia (*BS*). Offshore, humid tropical coasts are represented on the islands in Mozambique Channel, and in Madagascar on the east coast, and in the vicinity of Nossi Bé on the northwest coast. Diego Suarez in northern Madagascar and stations from Analava to Besalampy in the northwest, register *Aw*, while the west coast south from Maintirano is classified semi-arid (*BS*), with a short arid (*BW*) sector around Tulear. Port Victoria and Agalega Island have humid tropical coasts, but the outlying islands, including Mauritius, have more distinct dry seasons, and are classified *Aw*.

Arid and semi-arid coasts border the norhwestern Indian Ocean, including the Red Sea and the Arabian Gulf, and extend to Pakistan and northwest India. Dwarka and Veraval, on the Kathiawar Peninsula, are classified semi-arid (*BS*), but rainfall increases and the summer wet season lengthens around the Gulf of Cambay, where Surat is an *Aw* station. South of Bombay, annual rainfall increases, and *Am* conditions are typical, except at Mormugao in Goa, where very dry conditions in winter bring in an *Aw* classification. The transition to *Aw* occurs in Kerala, and the whole of the east coast is in this category. In Sri Lanka, Colombo is within the humid tropical southwestern sector, but parts of the east coast are subhumid, a dry June giving Trincomalee an *Aw* classification.

A transition from *Aw* to *Am* occurs along the Bangladesh coast, Chittagong being in the latter category, and humid tropical conditions are then maintained southwards through Burma and Western Thailand into Malaysia, the dry season becoming less marked southwards so that *Am* gives way to *Af* in the sector from Penang to Singapore, and on round to Kuala Trengganu on the east coast of Malaya. Revival of a winter dry season brings back *Aw* conditions on the shores of the Gulf of Siam, but tropical conditions occur on part of the Kampuchean and Vietnamese coasts (Fosberg *et al.*, 1961).

Towards the head of the Gulf of Tonkin, tropical climates give way to warm temperate *Cw* conditions, but parts of the Hainan coast are tropical, and although Yulin is *Aw*, some sectors may have a humid tropical classification.

The southern tip of Taiwan also registers humid tropical conditions (*Am* at Hengch'un), but Tainan (*Cw*) and Taipei (*Cf*) are both warm temperate stations.

Humid tropical coasts are extensive around the Indonesian archipelago and into Papua New Guinea. Most coastal stations are classified *Af* or *Am*, but Pasuran, south of Surabaya in eastern Java has a dry spell in late winter which gives it an *Aw* classification, and it is likely that this applies in a number of other sectors around the eastern islands : Dili and Koepang in Timor are both *Aw* stations, as is Port Moresby, and possibly also the southwest coast of Papua. Humid tropical coasts are similarly dominant in the Philippines (Flores and Balagot, 1969), although again there are a number of *Aw* stations (Manila, because of a dry February; Cuyo, because of a dry March; and Zamboanga on Mindanao because of dry conditions in late winter).

Coastal stations on inter-tropical Pacific Islands are generally *Af* or *Am*, exceptions being Ocean Island in the Gilbert and Ellice Group, and Canton Island, farther east, both of which have dry spells sufficient to render them *Aw*. In New Caledonia, Noumea and Pagouméne, both on the southwest coast, are *Aw* stations, but humid tropical conditions occur on parts of the wetter northeast coast and locally within the Loyalty Islands. Parts of the Hawaiian Islands are also humid tropical : Pepeekeo, on Hawaii, is *Af*, but Honolulu has an *Aw* classification.

Much of the north coast of Australia has a climate classified *Aw*, but Broome in the northwest and Burketown, south of the Gulf of Carpentaria, are sufficiently dry to be classified semi-arid (*BS*) stations. The humid tropical sector is confined to northeast Queensland, extending from south of Cooktown to north of Townsville (Bird and Hopley, 1969) : Cairns has an *Am* and Innisfail an *Af* classification, and these conditions are characteristic of the coastal fringe from north of Mossman to southeast of Ingham. Townsville is classified *Aw*, and a July mean of 17°C places Mackay just inside the warm temperate zone (*Cw*).

Before considering coastal features, it will be useful to review briefly some aspects of general geomorphology in humid tropical environments.

Geomorphology in the Humid Tropics

The characteristic features of geomorphology in the humid tropics have been summarised by Tricart (1972) and Thomas (1974), drawing upon the earlier review by Sapper (1935). It is well known that land surfaces in humid tropical environments are subject to the intense chemical and associated biological weathering of rock formations that proceeds under perennially warm and wet conditions. This has led to the development of deep mantles of decomposed rock material, in places more than 30 metres thick. The natural vegetation is rain forest, with a dense canopy and a thick organic litter that protects the land surface from the direct erosive effects of heavy rainfall; a sub-surface network of roots also binds and stabilises the upper part of the weathered mantle. Nevertheless, on steep slopes, runoff during heavy rain washes away surface material even where the vegetation is dense, and slope failure occurs frequently : landslides and mudflows scar the forested hillsides.

In general, runoff yields predominantly fine-grained sediment, silt and clay derived from the weathered mantle, and streams also carry large quantities of material in solution, acquired from ground-water that has dissolved out the soluble products of deep weathering. Where the weathered mantle is quartzose, the fluvial load includes sandy material, and where the streams have cut down into the underlying rock formations a gravelly load may be incorporated. The heavy and sustained rainfall produces a large runoff from humid tropical catchments, and the rivers carry an abundance of sediment down to their mouths. In consequence, depositional features are extensive, and steep sectors are relatively limited on humid tropical coasts.

Steep and Rocky Coasts

The steep coastal sectors show features similar to the slopes of incised valleys in the hinterland, usually with a dense vegetation cover extending down almost to high tide level, where wave action has washed away the weathered mantle to expose rocky outcrops on the shore. Actively-receding cliffs are rare, partly because of the great extent of coastal plains formed by deposition in front of hilly and mountainous terrain, partly because of the protection of fringing and barrier reefs of coral where these are present, and partly because of a general absence of strong wind and wave action in humid tropical regions.

Where high country extends to the coast, as in parts of northeastern Australia, Indonesia, Columbia, and the Ivory Coast in West Africa, steep promontories and coastal slopes descend to rocky shores. The slopes show slumping, which exposes the weathered mantle, but actively-receding cliffs are found only where rock formations that have proved resistant to chemical weathering outcrop on headlands washed by relatively strong wave action. Thus the cliffs between Mullins Harbour and Samarai in southeast Papua occur on a sector of steep coast reached by relatively strong waves, including ocean swell, that approaches through an area of deep, reef-less coastal waters (Löffler, 1977). Waves generated by onshore monsoonal winds have cut cliffs on parts of the east coast of Malaysia

(Nossin, 1961) and near Chittagong in Bangladesh (Ashfaque, 1967).

Steep slopes on basalt or granite, where the weathered profile often shows residual corestones as a result of decomposition along intervening joint planes, are fronted by bouldery foreshores where wave action has washed away the weathered matrix to leave an accumulation of the rocky corestones (Hopley, 1969). Once exposed on the shore, rock surfaces are quickly pitted and honeycombed as the result of intense weathering processes. These processes have been studied by Tricart (1972) on rocky shore outcrops at Cabo Frio and Ilheus in Brazil, and at Mamba Point in Liberia, but further research is necessary to distinguish between the effects of physical, chemical and biological processes, and especially to elucidate the parts played by wetting-and-drying, salt crystallization, and solution by rainwash and seaspray in the degradation and disintegration of these rock surfaces.

Deltaic Coasts and the Significance of Mangroves

Extensive coastal plains have been formed in the humid tropics by the deposition of large quantities of fine-grained sediment, supplied mainly by rivers, either directly, or after re-working by waves and currents in the nearshore zone. Much of the sediment is supplied during episodes of floodwater discharge, the supply diminishing during any dry season, on the north coasts of Java and Sumatra. In most cases the lowlands are essentially deltaic, built up by deposition around the mouths of rivers, but intervening swamp areas are also extensive. Progradation has been dominant, but there have been episodes of erosion, and the pattern of development of a deltaic coast can be related to fluctuations in the size, sediment yield, and outlet position of distributary channels that migrate to and fro as the delta grows.

Deltaic shorelines are generally fringed by intertidal mudflats, the upper zones of which are colonised by mangroves, which protect the coast from wave scour and may promote the accretion of sediment to build up new depositional terrain above high tide level. As accretion proceeds, the mangrove zone migrates seawards, followed by rain forest or freshwater swamp vegetation, which moves in to colonise the new depositional terrain. The extensive swampy lowlands bordering the Gulf of Papua developed in this way, with progradation continuing more rapidly around the mouths of rivers, as in the Fly estuary and on the Purari delta.

In recent centuries, deforestation and cultivation within the catchments of rivers in the humid tropics has led to soil erosion, which has accelerated runoff and increased fluvial sediment yield, resulting in rapid growth of deltas and progradation of adjacent lowland coasts. This was the case in the nineteenth and early twentieth centuries on the north coast of Java, where deltas enlarged and swampy shorelines advanced, particularly in Jakarta Bay (Verstappen, 1953), and along the low-lying coast to the east (Hollerwöger, 1964). On the Citarum River, dam construction has intercepted part of the fluvial sediment yield and initiated a new era of delta shoreline erosion.

Typically, sea areas off river mouths in the humid tropics are cloudy with fine-grained sediment in suspension, especially after episodes of river-flooding, or when strong wave action has disturbed the muddy sea floor. Dispersed along the coast, this sediment eventually settles in sheltered bays and inlets, the shores of which are colonised by mangroves. Tricart (1972) has suggested that the high temperatures of tropical sea water reduces its viscosity, permitting greater turbulence and wider dispersal of the suspended load of colloidal clay than would otherwise occur; Dey (1956) had invoked diminished viscosity to explain the calmness of muddy seas off the southwest coast of India.

Mangroves are of geomorpholcgical significance on humid tropical coasts, especially on the shores of deltas. The mangroves present are mainly *Sonneratia*, *Rhizophora* and *Bruguiera* species. In many sectors the mangrove fringe continues to advance seaward, especially in the vicinity of river mouths where there is an ample supply of silt and clay, mobile in the nearshore zone, and available to be colonised and stabilised by the mangrove pioneers. This has been documented on parts of the North Queensland coast, notably in Cairns Bay adjacent to the Barron delta (Bird, 1972a).

Elsewhere the mangrove fringe is being eroded away by wave and current action. In some places the swamp deposits end in a low, receding cliff and the shore is littered with undercut and fallen mangrove trees; in others, sand deposits have been washed up into the eroded mangrove fringe, conspicuous because of the exposed structure of close vertical trunks, which on an advancing margin are concealed by the green foliage of young or colonising mangroves. The pattern of advance and erosion is often intricate, and evidently relates to patterns of muddy sediment supply in some areas and a deficit in others : a mangrove-fringed deltaic shoreline commonly shows an alternation of advancing sectors around and downdrift from the sediment-yielding distributary mouths and eroding sectors no longer receiving muddy sediment because of the waning or deflection of the distributary that formerly fed them.

Detailed mapping of the extent of erosion on mangrove fringed coasts may indicate other relevant factors. For example it is possible that the slight rise of world sea levels during recent decades has impeded the spread of mangroves by deepening nearshore water, and increasing wave energy in sectors where the supply of sediment has been insufficient to compensate for the effects of modern submergence.

Sandy Depositional Features

Coastal plains on humid tropical coasts are typically bordered by sandy beaches, some of which have grown laterally to form spits, while others have prograded as beach-ridge plains, or sandy barriers. Sandy shores on humid tropical coasts are colonised by grasses such as *Spinifex* and creeping shrubs such as *Ipomoea* and *Canavalia*, with open woodland to the rear, including such trees as *Casuarina* and palm species, notably coconuts. Behind the beaches the terrain is typically swampy, with some residual lagoons. In most cases, lagoons that formed behind sandy barriers have been infilled by deltaic sedimentation or swamp encroachment, but some persist, usually with bordering swampy shorelines, as in western Nigeria, where it is possible that the persistence of the lagoons has been aided by coastal subsidence, and on the Pacific coast of El Salvador (Gierloff-Emden, 1969).

Coastal plains often incorporate the stranded relics of earlier sandy beaches or barriers, with intervening lagoon and swamp zones. These are well developed on the Ivory Coast in

West Africa, in eastern Nigeria (Allen, 1965), on the east coast of India (Dey, 1956), and on the oceanic shores of southwest Sumatra and southern Java (Bird and Ongkosongo, 1980). Cheniers, which are low sand ridges emplaced on a pre-existing plain of finer sediment, silt, clay and swamp peat, are extensive in Surinam and on the Guyana coast.

Nature and Sources of Beach Material

The origin of these sandy deposits varies. Beach ridges and cheniers may be produced where waves re-work fluvial sand delivered to river mouths and coastal waters during episodes of flooding; the Purari delta on the Gulf of Papua, exemplifies this, as do the chenier plains of the Surinam coast, north of the Amazon delta (Vann, 1969). Elsewhere, coastal plains include extensive barriers, built of sand that has evidently come from the sea floor, washed in by wave action. Le Bourdiec (1958) showed that the sandy barriers on the Ivory Coast were built of material brought up from the sea floor during and since the Holocene marine transgression, the world-wide eustatic rise of sea level that took place as climate ameliorated and glaciers and ice sheets in high latitudes melted away in late Quaternary times, and Allen (1964) recognised this shoreward contribution in the sandy facies of the Niger delta. Tricart (1959) invoked the same process of shoreward drifting to account for sandy beaches on the coast of Brazil, and there is little doubt that the tropical beaches lining the Pacific coast of the Americas (Gierloff-Emden, 1969), and those bordering the Indian Ocean, incorporate sandy material of this origin swept in by ocean swell of distant derivation. It is of course necessary to invoke this shoreward drifting of sand to explain the nourishment of beaches and barriers in situations where rivers draining to the coast carry only muddy sediment, and where there is little or no sand yield from the limited cliff sectors.

Sand swept shoreward must have been derived partly from the products of subaerial weathering of sea floor rock formations when they were exposed during low sea level phases of the Pleistocene, and partly from aeolian or fluvial sands deposited on the emerged sea floor during these phases. Prolonged lowering of sea level occurred at several stages of the Pleistocene during episodes of cooling of world climate, when the tropical zone contracted. Each time the sea level fell, land areas gained in elevation, and earlier shoreline features were left stranded far inland. In some areas the climate became sufficiently cooler and wetter during these low sea level phases to enable the rivers to derive large quantities of sandy sediment from headwater catchments that now carry tropical rain forest on deeply weathered mantles and yield only fine grained material. Moreover, since sea level fell by at least 100, and perhaps as much as 140 metres, during the Last Glacial phase of the Pleistocene, these rivers must have extended their courses and carried their sediment loads out across emerged sea floors that in places extended more than 100 kilometres seaward of the present coastline. Parts of the sea floor were thus exposed to atmospheric processes for up to 70,000 years, during which time the surface rock outcrops became weathered, and the emerged sea floor strewn with fluvial deposits and stranded beach sediments, some of which were probably re-worked by wind action to form dunes. As the Holocene marine transgression advanced across this terrain, sandy sediment was incorporated and carried landward, eventually to accumulate as beaches and barriers on oceanic sectors of the present humid tropical coastline, notably in Brazil and West Africa. The clay component of the transgressive sediment was more widely dispersed by waves and currents, and has not been retained in the energetic environment of these ocean beaches.

This sequence of processes has been much less effective within the shorter seas bordering Caribbean and Indonesian island shores, or in the lee of major coralline reef formations such as the Great Barrier Reef off Queensland and the similar structures off eastern Papua, Madagascar, and New Caledonia. In such areas the comparatively small waves, generated locally by onshore winds, have been less effective in bringing sand from the sea floor during and since the Holocene marine transgression. Where sandy beaches occur they are usually dominated by sediment derived arenaceous rocks exposed in adjacent eroding cliffs, from disintegration of coralline reef formations near the coast, or from rivers that supply sand as well as finer. sediment.

The emphasis given to the effects of intense weathering in humid tropical environments and the predominance of clay minerals in weathered mantles has tended to obscure the fact that some of the rivers also carry substantial quantities of gravel. Where the hinterlands are steep, as in much of Indonesia and New Guinea, and headstreams have entrenched valleys through the weathered mantle, the rivers are excavating gravelly material from the underlying formations. Coarse material thus derived is channelled downstream, and may be delivered to the coast. The Markham and the Ramu, in New Guinea, have substantial gravel loads derived from their mountainous catchments (and from gravelly terrace deposits which line their valleys), and some of this coarse material is swept down to the coast during floods : gravels from the Markham have been added to the beach at Lae, just north of the river mouth.

On the northern coast of New Guinea west of Wewak the beaches are supplied with sand and gravel brought down by rivers that drain the steep hinterland of the Torricelli Ranges, the load having been augmented after major earthquakes in this region (Bird, 1979). In Indonesia volcanic eruptions have deposited large quantities of ash in river valleys, and the beaches and barrier of southern Java have been episodically nourished by ash-laden runoff after such events (Verstappen, 1957). Beaches of granular sand and ferruginous gravel on the Ivory Coast close to the mouth of the Sassandra River are derived from coarse fluvial sediment brought down by flood-waters along a short, steep watercourse that drains a high granitic hinterland (Tricart and Cailleux, 1965). The significance of hinterland conditions is also illustrated in the Capibaribe estuary, near Recife, which receives sandy sediment imported by the river from the dry regions of the Brazilian plateau (Ottmann, 1965), and in a similar way the deltaic coastal plains of humid tropical Queensland include sandy beaches, beach ridges and cheniers that have been supplied with fluvial sediment by the Burdekin, the Barron, and other rivers with head-waters in the cooler and drier plateau hinterland of the Atherton Tableland, whence abundant sand supplies are derived from decomposed granitic outcrops. On this part of the Queensland coast the sandy features are interspersed with (and the cheniers emplaced on) alluvial and swampy terrain formed from the finer fraction of the fluvial sediment, notably the muds entrapped in mangrove swamps of the kind that fringe Cairns Bay

(Bird, 1972a). This varied terrain is the outcome of separation of the sandy fraction, deposited around the river mouths and spread along the shore by wave action, from the more widely dispersed silts and clays that settle gradually in the more sheltered environments of mangrove-fringed estuaries and embayments.

Beaches supplied with sand derived from nearby cliff erosion are comparatively rare in humid tropical environments, but on the east coast of Malaya Nossin (1965) reported that some beaches were nourished by sand eroded from cliffed sectors by the stronger wave action that occurs during the monsoon period.

Disintegration of coralline reef formations is another source of beach material on humid tropical coasts, notably in the vicinity of fringing and nearshore reefs. Reef growth in the humid tropics is inhibited by reduced salinity and by the turbidity of sediment-laden sea water off river mouths, so that fringing reefs and derived coralline beaches are generally restricted to rocky promontories, as on the southern shores of Huon Gulf along to Cap Nelson and Milne Bay in Papua New Guinea. Some fringing reefs have formed where corals and associated organisms have directly colonised a rocky shore; others formed offshore, usually on a rocky foundation, and became fringing reefs as the result of subsequent depositional progradation of the shoreline, as at Yule Point in Queensland (Bird, 1971). Coralline beaches have also developed on the shores of emerged reefs, as in Barbados, where wave attack has been strong enough to disintegrate the exposed coral, which is otherwise consumed mainly by solution processes to develop karstic 'notch-and-visor' profiles behind narrow shore platforms, well displayed on the coasts of northeastern New Guinea, New Ireland and New Britain (Lõffler, 1977). Once a fringing reef has developed, the beach to its rear receives sand and gravel of coralline origin, and this may be spread along the adjacent coast by wave action.

Other calcareous organisms living in coastal waters contribute to beach sediments. On sectors of the Brazilian coast between Salvador and Recife where the sea is not clouded by terrigenous sediment, the shell fraction is between 20 and 40 per cent of the beach sand (Tricart, 1959). Inshore from algal reef formations, calcareous algae form 50 to 80 per cent of beach sediments (Ottmann, 1965). Typically, the proportion of calcareous material diminishes towards sectors that have been nourished primarily by non-calcareous fluvial or cliff-derived sediment yield, but there are complex variations in grain size and sorting characteristics, related partly to distance from sources and partly to wave energy, which varies with aspect, coastal configuration, and nearshore bathymetry. This has been illustrated on the north Queensland coast by Bird (1972b).

The world-wide trend of sandy shoreline erosion during the past century (Bird, 1976) is evident on beaches and barriers on the coasts of the humid tropics. Sandy shorelines have here prograded in only a few sectors, mainly in the vicinity of deltas, where there is a continuing sand supply from rivers. Details of historical changes on humid tropical coasts are scanty, and the I.G.U. Commission on the Coastal Environment is now conducting additional surveys to determine the extent of sandy shoreline changes, particularly in South America, the Caribbean and south east Asia.

Beach rock develops on humid tropical coasts where beaches rich in calcium carbonate develop cemented sub-surface horizons (Russell, 1967). In some areas these horizons have been exposed as shore outcrops by subsequent wave erosion. The cementing carbonates may be derived either from percolating groundwater or from interstitial sea water (Stoddart and Cann, 1965). Their precipitation requires phases of drying out of beach sand, which occurs within the zone where the water-table rises and falls, provided there are sufficiently long episodes of dry weather. Beach rock has developed on the Ghana coast where there is a substantial dry season, but not on the permanently wet beaches of the Ivory Coast, to the west, or the coast of Nigeria to the east. In some sectors beach rock has been exposed by subsequent erosion as a nearshore reef, a feature that gives some protection to the adjacent shoreline, unprotected sectors often showing more rapid recession, marked by the excavation of lunate bays.

Coastal Dunes

Coastal dunes are poorly developed on humid tropical coasts, even where sand deposition has been sufficient to produce broad barriers and coastal plains with extensive beach ridge systems, as on the east coast of Malaya (Nossin, 1965) and parts of the north east Queensland coast (Bird and Hopley, 1969). The idea that dune development is impeded by the rapid invasion and stabilisation of backshore sands by luxuriant humid tropical vegetation (Jennings, 1964) appears unsound, because the presence of such vegetation would aid, rather than prevent, dune building if sand were being blown from the beach : any arriving sand would merely be trapped at the outer margin of the densely vegetated area as a precipitation ridge. The rarity of dunes on humid tropical coasts thus stems from a failure in sand supply from the beach to the backshore zone, evidently because beach sands become wind-borne far less readily than on coasts in cooler or drier environments.

Several hypotheses have been put forward to account for this. Morton (1957) suggested that beach sand deflation on the West African Coast was prevented by the formation of protective crusts of salt-cemented sand by evaporation of beach moisture in dry weather, but this process has not been substantiated on humid tropical coasts elsewhere, and its validity in West Africa requires further scrutiny. Beach sand in the humid tropics is generally too wet to be removed by wind action: strong winds can deflate wet sand, but the comparative rarity of strong winds in the humid tropics, and the fact that when they do occur they are usually accompanied by heavy soaking rains (Jennings, 1965) seems to go far towards explaining the poverty of coastal dune development in this environment. There are other, contributary, factors. Where the sands are of fluvial origin they may be too coarse to be deflated from a beach to build dunes; and where they are finer, they often incorporate a silty clay fraction which increases their cohesiveness and impedes their removal by wind action.

Coastal dunes are more likely to develop behind shores where the sand supply is sufficient to maintain a broad beach as a potential dune source area. In northeast Queensland, coastal dunes are generally rare, but they are present behind sectors where the beach widens as the result of longshore drift accumulation adjacent to headlands, and especially where the shore curves round to face into the prevailing southeasterly winds, as at Oombunghi, south of Cairns, where wind action coming directly onshore, has been strong enough to blow sand from the beach and build dunes up to 10 metres high at the outer edge of the tropical rain forest to the rear (Bird and Hopley, 1969). Ahmad (1972) has related dune sectors on the east coast of India to aspect in relation to the monsoonal winds. The importance of a sand source is illustrated by the dune features that have been described from Hood Bay, on the southeast coast of Papua, derived from sand supplied in abundance from the Kemp Welch River (Löffler, 1977), and from the southern coast of Java near Jogjakarta, where the rivers deliver large quantities of sand derived from the volcanic deposits of the Merapi volcano (Verstappen, 1957). In Sri Lanka, coastal dunes are extensive on sectors with a long dry season, especially where this is accompanied by strong onshore winds, and providing there is an adequate supply of sand to the beaches which are the source of dune material. Where strong winds coincide with the wet season, as in southwest Sri Lanka, dune development is impeded by the wetness of beach sands and because the beaches are eroded this season (Swan, 1979). Dunes have also been built on parts of the Bangladesh coast near Chittagong (Ashfaque, 1967; Siddiqi, 1967).

Effects of Monsoons and Cyclones

The rarity of strong wind action in humid tropical environments also means that wave energy is usually low (Cailleux, 1965), and the presence of coral and algal reefs in coastal waters can further reduce wave action approaching the shore. Reference has been made to the seasonally stronger wave action that accompanies an onshore monsoon and leads to cliff erosion and lowering of beach profiles on the east coast of Malaya (Nossin, 1965), and to the effects of transmitted moderate to high energy ocean swell in shaping beaches on the south-facing coasts of West Africa, Southern India and Sri Lanka, and the oceanic coasts of Sumatra and Java.

Occasional episodes of much stronger wave action occur on coasts subject to tropical cyclones. In Mauritius, McIntire and Walker (1964) documented the extensive coastal changes (notably cliff erosion, the cutting of platforms above calm weather sea level, the lowering of foreshore beach profiles, and the pilling-up of backshore beach ridges) which took place during the brief episodes of storm wave activity that accompanied tropical cyclones, when sea level was temporarily raised by the passage of the low pressure system. Similar sequences were described by Hopley (1974) as a consequence of cyclone Althea in the Townsville region of northeastern Australia. The geomorphological effects of occasional brief and localised storm surges may persist in the coastal topography, although subsequent phases of relatively calm weather are usually marked by the restoration of some of the eroded areas by deposition. Coastal vegetation is also damaged by cyclones and storm surges, but usually it regenerates quickly, and sometimes it is possible to find the tangled remains of windthrown vegetation overgrown by regenerated mangroves or rain forest, or incorporated in coastal peat deposits.

CONCLUSIONS

There are thus variation within the coastal geomorphological systems of the humid tropical zone that can be related to specific environmental factors. Perhaps the 'norm' should be taken as conditions found within the Indonesian archipelago, beyond the reach of ocean swell and transmitted storm waves. The equatorial zone is not subject to tropical cyclones, and so wave action is here perennially weak, and active cliffing very limited. Steep coasts are densely vegetated down to the littoral zone, where rocky shore outcrops exposed by wave sapping are irregularly pitted and honeycombed as the result of intense physical, chemical and biological weathering. Coastal sectors of emerged coral show a characteristic notch-and-visor profile, with narrow basal platforms, the outcome of weathering processes accompanying gentle wave action (Christiansen, 1963). Beaches are narrow and intermittent, and largely submerged at high tide; coastal dunes are very limited, but deltaic plains and mangrove swamps are extensive; coral reefs are patchy, with occasional fringing reefs on promontories. This is the equatorial environment, which is (and has long been) hot, wet, and relatively calm : elsewhere within the humid tropics the coastal geomorphology shows variations related to the incidence of cyclones or the incoming of monsoonal regimes with a dry season; variations related to exposure to oceanic wave action; and variations which are legacies from past conditions of climate and sea level.

SUGGESTED READINGS

Ahmad, E., 1972 : *Coastal geomorphology of India*, Orient Longman.

Allen, J.R.L., 1964 : Sedimentation in the modern delta of the River Niger, West Africa, in Van Straaten (ed.), *Developments in Geomorphology* 26-34.

Allen, J.R.L., 1965 : Coastal geomorphology of eastern Nigeria : beach ridge, barrier islands and vegetated tidal flats, *Geol. en Mijnbouw*, 44, 1-21.

Ashfaque, S.M., 1967 : Geomorphic features of southern Chittagong coast, *Oriental Geogr.*, 11, 12-22.

Aufrére, L., 1934 : Le role du climat dans l'activité morphologigue littorale, *Proc. 14th Int. Geog. Congress*, 329-31.

Axelrod, D.I., 1958 : *Map of coastal vegetation of the world*, University of California Press.

Bailey, H.P., 1959 : An analysis of coastal climates, with particular reference to humid midlatitudes, *Second Coastal Geography Conference*, 23-55.

Bird, E.C.F., 1971 : The fringing reefs neer Yule Point, North Queensland, *Australian Geogr. Studies*, 9, 107-15.

Bird, E.C.F., 1972a : Mangroves and coastal morphology in Cairns Bay, North Queensland, *J. Trop. Geogr.*, 35, 11-16.

Bird, E.C.F., 1972b : The origin of beach sediments on the North Queensland coast, *Earth Sci.*, 5, 95-105.

Bird, E.C.F., 1976 : *Shoreline changes during the past century*, I.G.U. Working Group on Shoreline Dynamics, Melbourne.

Bird, E.C.F., 1979 : Beach systems on the northern coast of New Guinea, *in preparation.*

Bird, E.C.F. and Hopley, D., 1969 : Geomorphological features on a humid tropical sector of the Australian coast, *Australian Geogr. Studies*, 7, 89-108.

Bird, E.C.F. and McArthur, J.A., 1978 : *The delimitation of humid tropical coasts*, UNESCO.

Bird, E.C.F. and Ongkosongo, O., 1980 : *Environmental changes on the coasts of Indonesia*, United Nations University.

Büdel, J., 1963 : Klima-genetische geomorphologie, *Geogr. Rundschau*, 15, 269-85.

Cailleux, A., 1965 : L'énérgie éolienne en France et dans le monde, *Annals de Géogr.*, 75, 257-70.

Christiansen S., 1963 : Morphology of some coral-cliffs, Bismark Archipelago, *Geografisk Tidjskrift*, 62, 1-23.

Derbyshire, E., 1973, (ed.) : *Climatic geomorphology*, Macmillan, London.

Dey, A.K., 1956 : The shore lines of India. *Quaternaria*, 3, 95-100.

Flores, J.E. and Balagot, V.F., 1969 : Climates of the Philippines, in Landsberg, H.E. (ed.) *World Survey of Climatology*, 8, 159-213.

Fosberg, F.R., 1966 : Vegetation as a geological agent in tropical deltas, in *Scientific problems of the humid tropical zone deltas and their implications*, UNESCO, 227-32.

Fosberg, F.R., Garnier, B.J. and Kuchler, A.W., 1961 : Delimitation of the humid tropics, *Geogr. Review*, 51. 337-47.

Gierloff-Emden., 1969 : *Die Küste van El Salvador* : Acta Humboldtiana, Series Geogr et Ethnogr., 2.

Hollerwöger, F., 1964 : The progress of the river deltas in Java, in *Scientific problems of the humid tropical zone deltas and their implications*, UNESCO, 347-54.

Hopley, D., 1969 : A coastal weathering sequence at Mount Douglas, north Queensland, *Rev, Géom. Dyn.*, 19, 9-14.

Hopley, D., 1974 : Coastal changes produced by tropical cyclone Althea in Queensland, December 1971, *Australian Geogr.*, 12, 445-56.

Jackson, I.J., 1977 : *Climate, water, and agriculture in the tropics*, Longmans, 44-47.

Jennings, J.N., 1964 : The question of coastal dunes in humid tropical climates, *F. für Geomorph*,. 8 (Sonderheft), 150-4.

Jennings, J.N., 1965 : Further discussion of factors affecting coastal dune formation in the tropics, *Australian J. Science*, 28, 166-7.

Köppen, W., 1936 : Das Geographischen der Klimate, in Koppen, W. and Geiger, R. (eds.) *Handbuch der Klimatologie*, I (C), 1-44.

Le Bourdiec, P., 1958 : Aspects de la morphogénese plio-quaternaire en basse cote d'Ivoire *Rev. de Geom. Dyn.*, 9, 33-42.

Loefflr, E., 1977 : *Geomorphology of Papua New Guinea.* A.N.U. Prees, Canberra.

McIntire, W.G. and Walker, H.J., 1964 : Tropical cyclones and coastal morphology in Mauritius, *Ann. Ass. Amer. Geogr.* 54, 582-96.

Morton, J.K., 1957 : Sand-dune formation on a tropical shore, *J. Ecol.*, 45, 495-7.

Nossin, J.J., 1961 : Relief and coastal development in north-eastern Johore (Malaya), *J. Trop. Geogr.* 15, 27-38.

Nossin, J.J., 1965 : Analysis of younger beach ridge deposits in eastern Malaya, *Z. für Geomorph.*, 9, 186-208.

Ottmann, F., 1965 : *Introduction á lo géologie marine et littorale*, Paris.

Richards, P.W., 1952 : *The Tropical Rain Forest*, Cambridge.

Russell, R.J., 1967 : *River plains and sea coasts*, University of California.

Sapper, K., 1935 : *Geomorphologie der feuchten tropen*, Leipzig.

Siddiqi, J.A., 1967 : Geomorphic features of northen Chittagong coast, *Oriental Geogr.*, 11, 1-11.

Stoddart, D.R., 1969 : Climatic geomorphology : review and assessment, *Progress in Geography*, 1, 159-222.

Stoddart, D.R. and Cann, J.R., 1965 : Nature and origin of beach rock, *J. Sedim. Petrol*, 35, 243-7.

Swan, S.B., 1979 : Sand dunes in the humid tropics : Sri Lanka, *Z. für Geomorph.*, 23, 152-71.

Thomas, M.F., 1974 : *Tropical Geomorphology, Macmillan.*

Tricart, J., 1959 : Problémes géomorphclogiques du littoral oriental du Bresil, *Cahiers Océanogr.*, 11, 276-308.

Tricart, J., 1972 : *Landforms of the humid tropics, forests and savannas*, Longmans.

Tricart, J. and Cailleux, A., 1965 : Particularités de la morphogénes littorale, *Le modele des regions chauds forests et savanes*, 106-49.

Tricart, J., 1972 : *Introduction to climatic geomorphology* (trans. C.J.K. de Jonge), Longmans, London.

Vann, J. H., 1969 : *Landforms, vegetation and sea level change along the Guinea coast of South America*, Buffalo.

Verstappén, H. T., 1953 : *Djakarta Bay : a geomorphological study on shoreline development*, The Hague.

Verstappén, H. T., 1957 : Short note on the dunes near Parangtritis (Java), *Tijdschr. Kon. Nederl. Aardrijks. Gen* , 74, 1-6.

Wilcock, A. A., 1968 : Koppen after 50 years, *Annals. Assoc. Amer. Geogr.*, 58, 12-28

EIGHTEEN

Jag Mohan Agrawal

A NEW CONCEPT—SUBMARINE TERRAIN EVALUATION

THE CONCEPT

IT IS well known that the land part of the earth's surface comprises of various features which by origin could be grouped into arid, glacial, fluvial and coastal, etc. These features when studied minutely reveals that they themselves are the result of various subfeatures which has got a broad resemblance in formation and the manner of their associations, hence their explorations, if based on some special methods and study, gives a better output and avoids various problems, at the same time it helps to predict various information about a similar feature with similar atmospheric conditions.

The terrain of sea bed is also comprised of various geomorphic units which must have come into being after various geomorphic processes with a bearing on geology, geomorphology and geophysics and various other branches of Modern Science.

The practical approach for classifying the submarine terrain with an aim to evaluate its uses for various requirements after their special study with a flexibility to accommodate refinements, may be called submarine terrain evaluation.

SUBMARINE TERRAIN CLASSIFICATION

The following features affect the evaluation of a given submarine landscape :

(i) Climate.
(ii) Geology: Rock types and structure including tectonic set-up.
(iii) Geomorphological Processes—Operating since history and the present stage of the pattern.
(iv) Location—Bay, gulf and the fetch area, etc.
(v) Effect of land born forces and their contribution.

Method of study

To prepare pattern and facet charts for any given area following procedure may be adopted :

Literature

On geology, geomorphology, Fetch areas amount of discharge received from land, tonnage, sailing directions, Historical background, etc.

Photography

Under-water photography of specific areas.

Bathymetric Charts

Based on all existing sources on a small scale for patterns and on a large scale for facets. Bathymetric charts should be able to depict topography and its three-dimensional effects.

Field Checks

At selected locations to confirm and amend observations, collection of sea bottoms and taking field photographs.

Laboratory

Studies of sea bottom, samples and their economic importance.

A pattern will thus be prepared followed by its facet maps The facet maps will have boundaries of two types. The thicker boundary will be of a pattern and the thinner lines will be for the facets. Patterns having a two digit while the facet with a three digit number. For easy references, the facet may have a five digits, i.e., from left to right two digits for pattern and the last three for the facet concerned. Thus 15,063 means 15 is the pattern number and 063 is the facet number.

Storage of information:— The facet and pattern information will be stored in cards and can also be computerired. A facet card and its required information is explained below.

Facet

A facet will at least have one character inherent in it which will define it with respect to other facets and thus small sections of Submarine Terrain will be carefully compared with a view to determine precisely in what ways each section a "Facet" thus named, differs from other sections.

Few salient character of Facet can be described as (1) Slope, (2) Overlying Material, (3) Associations, and (4) Extent. Demarcation of a Facet will actually depend on various scientific processes and evaluation and may or may not differ on the basis of their uses. Similarly a facet can be subdivided in various sub-numbers/sub-Facets.

Slope

It will refer to the inclination of the seabed at specific spots with the help of existing depths. The slope can be termed as— Gentle Slope, Steep Slope or Moderate Slope as the case may be. A Facet will have either one of the type of slopes or a combination of the slopes with a difference in frequency as compared to other Facets.

Overlying Material

If a close study of seabed is done it will be observed that nature of material overlying an area will be predominant enough to give it a separate name as compared to its surroundings. Rock type of Facet, origin of overlying material and its processes of its accumulation on the facet will be an added attraction to define a particular facet.

Associations

Arrangement of different facet at a certain level will give a clue for individual number. Underwater photography, maps/ charts prepared by relief contour and other methods showing the sea terrain in its real shape will add to signify the importance of arrangement and association character. Arrangement of Facet can be of significant clue to the geological history of the Facet and the Pattern as well. Association in horizontal and vertical plane with the help of various graphs and curves will help to differentiate a facet and add to its special characters which will be of great use in various economic and scientific fields.

Extent

Extent of Facet though with the same characters can result in to a separate Facet number depending on its use and scientific values. The horizontal and vertical layout of subgeomorphic features such as shoals, deeps, valleys, small undulations, small deep patches will strengthen the reasoning of a separate Facet.

The study of Facet will not only be based on submarine geomorphology but it will require cooperation from Geologists, Oceanographers, geophysicists, Biologists, Mariners and various other groups for collection of following information to fill in the Facet Card.

FACET CARD

1. Number——————————
2. General location——————
3. General depths from——————to——————
4. Type of marine life predominant——————
5. Guest creatures who oftenly comes to Facet————
6. Types of bottoms——————
7. The bedrock——————
8. Amount of river discharge received from land if any——————
9. B.T. Data at various layers——————
10. Sigma T values——————————
11. No. of wrecks—————
12. Size of wrecks——————
13. History of wrecks——————————
14. Tidal streams——————————
15. Currents——————————————
16. Surface winds, their speed and direction————
17. Range of Tides——————————————

18. *General description* : A facet card will be accompanied by a jacket containing the information such as the name of the Scientists and the Project/Ships' name engaged-in while contributing for the description and data for the facet. The collection of data whether it is exclusively carried out for facet analysis or as a subsidiary job during other research work should be specific and accurate with a brief description of the instruments used, and process of analysis adopted for. General description will include the details of the surrounding locality, type and way of sediments brought from land/sea. Any historical part if played by the facet in previous wars or economic developments. The geological history of the facet and its related facets and patterns. The navigational charts available for the area including surface navigational/submarine navigational aids if available. The under-water caves, spurs, and other geomorphic features. The general shipping over the facet and the type of nearest ports and its coastal attributes. The submarine navigation known to have existed on the facet. Various profiles, viz., Superimposed Profiles, Composite Profiles, Geologic sections and Hypsographic curves, and their brief descriptions. Hazards to navigation on the surface of the facet or above it, if any, will be specified. Block diagrams of the entire pattern showing its various facets will be discussed in light of respective facet. The development of mineral, oil industry and the results if obtained. The type of fishing practised and the development of Fishing Industry on the facet. The general distance from the main coastline and relation with the international sea boundaries and oceanic geomorphic features.

Utility of Submarine Terrain Evaluation

Like terrain evaluation on land submarine Terrain Evaluation can be of high assistance for Submarine navigation. Surface navigation, Fisheries, Mineral explorations, Oil exploration and will be an added help for deciding Median lines and International boundaries.

Navigation

The navigator knows in advance the trend of sea bottom, the general life of the sea and the deposits of the bottom. He therefore, will not be misled by false soundings and phantom layers, at places he will be then able to calibrate his sounding equipment with the help of various facets and their description.

Submarine Navigation

The three-dimensional diagrams and charts of various facets of a pattern will help a submarine to decide its routes, and subroutes if it is forced to adopt during emergency.

Fisheries

A facet is numbered and will be the same wherever it is found, hence lot of field work can be estimated on the table before actually going to the field. The type of life predominant on a facet will be common wherever the facet is found and accordingly methods of exploration will be developed.

Mineral Explorations

A facet and its line, joining the other facet will give the amount and type of mineral wherever it is found "as development of a facet depends on geology, and marine environment", and will help in mineral exploration including oil exploration.

Median Lines

Knowledge of general slope, type of bottom and benefits of certain area may attract the nation to opt for a particular area even at the cost of unused and unimportant area which to other country might look important due to its area.

Another advantage of the system is that once the submarine terrain classification of the area is done and confirmed by the field checks, the readymade information on various attributes of the studied pattern can be very well utilised to infer terrain conditions of the given area.

The system will be flexible to accommodate changes or additions in the classification and information stored. If required two similar facets may be grouped as one. The system thus has high potentials for varied tasks at sea.

NINETEEN

Satya Prakash and *J. M. Agrawal*

CARTOGRAPHIC PRACTICES USED IN REPRESENTATION OF SUBMARINE TERRAIN*

INTRODUCTION

Our knowledge about the sea was earlier confined to navigation and fishing in the coastal margins but now it extends to a wider zone. Users needs have broadened enormously in recent years for a larger variety of alternative methods of graphic representation, including a scientific study of the ocean basin, terrain features, location of trenches, faults, canyons, depth and structure of sedimentation for mineral reserves, water masses alongwith their physical character and all related Oceanographic, Geological, Geophysical, Biological and Meteorological parameters. With the increasing population and diminishing resources on land, man is diverting his attention to ocean and scientists will be demanding charts for spatial distribution, quantification and comparative analytic studies. The scientists of different fields will though require specific charts in their specialised fields, the basic needs of everybody is sea floor relief charts.

The crux of the problem has been as how to present the three-dimensional terrain onto a plane surface in a natural but useful shape without much sacrifice of natural scale and true positions. The land is accessible and can be well mapped through the aerial photography and Landsat Imagery but sea floor is remote, alien and almost invisible. The data is expensive

* Paper presented at the 2nd National Cartographic Conference held in Pune from 12 to 14 Dec. 1980.

and scarce. Moreover knowledge of various fields viz. Geology, Geomorphology and Cartography is required to analyse and chart this data. Recent development in colour photography has been of some help. However, the new technique of Photobathymetric charts is possible only in shallow waters.

The many graphic techniques for the representation of terrain, which cartographers have developed to represent land areas are directly or with some modifications applicable for submarine terrain also. The Oceanographic Cartographers have been using various graphics at times with slight modification and at times with combination of graphics according to the need of the users.

Since the complex nature of processes and characteristics of the sea, which vary through both the vertical and horizontal dimensions, are extremely difficult, the imaginative and innovative cartographers have to design and develop new graphics.

Graphics In Representation of Submarine Terrain

The functions of submarine tarrain charts are undoubtedly the same in analytical context as for land areas. The authors who are engaged in 'Nautical Charting,' have taken up the task of development of submarine terrain representation techniques for quite some time and have observed that the navigators are not the only workers in the marine environment. New types of carto-techniques are needed to fulfil the requirement of specialist users. A basic relief chart of submarine terrain is the prime requirement of each class of users, viz., Physical Oceanographers need it for depicting physical oceanographic parameters, Geologists, require Seafloor shape and distribution of physiographic features, preferably with 3-D view; Biologists want delineation and distribution of physiography affecting movement of marine life; Marine Engineers, Miners, and Explorers need it for planning, prospecting and exploitation of sub-Marine projects whereas Marine legal experts need it for delineation of continental margins before finalizing International Maritime Boundaries. The authors have made a critical study of function and design of various graphic techniques used for representation of submarine terrain and a brief resume of the same alongwith the authors experimentation in the modern graphic techniques for representation of submarine terrain is appended below :

Result of Hydrographic Surveys and Nautical Charts

Result of survey is an end product of a hydrographic survey which portrays a part of earth's marine surface on a flat sheet using T.M. Projection or UTM and indicates submarine relief by means of dense continuous pattern of soundings (Density depending on the Scale of the Survey) as extracted from the echo sounding traces. These surveys are the prime sources for the compilation of nautical/bathymetric charts of various categories (Figs. 19.1 & 19.2).

A nautical chart is the chief nautical instrument for safe navigation and marine exploration. It portrays a part of earth's marine surface on a flat sheet using appropriate projection (Mercator, Gnomonic, or Transverse Mercator) and standard

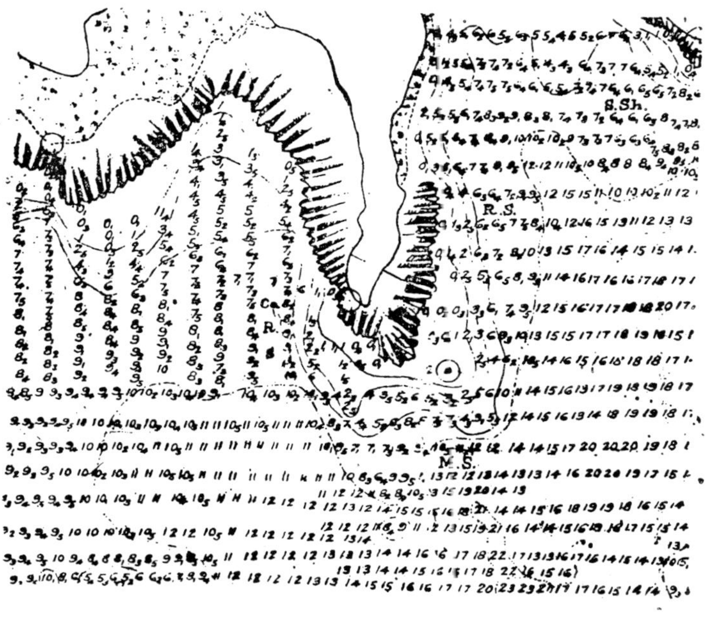

FIG. : 19.1 *Portion of Hydrographic Survey*

scales and indicates irregularities of earth's surface both from above and below the sea by means of selected soundings, contours, spot heights alongwith data for position fixing, route finding and safety of navigation. It shows the mariners, how the landmarks and sea surface look like and tells him where to go, what to stay away from, where ship is in the ocean and where lies his destination on shore. A mariner places boundless confidence in a chart and therefore, it must be most accurate, intelligible, up-to-date and should cater for present day needs. It must respond to the fast changing requirements of the navigators and marine explorers. They give sufficient information in shallow and coastal waters and provide selected and sometimes approximate information in deep waters. It is the elementry tool for every Marine Specialist User, viz., Oceanographers, Geologist, Biologist, Engineers, Miners, Explorers etc. It is though used by Marine Legal Experts as basic docu-

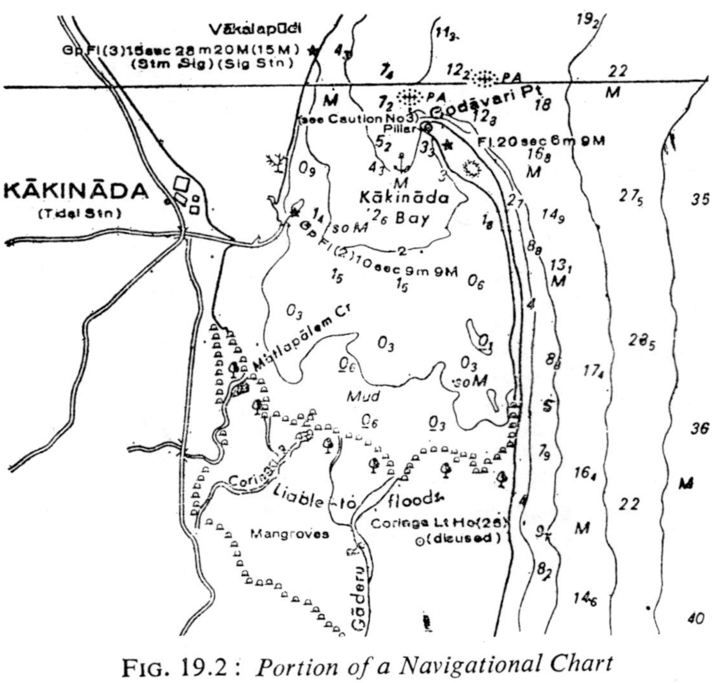

FIG. 19.2 : *Portion of a Navigational Chart*

ment yet it has an inherent drawback of being prepared on non-equal area projection.

Ocean Sounding Bathymetric Contour and Bathymetric Charts
Ocean sounding charts contain track soundings generally beyond 200 metre line (Continental Shelf) taken by various oceanographic, surveying, research and exploration cruises and are the only sources so far in offshore areas. These charts are generally available on smaller scale (1:1 M) on Mercator Projection and form the basic document for the preparation of various graphic charts for representation of Submarine terrain.

The bathymetric contour charts (BC's) contain the interpolated depth contours beyond 200 metre line at selected intervals from the above mentioned track sounding charts. These are generally available on smaller scale (1:1 M) on mercator projection. (Figs. 19.3 & 19.4).

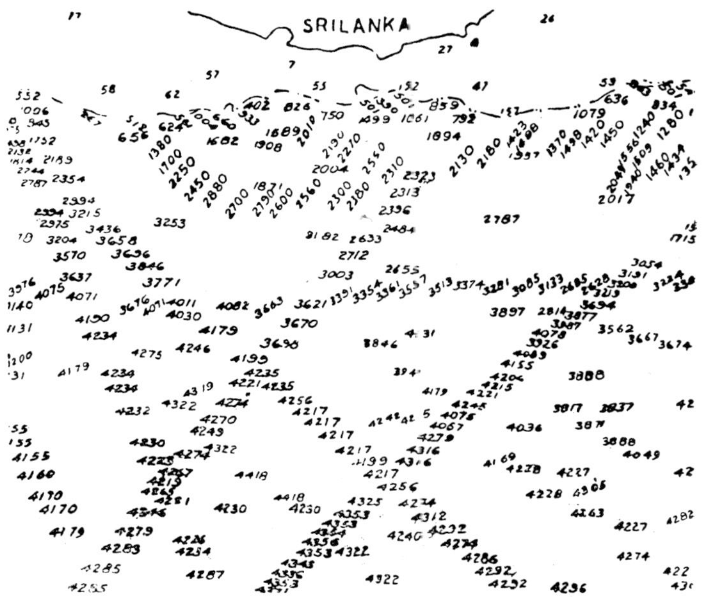

FIG. 19.3 : *Portion of an Oceanic Plotting Sheet*

The Bathymetric Charts show the physiography of Submarine terrain with the help of various selected depth contours supported by controlling important depths which are though scattered yet they bring out the actual undulation of the seabed. These Charts could be categorized as the most useful charts to represent the seabottom on a plane surface. They are generally prepared on a smallar (1:10 M) scale and Mercator Projection. Polar stereographic projection is however used for polar areas.

A gamut of symbology has been introduced by Cartographers for better representation of submarine relief through these Bathymetric Charts. Layer colouring or hypsometric tints is the most prevalent method. The use of blue colour for Ocean floor has been universally accepted and differentiation produced for varying depths by changes in tone, is quite appreciable. The deeper having a heavier tone. This conti-

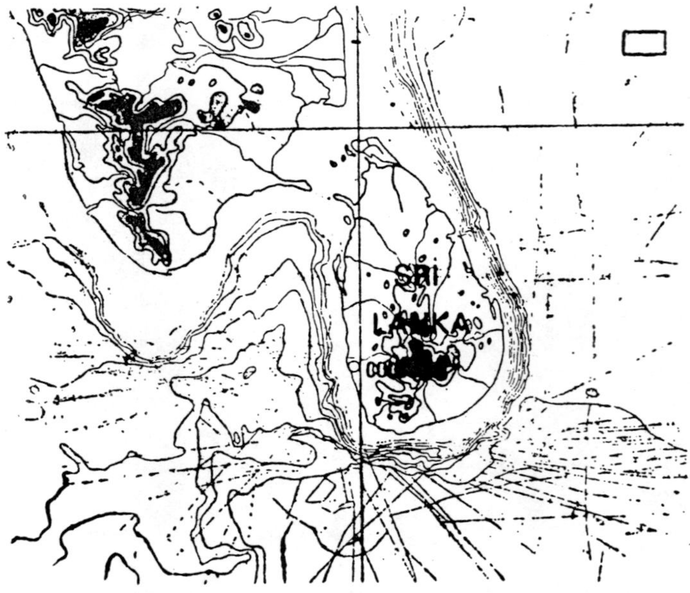

FIG. 19.4 : *Portion of a General Bathymetric Chart of Ocean*

nuous increase in tone offers a perceptible pictorial interpretation in areas of smoother slope. However, tone identification becomes difficult in undulating terrain. At times deeper tone also becomes so dark that other information is obscured.

A world series of General Bathymetric charts of the Oceans (GEBCO) at a scale 1:10 M covering the World Oceans in 18 sheets has been developed under International Hydrographic Organisation (IHO) on Mercator/Polar Stereographic (Polar areas) Projection. Under this programme the basic collection of the bathymetric data in the form of 600 Oceanographic Plotting Sheets on 1:1 M Scale is carried out by 18 volunteering hydrographic countries, for final preparation of 1:10 M GEBCO's. The fifth edition of this GEBCO Series is being prepared under the Joint IOC/IHO GEBCO Guiding Committee (of which India is a member). Canadian Hydrographic Office has volunteered to publish this series which is likely to be completed by the end of 1982.

Physiographic or Shaded Relief or Pictorial Charts

Hachuring, first used by Lehman, is the oldest well-known device of depicting elevated terrain. It is an artistic way of presenting relief in a three dimensional fashion as viewed from normal plane without sacrifice of natural positions.

Cartographers made further use of this graphic and produced shaded relief charts as viewed from oblique plane. The features are shown by line shading running parallel to slope. In pictorial graphic the terrain is drawn so artistically with natural shapes as if an oblique photograph of ocean floor is taken with imaginary light falling from a side, preferably North-West. These charts give true pictorial representation and offer good general knowledge of ocean floor for a layman but they are of little use to scientists because they are not on natural scale and they do not show true geographic positions. National Geographic Society has published such maps of oceans (Fig. 19.5).

Illuminated and Shaded Contour (Relief Contour) Charts

This graphic is slight modification of contoured bathymetric charts. The contours enhance the three-dimensional effect and quantitative analysis is quite accurate and easy. In this method,

Fig. 19.5 : *Portion of a Shaded Relief Chart (After Heezen and Tharp)*

it is imagined that light is falling on relief from an oblique plane from one direction (preferably North-West). Thus a principle of illuminated and shaded contours is adopted, the contours falling in forwarded side, are illuminated (printed white), the contours falling in backward side are shaded (printed dark blue or black). The contours which are slightly thick near the tangential plane, gradually thins out towards tapering end to give a natural effect. The general blue wash is kept clear from contours which is achieved by photomechanical masking. This method solves partially, the problem of presentation of three-dimensional submarine terrain and offers good representation of physiographic features but cannot depict vertical distribution of oceanographic properties. One of such charts produced by Naval Hydrographic Office, Dehra Dun for Bay of Bengal (7013) has been widely appreciated by marine scientists in India and abroad.

Block Diagram Technique

In this graphic, view of terrain is drawn by taking one profile or layer for one contour. All shapes are projected onto an oblique plane. Block diagrams can be prepared one or two point perspective. The shapes are altered by being foreshortened in the direction of viewing. In certain cases the dimension parallel of viewing is shortened by cos θ (being the mapping plane) while dimension perpendicular to the direction of viewing remain unaltered and dimensions intermediate between the two are altered by proportionate amount. This means that the map has a varying scale. Further developments have been made by applying mathematical correction to the scale factor. However, the block diagrams can only depict general physiographic configuration but will be of little use to depict distribution of any kind of Oceanographic property. Canadian Hydrographic Department has published such charts. (Fig. 19.6)

Three-Dimensional Relief Models and Plastic Relief Charts

The sea bed is covered by a thick sheet of water and thus makes direct photography rather impossible as against land areas. Underwater cameras, though proved to be of great assistance, can only portray a small areas and are quite costly, while modern colour photography can cover only shallow and

FIG. 19.6 : *From Canadian Hydrographic Department*

clearwater areas. Therefore the oceanic cartographer have to construct the three-dimensional models for depiction of 3-D view of submarine terrain. However, the scanty data creates many difficulties in this work. Accordingly he has to apply a great deal of imagination and judgement for preparing suitable graphics for the specific tasks.

A 3-D Model can be prepared by 'Contour layer and Tint Method' wherein contours drawn on transparent sheets (One contour on one sheet) and tinted with a light blue colour in the deeper side beyond each contour is superimposed over each other, which presents a wonderful 3-D view of submarine terrain when seen through light.

The relief models of submarine terrain can also be prepared directly with plaster of paris. Alternatively a plastic sheet model can be made. In this method the contours are printed on several plastic sheets (Number of sheets depending on the number of contours). The plastic sheet is cut beyond the limit of each contour and pasted over each other to form the model. The gaps between the steps are then smoothened before applying the tints. A negative cast of above finished plastic sheet is then taken onto the plaster of paris to obtain a positive epoxy cast. The basic coloured chart of the area is then printed on the special HIP (High Impact Polyster) Sheets which can be vacuum formed over the epoxy model to obtain a plastic relief model.

Anaglyphs Prepared from the Models of Submarine Terrain

In Anaglyph method a common portion of two photographs of the Relief Model taken consecutively with 60 per cent overlap, is superimposed by printing in complementary colours (viz., red, in right; and green, towards left) with a registration gap proportional to the vertical scale. This Anaglyph of the area provides the 3-D view when seen through a special double colour (Red and Green) spectacles. Alternately 'Contour Anaglyphs' can also be prepared in similar way by drawing all the contours in red and green, the red contours being displaced to the right and the green to the left. The amount of displacement is proportional to the actual depth of the contour and when viewed through red and green spectacles, the impression is one of a 3-D model.

However, the authors feel that the anaglyph can be further improved by applying the shading effect on the features of the model before taking the photograph. They have also developed the possibility of introducing other oceanographic parameters through appropriate symbology. Thus solid three-dimensional parameters of the sea (Spatial relationship) are depicted on a plane sheet effectively and accurately.

Computer Generated Graphics through CRT Graph

With the introduction of automation in the oceanographic data aquisition, analysis and display on the graphic terminal, it is possible now to portray the spatial changes for the oceans through the time. As a numerical data base, the computer can be used in the study of correlation of data either numerically or graphically. The advantage of drawn 3-D objects on the graphic terminal is that many different views of certain aspects can be displayed quickly. It would be possible to rotate the graphic display into the position that best reveals some aspects of particular interest through change of data. Furthermore 'hard' copies can be obtained at each change.

The information can be further displayed through varying time by utilizing 'electronic display' after the time slices are made very small in a quick succession. This will develop a system quite close to cinematography. Thus the cartographer will be able to 'Freeze' the map at any instant of time.

CONCLUSION

The study of various graphics described above reveals that it is difficult to achieve all the properties on a single graphic out of the non-computer based graphics. However, the anaglyphs prove to fulfil the maximum requirements with some limitations. It may, therefore, be concluded that the computer generated graphics can achieve all the required properties and prove to be the best graphics of the time to suit the needs of the marine specialists and geoscientists.

Comparative Analysis of the Properties of the Various graphics used in Representation of Submarine Terrain

	Hydrographic charts	Bathymetric Charts	Shaded pictorial charts	Illuminated shaded contours (Relief-Contour Charts)	Block Diagrams	3-D Relief Models and Plastic Relief Charts	ANAGLYPHS (Seen through Spectacles)	Computer based Graphics
	(1)	(2)	(3)	(4)	(5)	(6)	(7)	(8)
Horizontal Scale	Yes	Yes	Not very accurately	Yes	Varying	Yes	Yes	Yes
Vertical Scale	No	No	No	No	Yes Spatial Relationship	Yes	Yes	Yes
Projection (Geog. Coordinates)	Yes For navigation purposes only	Yes	No	Yes	No	Yes Can also be in true perspective	Yes Can also be in true perspective	Yes

(*Contd.*)

	(1)	(2)	(3)	(4)	(5)	(6)	(7)	(8)
Three-dimensional Image	No	Yes, if colour layer and tint method is used	Yes, (not in true perspective)	Yes	Yes, Not in true perspective	Yes	Yes	Yes
Depiction of other spatial relationship between Oceano-parameters	Yes, Hori-zontal distribution only	Yes, Hori-zontal distri-bution only	Not, suitable being an arti-stic depiction only	Yes, Hori-zontal distri-bution only. However, excess of data may obscure the 3-D quality	Not, suitable being a dia-gram only.	Yes, to a limited extent.	Yes, on develop-ment of proper symbology.	Yes

Note :— However, certain properties can be achieved on above graphics by introducing appropriate symbology.

[FIG. 19.6A : *Portion of Shaded Relief Chart (from National Geographic Magazine)*

SECTION FIVE

REMOTE SENSING AND GEOMORPHOLOGY

INTRODUCTION

PROBABLY the most significant geomorphic tools that have become available during the present century are the aerial photographs and Landsat imageries The earliest form of remote sensing is aerial photographs which has been in vogue since the beginning of the 20th century. It was used in the First World War for terrain and other studies. Since then the improvement of cameras and the films have taken place rapidly. Aerial photography proved the main source of collecting reconnaissance data during the Second World War. Since 1965, multi-spectral scanners, thermal scanners and radiometers have been used on aircraft. Since 1970, satellites have been used to collect the data.

The paper by K.R. Rao and Naresh Bedi illustrates the use of remote sensing technique in geomorphology. This study indicates that regional geomorphic remote sensing studies can be very conveniently conducted on the Landsat imagery. This is more so in case of areas which have high relief hampering the accessibility and have thick vegetation and weathering mantle affecting the mobility and fixed observations. In such cases the saving in time and cost factor is an important gain. The geomorphic map prepared by the technique can be further utilized for land systems, land face and land use studies.

TWENTY

K. R. Rao and *Naresh Bedi*

REMOTE SENSING FOR GEOMORPHOLOGY

INTRODUCTION

WITH the advent of Remote Sensing techniques, geomorphic studies have got an impetus. Remote sensing techniques when applied to geomorphic studies have two aspects :

(a) acquisition of data and processing it into suitable formats; and
(b) interpretation of data for geomorphology, i.e., study of landforms and processes leading to their development.

Acquisition of data is carried out by sensors installed in aircraft or satellites. The sensors usually operate in the electromagnetic spectrum. They measure the reflected/scattered energy impinging on the earth's surface. Examples of sensors are aerial and space photographic cameras, thermal scanners, aerial and space borne multispectral scanners, vidicon cameras, radiometers, microwave radars etc. The source of energy is the sun or an artificial source (as in radar). Processing of data is done to organise the data into acceptable and easily intelligible formats with annotations about the conditions under which the data is collected. Interpretation takes place afterwards either visually or through interactive computers or through a combination of both (Plate 1 & 2). The remote sensing data is highly versatile and is used for a variety of fields of application:

Geomorphology, Geology, Agriculture, Forests, Soils, Hydrology, Oceanography, Cartography, etc. The term "Remote Sensing" is used to emphasise the use of instruments (sensors) placed in remote platforms like aircraft or more recently satellites. It is used in direct contrast to local "*in situ*" measurements.

EVOLUTION OF REMOTE SENSING

The earliest form of remote sensing is aerial photography which has been in vogue since the beginning of the 20th Century. It was used in the First World War for terrain and other studies. Since then, the improvements in cameras and the films have taken place rapidly. Aerial photography proved the main source of collecting reconnaissance data during the Second World War. Stereo-aerial photography has extended its use in topographic mapping and thematic mapping (covering themes like geology, hydrology, soils, forests, hydrology, etc.). Types of photography used in aerial photo interpretation include black and white, infra-red, black and white as well as colour, true colour, false colour (for portraying data from the visible band), multiband photography, etc.

Since 1965, multispectral scanners, thermal scanners and radiometers have been used on aircraft. Since 1970, satellites have been used to collect the data. An example is the Landsat series of satellites launched by the USA. The Multispectral Scanner (MSS) carried in these satellites has a resolution of about 70 metres. By 1990, it is anticipated that the resolution of 10 metres will be available for civilian use on a wide scale.

India entered the field of aerial photography during the year 1940. The Indian Photo-Interpretation Institute (IPI) at Dehra Dun (U P.) has been functioning since 1966. Considerable work on aerial photographs has been carried out here. It has become an important centre for imparting training to several scientists from India and abroad. The National Remote Sensing Agency, (NRSA) was established in Hyderabad in 1975. The computer aided interpretation facility was established here in 1976. To acquire aerial remote sensing data, the NRSA aircraft Research Flight Facility (RFF) were established in 1975 at

Bangalore. A great boost to remote sensing activity in the country was given with the establishment of NRSA Satellite receiving station at Shadnagar (a place about 60 km south of Hyderabad) in 1979. The activity was further integrated with the merger of IPI with NRSA in 1976. Through these units the NRSA has conducted several operational programmes in the field of geomorphology using both aerial and satellite data. An example of these studies, from one of the projects, will be given later.

Nature of Remote Sensing Data used in Geomorphic Studies

A wide range of remote sensing data in different formats is being used in the geomorphic and related studies. Important of them include the following :

Aerial and Space Photography

1. (a) Conventional films (0.4 μm—0.7 μm)—Black and white, true and false colour.
   (b) Infra-red, films (0.6 μm - 0.9 μm)—Black and white and colour.
   (c) Multispectral bands (0.3 μm—0.9 μm)—Black and white 4 bands simultaneously.

Scanners

2. (a) Airborne multispectral scanner (MSS) imagery, 0.4—1.1 m in visible and near infra-red region and 0.5 m to 12.5 m in thermal scanning.
   (b) Space borne MSS imagery—Landsat imagery in four bands—Band 4 from 0.5 to 0.6 m, band 5 from 0.6 to 0.7 m, band 6 from 0.7 to 0.8 m, and band 7 from 0.8 to 1.1 m.

The Landsat data is available in the following types of products:

(i) Black and white imagery—hard prints, diapositives, 70 mm negatives (four bands) etc.
(ii) False colour composites.
(iii) Computer compatible tapes (CCTs)

(c) Meteorological sensors—Advanced very high resolution

Radiometer (AVHRR)—5 channels (0.55-12μm) with 1.1 km resolution.

(d) Thematic mapper, 7 channels (0.45-12.5μm) with 30 m resolution to be flown in LANDSAT-D.

(e) Charged coupled Device (CCD) Array in SPOT 0.5—0.75μm) in two channels, to be launched.

3. Return Beam Vidicon (RBV) Camera (0.5 to 0.75μm) in Landsat-C.
4. Side Looking Radar (SLAR)—Air borne versions in use now (1.5—35 GHz range). One space borne version was flown in sea sat. Operates on 9 GHz band (Plate 3, 4, 5).

Data Enhancement

Quality of raw data is improved by data-enhancement techniques. The data enhancement involves manipulation of raw data to make certain features more prominent in comparison to others. It facilitates the data interpretation. Two types of data enhancement techniques are in practice—optical (photographic) and computer.

The computer techniques are used for the digital data whereas the optical techniques are used for photographic imagery. The latter are being mostly used for enhancing the data visually for geomorphic interpretation purposes. Prominent of these is improvement of tonal contrasts.

Though not falling exactly under the perview of data enhancement the false colour composites, prepared from Landsat image bands 4, 5 and 7 have proved to be highly suitable for geomorphic studies. Similarly, the interpretability of Landsat imagery and multiband photography is increased by the use of additive colour viewer.

Advantages of Remote Sensing Techniques

Some of the operational advantages of using remote sensing techniques are discussed below :

1. Landforms are the most conspicuous phenomena appearing in the imagery obtained through remote

1. Multispectral data analysis M Das—A computer system for analysing land data in CCT form

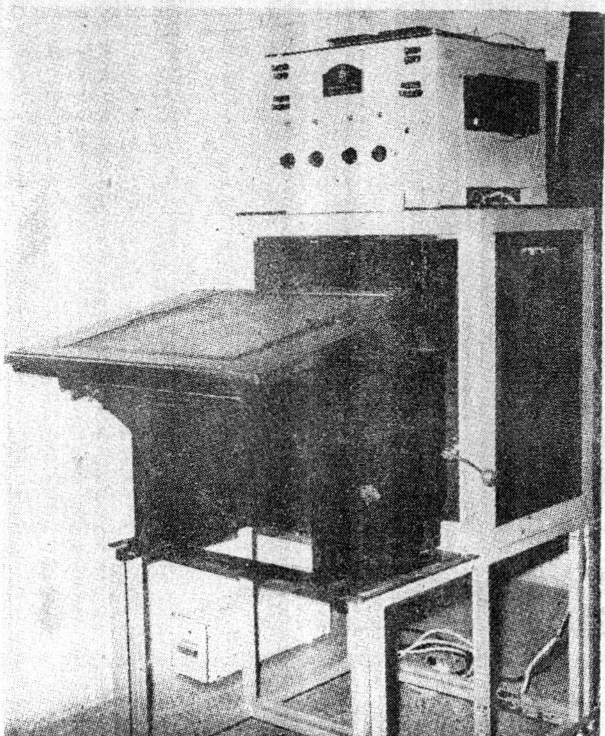

2. Multispectral additive colour viewes

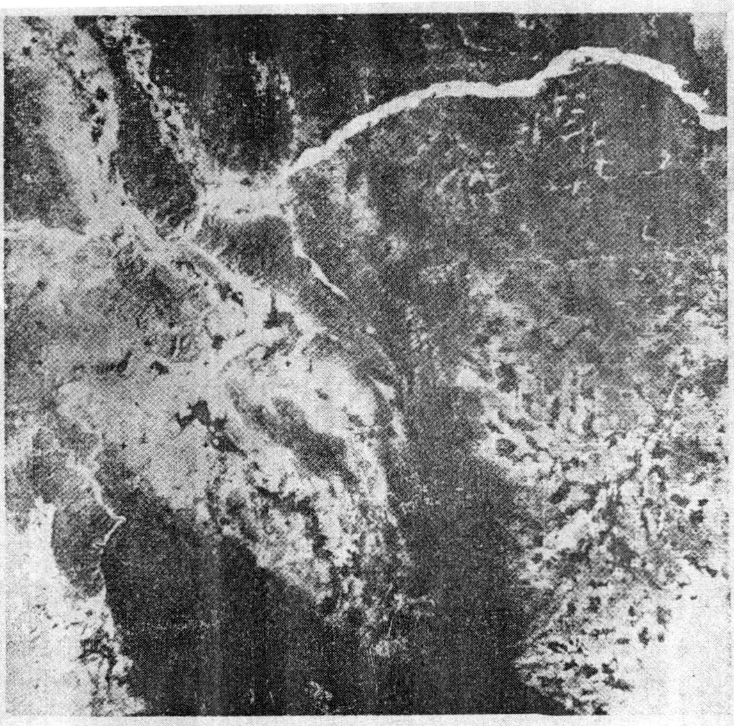

3. Landsat RBV picture of a Part of Andhra Pradesh, India dt. 15.3.80

sensing techniques. Relief or landforms make themselves felt in three dimensions in case of stereoscopic imagery and through shadow or drainage pattern in case of non-stereoscopic imagery.

2. The ever improving quality of imagery and their diversity in type and scale now enable a geomorphologist to map measure and study a wealth of geomorphological phenomena singularly as well as sequentially.
3. Delineation of landforms on aerospace imagery is a very direct type of interpretation as the landforms are visible directly on the image. This brings in a high degree of reliability. Subsequent deductions about the morphogenesis or processes can be made keeping in view the objectives of the survey.
4. The remote sensing data provides a synoptic (at a glance) view over a large area. The areal coverage depends upon the scale and type of imagery, e.g., a Landsat image covers about 32,900 sq km. of area. This helps in :

   (a) interpolation and extrapolation of local observations to regional scale and vice-versa;
   (b) direct comparison of landforms found in otherwise distantly located areas; and
   (c) direct comparison of geomorphic processes operating in different areas like sand movement, floods, coastal erosion etc.

5. The capacity of remote sensing data to provide a synoptic and repetitive coverage facilitates the study and monitoring of static as well as dynamic geomorphic phenomena, e.g., changes in the river channels, processes of erosion and sedimentation, coastal dynamics, changes in glacier position, displacement of active dunes—disertification, accelerated erosion, land slides, etc.
6. The spatially as well as spectrally (visible and non-visible range) registered data collected at regular intervals offers more scope for identifying landforms

which are fully or partially covered by vegetation in certain seasons. For example in a thickly forested area, aerial photography or Landsat imagery obtained in autumn would definitely show landforms which would be covered in other seasons. A band 7 Landsat image (near infra-red) also serves a similar purpose.

7. The geomorphic survey can be carried out directly on aerospace imagery even if no reliable topographic maps of the area exist. This is done in a less time and cost.
8. Morphometric details like relative heights, the slope steepness and classes, drainage density, etc., can be easily measured on certain types of aerospace imagery, eg., vertical aerial photographs. These details are of great importance in analytical as well as applied geomorphological studies.

Geomorphic Interpretation of Remote Sensing Data

The geomorphic interpretation of aerospace imagery involves deduction of information about landscape elements, from landforms as seen on the imagery. The outcome of interpretation is affected by the suitability of the imagery, equipment and methods used and particularly by the capability of the interpreter. Depending upon the purpose the geomorphic interpretation of aerospace imagery involves the following four steps.

(1) To begin with the geomorphic landforms are detected and delineated on the aerospace imagery. This is done by using the interpretation elements of objects or landforms appearing on the aerospace imagery. They include shape, size, shadow, colour and tone, texture, pattern, association of objects, resolution of the imagery, etc. All these elements are controlled by the size of the object and nature of the imagery. In case of different imagery (sensors) the resolution is different at different flying heights, e.g., aerial photography and Landsat imagery. Here to increase the interpretability of the imagery enhancement techniques are also used.

(2) In the second step delineated landforms are identified and classified into known categories. Interpretation keys are used. Important factors involved in this step are the professional knowledge and the personal knowledge of the interpreter about the area.

(3) The third step includes the delineation of categories of landforms into zones.

(4) These demarcated categories can now be classified with respect to the objective of the interpretation or geomorphic mapping. They can be synthesised with the other landscape elements like soil, land cover etc. Here, a considerable amount of 'induction' and 'deduction' is involved. Sometimes inferences have to be drawn about the sub-surface conditions. Based upon the features delineated at the surface a three-dimensional model has to be conceived which helps in drawing inferences about the conditions and processes going on underground which are not visible at the surface. The linkage between visible and invisible landscape elements is usually rooted in an evaluation of the ecological situation of the landscape in the area of study. The features observed on the imagery, conception of three dimensional model and deductions to which they give rise vary to some extent with the physiological, psychological and professional characteristics of an interpreter. The 'inductions' and 'deductions' made on the strength of systematic analysis are the most difficult but essential aspects of geomorphological image interpretation. Without them it is not possible to analyse the landscape and conceive a surface to sub-surface model, to complete the interpretation. Keeping in view the objective of the study, the process of induction and deduction can be induced towards either the morphogenetic or general environmental geomorphic interpretation. In the first, the landforms are classified on their genetic characters so that it is easier to deduce present as well as past processes. This is essential because genesis of a landform effects its economic potential.

In the environmental approach, the connection of landforms with other landscape elements like geology, soils, vegetation, hydrology, etc. is correlated and inferences drawn. Both approaches are however complementary and can be put to advantage conjunctively. Thus the efficacy of the whole process of interpretation depends on how well the landforms are detected, delineated, classified and the inferences drawn from them. As already mentioned, different remote sensing data require different modes of interpretation. Aerial photographs are best studied under the stereoscope as the relief factor important for geomorphic studies is best brought out. Certain quantitative measurements can also be made here. Use of stereoscopic techniques is not yet possible in case of MSS, thermographic or single radar imagery. In such cases the interpretation is either done visually or through automatic techniques like computer or density analysis. The MSS data is available in two forms—photographic imagery and the digital computer compatible tapes (CCTs). For the areas with high relief the photo imagery can be studied visually. A very convenient way of studying the MSS imagery or multiband photography is by the additive colour viewer. In case of the low relief areas automated computer analysis is best suited for the pattern recognition. The computer analysis is also very convenient for quickly recognising the geomorphic hazards like landslides or accelerated erosion.

Sequence of Remote Sensing Geomorphic Mapping

Remote sensing technology has reversed the conventional pattern of geomorphic surveys where studies were carried out locally and then a regional picture was construed. Now it is possible to do regional studies straight away on aerospace imagery and then the smaller areas selected for detailed work. The Table 20.1 gives the type of features observable on different aerospace imagery.

TABLE 20.1 : **Geomorphic Units Observable on Aerospace Imagery**

Land Unit	Description
A. *Macro Units (Delineated on Landsat Imagery)*	
1. Geomorphic/ Land Province	: Major physiographic unit with distinct assemblage of rock types, land forms of second order and almost uniform climatic conditions.
2. Geomorphic/Land sub-province	: A major sub-division of a province with a distinct land system assemblage—complex or compound.
3. Landform unit	: Mostly third order landforms having distinct shapes and covers.
B. Micro units : *Occasionally identified on Landsat imagery, mostly delineated on Aerial photographs)*	
1. Land system	: Area with a definite landforms of 3rd or 4th order, soils and vegetation association.
2. Land Catena	: Recurring group of smaller (4th order) landforms with a characteristic soil association.
3. Land facet	: A distinct landform unit with which a distinct vegetal cover is associated.

Modified after Howard, J.A., 1976).

The following sequence of geomorphic mapping is generally followed through remote sensing techniques.

(1) Regional maps on scales ranging from 1:1 million to 1:250,000 are prepared from small scale imagery like Landsat, MSS and RBV imagery.
Landforms are divided hierarchically into "macro-units". (Fig 20.1 and 20.2)

(2) Detailed geomorphic maps covering small areas, on scales varying from 1:50,000 to 1:15,000 are prepared by using airborne data like photography. The "Micro-units" can be demarcated in this stage. (Fig. 20.1 and 20.2)

An example of Geomorphic Interpretations of Landsat image

This Landsat scene (Plate-2) Path Row No. 156-051, covers a part of western Karnataka in India. The area is bound by latitudes 12°05′00″ and 13°45′00″N and longitudes 74° 30′00″ and 76°05′00″E approximately. The remote sensing data used included :

1. Landsat image—Black and White band 5 & 7, on scale 1:1 million and 1:250,000.
2. False colour composite of the above.

These imagery were interpreted visually. Planimetric details were taken from the standard Survey of India toposheets on scale 1:250,000. Geological background was obtained from the Geological Survey of India map of Karnataka on 1:1 m scale. Auxiliary data was collected from other sources. Before classifying the geomorphic units general attributes of the area were kept in view. The interpreted units were confirmed by the selective field work.

General Attributes of the Area

(1) Physiographically the area can be divided into three main divisions.

(a) The western ghats upland including the great ghat scarp. Here again there are three relief zones, viz:

(i) High relief with steep slopes near the ghat scarp,

(ii) Moderate relief, medium slopes and with a few steep longitudinal ridges,

(iii) Low relief with rolling to undulating surfaces in the far east.

(b) The coastal low and hilly belt with moderate relief.

(c) The narrow coastal plain.

(2) There are distinct climatic zones with consistent rainfall and temperature distribution in a linear fashion. These zones run parallel to the ghat's disposition with maximum rainfall zone restricted to the high relief area. The rainfall decreases towards east where the climatic transition from hot moist to almost sub-humid,

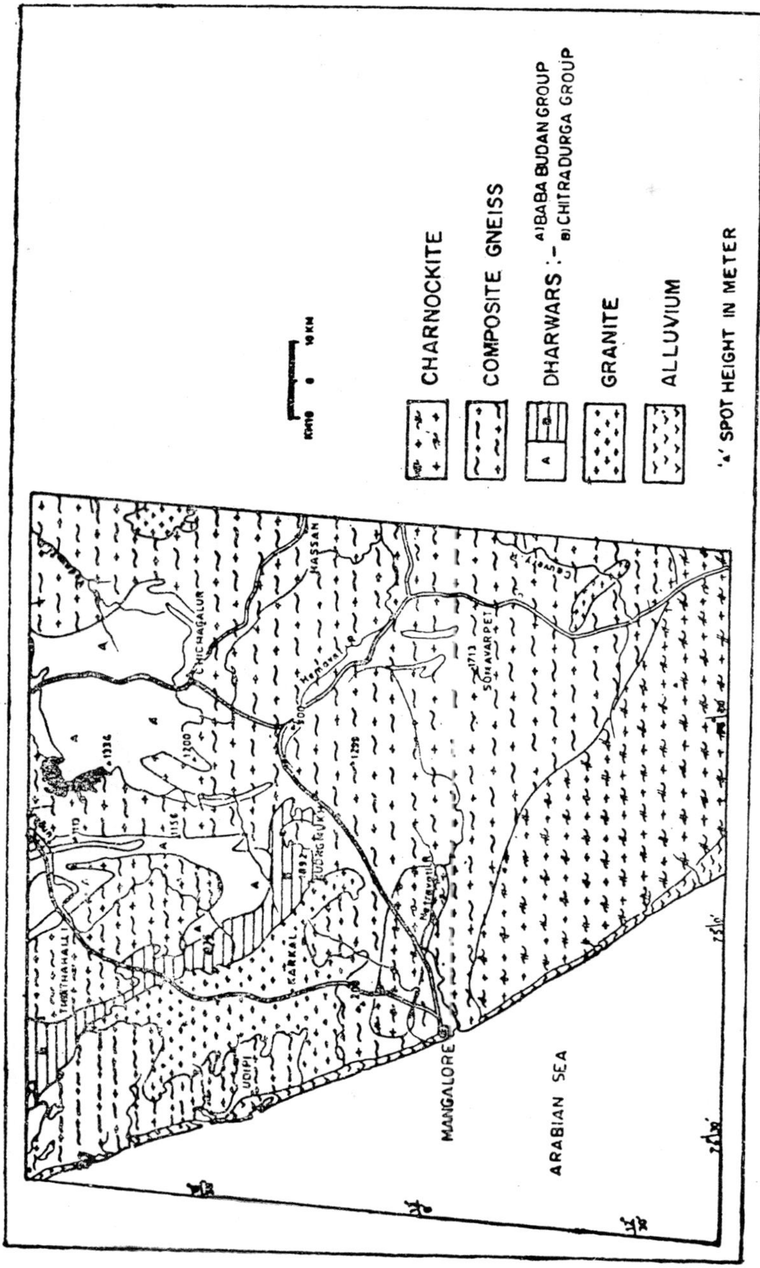

FIG. 20.1 : *Regional Geological Map of Parts of Karnataka and Kerala.*

even to semi-arid, also takes place. Most of the area however, gets fairly high, 250 mm, rainfall.

(3) The area is occupied mostly by Pre-cambrian rocks (Fig 20.1) comprising charnockites in the southern parts; composite gneisses, migmatite complex, granite and granodiorites in most of the parts; Dharwar super group of rocks constituting Babahudan group of metasedimentary and metavolcanic rocks in the upper central parts and Chitradurga group of metasedimentary rocks in upper western ghats areas. The unconsolidated fluvio-marine and fluvial sediments of recent age are found in the coastal plain and the river valleys respectively.

4. Most of the afore mentioned rocks exhibit megalineaments in NE-SW and NW-SE directions, as well as several groups of other local fractures, and faults. The tectonic frame work has changed from time to time and this had an influence on the morphogenesis.
5. In the western ghat area and its vicinities there is a continuous ground cover of living plants, trees, grasses, scrubs, etc., occupying the interface between ground and air. In the thickly vegetated areas there is a humus material occupying the ground surface and mixing with soil.
6. Physico-chemical activity is one of the most important and widespread process currently going on in this area. Soils show equilibrium with the present atmospheric conditions and are mostly products of chemical decomposition.
7. More water is entering the vegetated area as precipitation than is utilised there, so that a water surplus moves down slope out of the vegetated setting, especially on the western ghat scarp slopes.

GEOMORPHOLOGY

From the synoptic view and the general distribution of landforms, as appearing on the Landsat image the area can be divided into the following geomorphic units (Fig. 20.2)

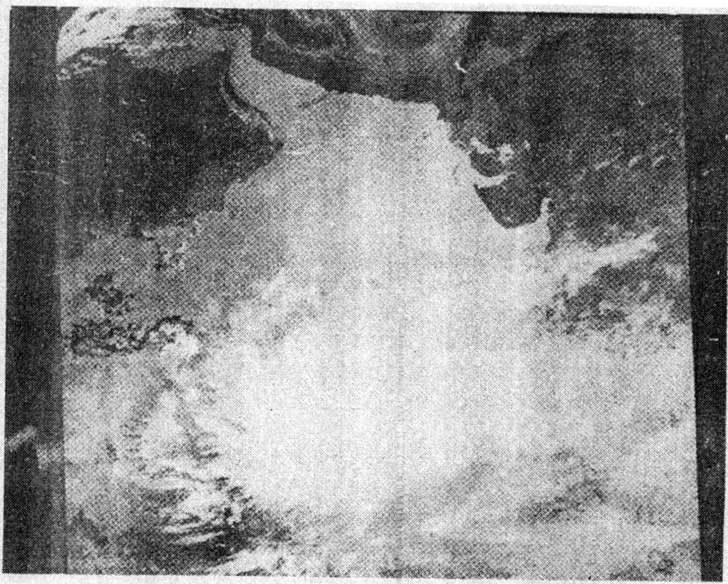

4. Metsat (AVHRR) picture of India, Middle-East and Arabian sea dt. 13.11.79, shows generation of cyclone

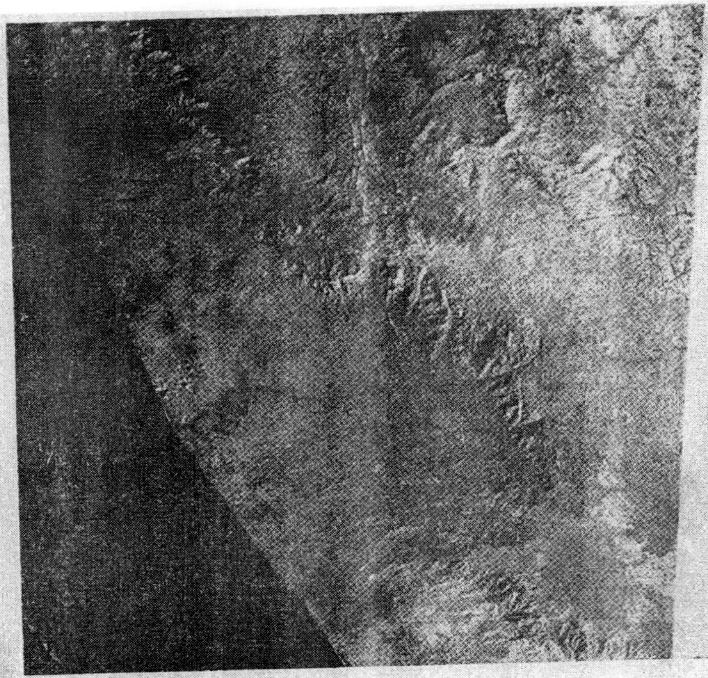

5. Landsat MSS image, Band 7, of a Part of Karnataka India, dt. 6.1.73

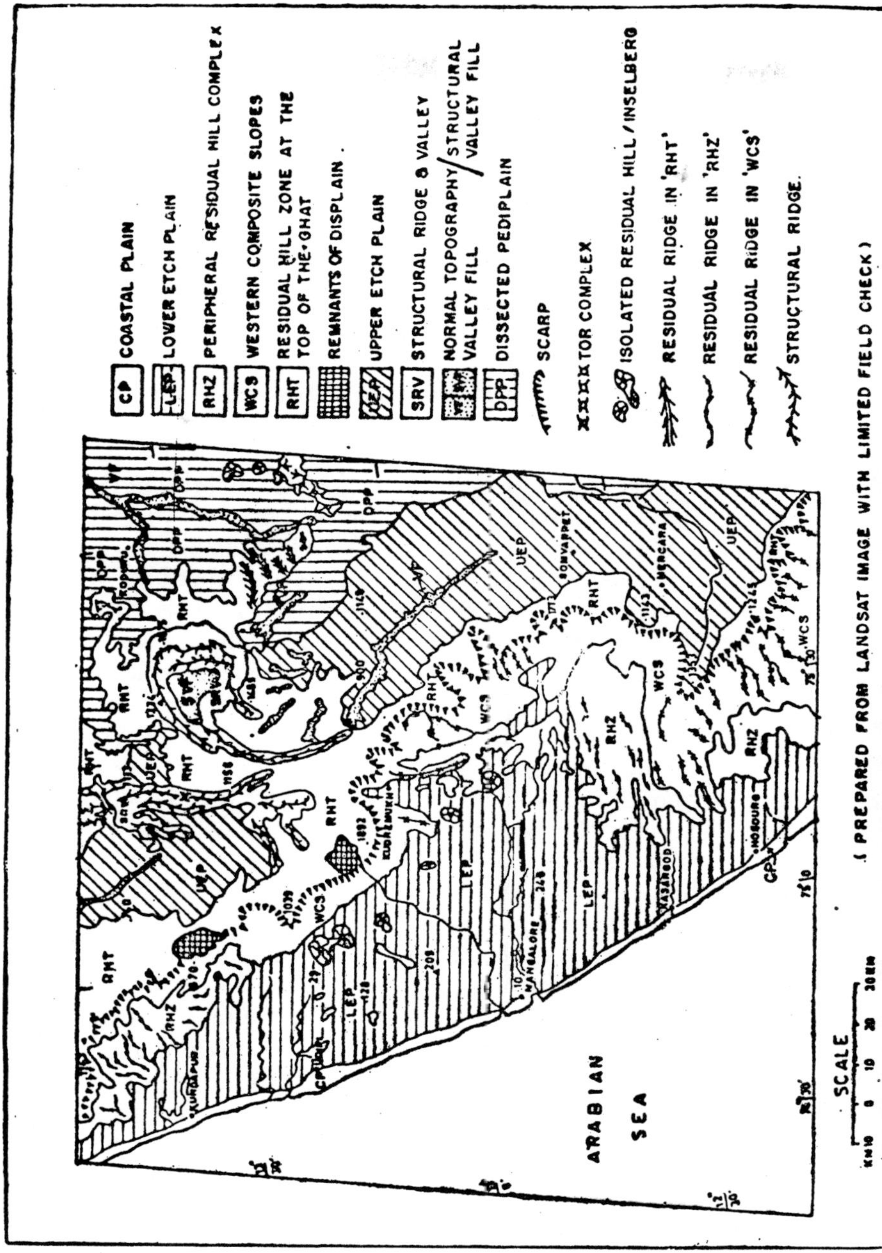

FIG. 20.2 : *Regional Geomorphological Map of Karnataka and Kerala*

1. Coastal plain.
2. Coastal low land etch plain.
3. Peripheral residual hills near the western ghat scarp.
4. Western composite slopes including the great ghat scarp.
5. Residual hill zone at the top of the ghat scarp.
6. Upper etch plain.
7. Structural hills/ridge with filled in valleys.
8. Dissected pediplain.

The coastal plain (CP) ranging in width from 1 to 12 km. is active comprising mostly a single berm beach and beach ridges with intervening backwaters. These indicate that the coast is of emergent nature. The coastal low land etch plain (LEP) extends between coastal plain and the peripheral residual hills complex with a width ranging from 4 km in north, to 50 km. in the central part. It is constituted by the bun shaped polyconvex hills and flat topped mesa-like hills punctuated by open trapazoidal valleys. The maturity of this landform unit increases towards coast. Round hills have truncated lateritic profiles whereas flat hills have hard silerete curraise at the top. Valleys contain mixed colluvio-fluvial material near to the ghats and sandy clayey-material near to the coast. At places one to two terraces also have been cut, e.g., near Udipi. Valleys are under cultivation whereas hill slopes have sporadic plantations. On the whole thus, two types of etch plains are found—lateritised and dissected formed by intensive chemical weathering, and wash and creep processes.

The peripheral residual hills (RHZ) represent the detatched portions of the main western ghat plateau mass. Since the degree of denudation is higher away from the main mass the size of the hills decreases outwards. Their orientation is controlled by the strike of the rocks. The western composite slopes (wes) comprise the precipitous main ghat scarp and the radiating ridges and spurs from it with steep slopes. Mostly they are under a thick forest cover. Slopes as well as scarps are basically of erosional nature. Highly sinuous outline of the main ghat scarp has resulted due to the deep indentations created by headward advancing west flowing streams. Width of this zone varies from 3 km in the central portion to 30 km near Mercara. Bordering the upland in the west, standing immediately above the ghat scarp, there is a linear zone of residual hills (RHT) whose width varies from less than 1 km. to 12 km. The ridges are mostly strike ridges and have knife edge sharp crests and V-shaped intervening valleys. Slopes are mostly convex and covered by a thick vegetation. Perched at the top of these are a few remnants of old planar surfaces (RDP). These have further been dissected to form bowl type enclosures. These represent the surfaces which were uplifted in the past truncating the then on-going cycle of erosion.

Towards east the 'RHT' gives way to the Upper Etch Plain (UEP). The transition is abrupt and the fall in elevation is from 1890 m to a steady 800 to 900 m of hill top accordance. This is constituted by interconnected polyconvex hills and flat bottomed valleys. Intensive, chemical activity is consuming different rock types uniformly leaving a thick mantle of weathering. Fine weathered material is carried down the hill slopes by water and deposited in the valleys which have thick loamy clayey soil of mostly reddish colour. These valleys are intensively cultivated where as the hill slopes are cut into benches and put under plantations of coffee, tea, etc. Maturity of this etch plain decreases towards east.

In the upper central portion, there are linear and semi-circular zones of structural hills (SRV) of ferruginous quartzites and intervening filled in valleys. Enclosed within the Baba-Budan semi-circular ridge complex is a thickly filled-in valley (VF/SVF). The ridges stand out to their structural control and hard resistant lithologic content.

The eastern most zone is the dissected pediplain (DDP) having a sub-humid to semi-arid climatic variations towards east. A dendritic net of ephemeral drainage is dissecting it. There are residual hills here and there surrounded by pediments of different categories. The stream channels are cultivated.

The above described landforms classified on the morphogenetic basis now can be analysed for either constructing the geomorphic evolution of the area or towards assessing the environmental geomorphic picture whereby the economic potential of each zone can also be evaluated.

This case study indicates that the regional geomorphic remote sensing studies can be very conveniently conducted on the Landsat imagery. This is more so in case of areas which have high relief hampering the accessibility and have thick vegetation and weathering mantle affecting the mobility and field observations. In such cases the saving in time and cost factors is an important gain. The geomorphic maps thus prepared can be further utilised for land system, land face and landuse studies. The landforms with genetic classification are helpful in soil studies and even in the exploration of minerals produced by weathering processes.

SUGGESTED READINGS

Bedi, Naresh, 1978 : Geomorphic evaluation of Andhra Pradesh on Landsat imagery. *Proc. Symp. Morphology and evolution of Landforms*, Dept. of Geol. Delhi Univ. (NRSA Technical Report No. 0728).

Bedi, Naresh, 1979a : Regional hydrological studies for ground water prospecting using Landsat data in two different parts of Southern India. (*NRSA Technical Report* No. 0743).

Bedi, Naresh, 1979b : Larger perspective for geomorphic studies on Landsat imagery, *Proc. 13th Inter Symp. on Remote Sensing of Environment*, ERDM, An Arbour, Michigan USA. (NRSA Technical Report No. 0745).

Bedi, Naresh, 1980 : Chapter on geomorphology in NRSA report on satellite remote sensing survey of part of west coast region (from Ratnagiri to Kanyakumari) prepared for CWC (Govt. of India) Vol. I part II pp. 12-36. (NRSA Project Report prepared for CWC).

Fairbridge, Rhods W., 1967 : *Encyclopaedia of Geomorphology*, Reinhold Publ. Corp. New York.

Lueder, D.R., 1959 : *Aerial photographic interpretation principles and applications*. McGraw-Hill Book Co. New York.

Rao, D.P., 1979 : Utility of Landsat coverage in small scale geomorphological mapping—some examples from India, *Int. Symp. on Res. Eng. and Tech.* IIT, Bombay. (NRSA Technical Report No. 0727).

Rao, K.R., 1976 : Utilisation and applications of remote sensing techniques—key note address at *IETE Symp.* on remote sensing in India, Ahmedabad. (NRSA Technical Report No. 0708).

Rao, K.R., 1977 : Remote sensing for natural resources Key note address at the state convention on natural resources, A.P. Academy of Sciences, Hyderabad. (*NRSA Technical Report* No. 0731).

Rao, K.R., 1978a : Remote sensing in arid zone surveys. *Int. Sym.* on Arid zone Research and Development held at Jodhpur, (CAZRI) (NRSA Technical Report No. 0741).

Rao, K.R., 1978b : Earth resource surveys by satellites. Seminar on Satellite Communication (P & T) New Delhi. (*NRSA Technical Report* No. 0746).

Verstappen, Th., 1977 : Remote sensing in geomorphology, Elsevier Scientific Pub. Amsterdam.

INDEX

Abich, H., 281
Abikh, G. (geologist), 256
Ackermann, H.D., 72
Aerial photographs, 399, 402, 408
Africa, 55
Agalega Island, 356
Aggradation plain, 184
"Aggregates stability" (mechanical resistance of soil), 33
Agrawal, Jag Mohan, 326
Ahmad, E., 102, 111, 120, 121, 125, 369
Ahnert, F., 3
Ajara-Imereti Range, 287, 288, 293
Akagi, Y., 73
Akhuni Ridge, 92; erosion steps of, 93
Alabama, 50
Alases, 306; coalescence of, 308
Alaska, 72, 313
Albany, Western Australia, 156
Alexander, L.T., 187
Alice Springs, N.T., 152, 153
Allen, E., 363
Allen, J.R.L., 363
Alpine karst, 256; massives, 258; paleoglacial schemes, 286; plains (plateaus), 84; tradition, 281
Alpine mountain system, 214
Althore peaks, 200
Alwar quartzites, 171
Amazon delta, 363
Ambrose, J.W., 75
America, Pacific coast of, 354
Anaglyph method, 391
Ancient glaciation, regional survey of, 290
Anfrere, L., 351
Antarctica Ocean, 330
Anthropogenic history, 84
Anticaucasus upland, 83
Arabian Gulf, 356
Aragats massif, 285
Aravalli range, 111; region, 104ff, 110
Archaeans, 107
Arcoona plateau, 190, 198
Arctic ocean, 330
Area (in remote sensing), general attributes of, 410
Arsena Okrojana Shvilis Cave, 264
Ashfaque, S.M., 360, 369
Asia, 40, 214
Astakhov, N.E., 85, 96
Astronomical data, 15ff
Atkinson, T.C., 208
Augite-enstatite granite, 120
Australia, 6, 47, 76, 134, 161, 187, 188, 249, 337, 339, 357; coast of, 351
Australian aborigines, 189
Austria, Alps in, 211, 214; Eirienseswelt cave, 231; Gross Badlhohle in, 230
Axelrod, D.I., 352
Ayyar, N.P., 112

Baffin Island, 39
Bagchi, K., 102
Baghelkand plateau, 102
Bailey, H.P., 352
Bain, A.D.N., 131, 152, 157
Balagot, V.F., 357

Balk, R., 153
Balazs, D., 212
Banda Pahar, 122
Bangalore, 403
Bangladesh, 360; coast of, 357, 369
Barakar basin, 115, 118, 122, 123, 125
Barbier, R., 130
Barrell, J., 63
Barrére, P., 72
Barrier islands, 334, 339
Barron delta, 361
Barton, D.C., 131, 152, 156
Bascom, F., 63
Bascom, W.N., 341
Bathymetric data, 387
Bauer, F., 214
Baulig, H., 67
Baydzharakhs, 308
Bazárdjuzi mountain, 293
Beach crest, 337; material, nature and sources of, 363; morphology, 337; ridge plains, 362
Bedi, Naresh, 399
Belgium, R. Ourthe's cave, 238
Bengal, Bay of, 169, 389
Benson, W.N., 195
Bereznjakov, A.I., 292
Bermo Carboniferous earth movements, 125
Bihar, 119
Binjh range, 109
Biocenosis, influence of, 32
Bird, E.C.F., 325, 334, 351, 354, 358, 361, 363, 365-369
Birot, P., 5, 57, 94, 130, 150
Birot, P.A., 74
Black Sea, 281, basin, 84; shore, Gudauta, 263
Blackingstone rock, 128
Blackwelder, E., 72
Bleahu, D.M., 212
Bodra Pahar, 124
Boer, G. de, 333
Bögli, A., 211, 226, 227, 238, 242
Bohemian massif, 159
Bombay, 357
Bornhardts, W., 127; age of, 144 ff; distribution of, 132 ff; modifications of, 151ff
Bose, S.C., 119, 120, 122, 123
Botswana, 160, 161
Bourne, J.A., 72, 75, 127, 129, 135, 138, 140, 141, 146, 149, 152, 153, 156, 157, 193, 198, 199, 201
Bowen, A.J., 337
Boyé, M., 140, 141
Bradly, W.C., 159
Bradshaw, M.P., 337
Bray, L G., 224, 226
Brazil, 360, 364; coast of, 354, 366
Brazilian plateau, 365
Bretz, J.H., 76, 235, 236, 244
Brook, D., 244, 248
Brook, G.A., 134
Bryan, K., 72
Buchanan, F., 188
Budagov, B.A., 296
Büdel, J., 37, 43, 44, 53, 55, 132, 314, 351
Burma, 357
Butt, C.M., 192
Bzibi range, 262, 263

Cady, J.G., 187
Cailleux, Andre, 4, 5, 19, 29, 57, 58, 243, 351, 365, 369
Caine, N., 159
Cairns Bay, 365
Calcrete, 191
California, 128, 135, 144, 146
Callen R.A., 150, 199
Canada, 39, 75, 304, 317
Canadian Hydrographic Department, 389
Canadian Lead, New South Wales, 245
Cann, J.R., 367
Canton Island, 358
Cape Province, Bushmanland Surface of, 74
Caribbean Island shores, 364
Carr, A.P., 339
Cartwright, D.E., 345

Castle Koppies, 153 ff, 157, 160; rock, 156
Catastrophic erosion, 42
Caucasian glaciation Chronology 279, 297, 298, problem of, 294 ff; Isthmus, 255; study of, 279, 281
Caucasus, glaciers of, 282ff; minor, 90; slope of, 284
Cave, G.B., 241
Cave Breakdown, 229ff
Caves, 225, 234; epiphreatic types of, 239; evolution of, 244; evaluation, foctors in, 242 ff; types of : Dolomite, 249; Gypsum, 249; Epiphreatic, 245; Limestone, 224, 249
Caves, Abrskil's, 263, Castle-guard, 247; Demanovskè, 245; El Abra, 240; Garakha, 271 Kartskheli, 271; Mendip, 237; Mothena, 264, 273; Mullamullang, 231; Nazodelavo, 271; Sakazhia, 270, Sathaplia, 263; Sovekuo (conglomerates), 271; Tekenteri, 270; Tskhrajvari, 268ff
Cenozoic periods, 88
Central Australia, 144, 150, 193, 199
Central Europe, 22, 41, 47, 313
Central Highland, 101, 105
Central Norway, 39
Chapman, C.A., 139
Chappel, J., 337
Chatterjee, S.C., 123
Chatterjee, S.P., 102
Chesil Beach, 339, 341
Chikishev, A.G., 247
Chorley, R.J., 3, 50
Chota Nagpur plateau, 102, 110, 111, 118, 124
Choubey, V.D., 75
Christiansen, S., 370
Cimea, 256
Classical karst contact, 216
Clayton, R.W., 141
Climate, ecological aspects of influence, 26, 27
Climate and process (relationship), 44ff
Climate process system, 45, 46
Climate related to sediments, 49t
"Climatic accidents", 21; geomorphology, 4, 24, 29, 37, 57; difficulties of, 28; parameters, correlation of, 29; range overtime, 41 ff; zones, 43t
Climatogenetic geomorphology, 37, 57
Climatomorphological zones, 55
Cline, M.G., 187
Cloos, H., 153
Coast, 327; classification of, 328; hardrock, 332; low marine, 332; mangrove fringed, 362; neo-trailing edge, 329; soft rock, 332; steep and rocky, 359; storm wave, 330; trailing edge, 329
Coastal dunes, 367ff, 368; geomorphology, 351; land forms, 325, 327, 344, 352; *Biotic process of*, 346; process of, 342; plains, 360, 362, 414, processes, 325
Coasts, geometrical and rhythmic forms, 335 ff; monsoon, 338
Colchis flora, 257
Colombia, 30, 355
Colombian Andes, 33
Colorado, 72; Rockies, 135, 139, 159; valleys of, 149; stand, 160
Contemporary glaciation, 291; glaciers, dimensions of, 284t; distribution of, 284t
Contour layer and tint method, 390
Cooling Corrosion, 227
Cooleman plain, New South Wales, 244
Corbel, J., 48, 50
Cotton, C.A., 21, 22, 67, 141
Cowell, P., 337
Cowie, J.W., 75
Cretaceous-Paleogue limestones, 256

Crickmay, C.H., 67, 68
Crimea-Caucasion province, 94
Curray, J.R., 340
Czudek, T., 305, 316

Daily, B., 129, 150, 156, 198, 199
Dale, T.M., 128
Damodar trough, 125
Darwin, N.T., 151
Dave, V.K.S., 121
David, T.W.E., 76
Davies, J.L., 329, 336
Davies, S.N., 235
Davies, W.F., 238
Davis, W.M., 21, 35, 38, 64, 66, 67, 230, 239, 280, 301
Davision Geomorphology, 22
De Saussure, H.B., 25, 131
Deccan, 55; Foreland, 6, 101, 102, 105, 110, 112; plateau, 101; trap, 75, 102
Degradational and dissected plain, 123ff
Deimos, 15; morphology of, 18
Delhi, 169, 174
Demek, J., 159, 280, 305, 313, 316
Derbyshire, E., 351
Development, theories of, 130ff
Devil's Marbles, N.T., 129
Dey, A.K., 361, 363
Dhanbad, 119, 124
Dharwar, 102
Didgverdi massif, 290
Dill, R.F., 244
Dikhtau, summit of, 291
Dinaric karst, 213
Dissected pedi plain, 415
Dolidze, J., 94
Donghuzorum, summit of, 291
Dogleg caves system, New South Wales, 238
Douglas, I., 47, 212
Dowling, R.K., 225
Dresch, J., 94
Drew, D.P., 208
Dube, R.S., 109
Dumanowski, B., 141
Dumitrashko, N.V., 85
Dunkerley, D., 24, 26
Dunn, J.A., 102, 111
Durgapur, 124
Duricrusts, 75, 187ff
Duyoda, 308
Dynamic equilibrium, 38; geomorphology, 37; phreas, 237ff
Dzangitan, summit of, 291
Dzirula horst, 86; massif, 87, 89; peneplane, 87
Dzulynski, S.T., 159

Early Cretaceous planation surface, 137
Earth, climatic morphological zones of, 53t
East Chkhorotskhu, 271
East African inselbergs, 150
Eastern Germany, 22
Eastern Siberia, 316
Ecosystem (ecodynamics) biocenosis of, 33; dynamic of, 34
Ede, D.P., 212
Eggler, D.H., 159
Egypt, 141
Ek, C., 213, 233
El Salvador, pacific coast of, 362
Embleton, C., 302
Emilian, C., 42
Energy, source of, 11
England, 23; east coast of, 340; Mendip caves in, 229
Enguri basin, 291
Epigenetic permafroast, cryogenic textures, 312
Eraso, A., 211
Erosional surface, 109, 115ff
Etch plains, 75; *surfaces*, 75
Europe. 5, 21, 46, 57, 223
Ewers, R.D., 237, 239, 240, 247-249
Exhumed plains, 75
Extensive cave systems, 223
Eyre Peninsula, 73, 139, 141, 142, 193, 199, 201

Facet, salient characters of, 377

Fairbridge, R.W., 150, 198, 199, 314
Fenner, C., 195
Ferricrete, 188
Finkl, C.W., 150, 198, 199
Fish, J.E., 240
Fisher, O., 64
Flandrian transgression, 339
Flinders ranges, 137
Flint Mammoth caves (world's longest), 243; system, 223, 245
Flint ridge, 243
Flohn, H., 41, 46
Flores, J.E., 357
Fluvial erosion, 45; sediments, 318
Fontaine de la Vauciuse, South France, 239
Ford, D.C., 215, 224, 237, 239-240, 247, 249
Fosberg, F.R., 352, 357
Fournier, F., 47-50
France, 213, 256; Les Baux in, 188
Frankel, J.J., 189, 191
French, H.M., 316
Fritsch, P., 140, 141

Gams, Ivan, 208, 211, 212, 215, 216, 218
Gascoyne, M., 211
General systems theory, 3
Geomorphic processes, 29, 31, 33, 34
Geomorphological analysis, 38; research, 3
Geomorphology, quantitative tendency of, 9
Georgia, 83, 89, 92, 255; Kalichona cave in, 27
Georgian karst, 255, 257, 261, 272, 274
Gerassimov, I.P., 88
Gerson, R., 214
Gerstenhauer, A., 211
Ghana coast, 367
Gierloff-Emden, H.G., 333, 362, 363
Gigineishvili, G., 87
Gilbert, G.K., 128, 334
Gilluly, J., 68, 72, 146
Giridih, 122; coal basin, 122
Glacial processes, 29
Glaciated zone, 55
Glacier, Abudelauri, 295; contemporary, 289, 292; Didgverdi, 290; Mulakhi, 289; Namkvami, 285
Glaessner, M.F,, 196
Glennie, E.A., 237, 239, 243
Goa, 357
Godard, P.A., 74, 139, 154
Gombori ridge, 97
Gondwana, 87, 102
Goudie, A., 188
Gouffre Jean Barnard (French Alps) (World's deepest wave), 223
Granite inselbergs, types of, 161
Great Artesian basin, 75
Great Caucasian range, 255
Great October Revolution (1917), 256
Greece, 244
Griggs, D.T., 152, 157
Gromov, V.I., 296
Groom, G.E., 212
Ground surfaces, mechanical properties of 33
Guyana, 55; coast of, 363
Guza, R.T., 337
Gvozdetskiya, N.A., 247, 256

Habic, P., 212
Hack, J.T., 76, 138
Hadley, R.F., 71
Halliday, W.R., 224
Hamersley ranges, 198
Handley, J.R.F., 132
Harmon, R., 211
Harris, W.K., 150, 199
Hassenfratz, J.H., 132
Hawaiian islands, 358
Hays, J., 157, 198
Heron, A.M., 110, 174
Hills, Bundi, 104; Carappee, 201; Chepo, 122; Chiltern, 52; Christian, 122; Corrobinnie, 134, 152; Dalma, 104; Delhi University,

174; Dhangi, 124; Jawaharlal Nehru University, 174; Kenja, 124; Khandauli, 122; Kokerbin, 134, 201; Moradabad, 174; Parasnath, 118, 119, 120, 124; peripheral residual, 414; residual, 121; zone of, 415; Qutab, 174; Teharkuldu, 153; thousands, valley of, 134, 149; Tundi, 124; Ucontitchie, 139; Waulkinna, 154
Hills, E.S., 74, 198
Himalayan limestones, 249; Orogeny, 118
Historic-genetic approach, 3
Holland, coast of, 339
Hollerwoger, F., 361
Holloch cave (Switzerland), 241, 245
Holmes, A., 128, 130
Holocene, 26
Hong Kong, 151
Hood Bay, 369
Hopley, D., 351, 358, 360, 368-370
Horwitz, R.C., 191, 192, 199
Howard, A.D., 71, 130, 235
Howchin, W., 76
Hoyt, J.H., 334
Hugo, M.L., 134
Humid tropical catchments, 359; climate, 214; coasts, 352, 356, 357, 359, 366, 367; environments, 369; zone, 370
Humps, 200
Huon Gulf, 366
Hutton, J.F., 150
Hutton, J.T., 189-191
Hyden rock, 134
Hydrodynamic domains, 232ff
Hydrographic survey, 383
Hydrographic surveys and nautical charts, result of, 383ff

Ialbuzi, summit of, 291; mountain, 282, 285
Ice, thermal conductivity of, 27
In situ measurements, 402
India, 249, 338, 389, 402; coast of, 329, 340; east coast of, 331; south-west coast of, 361; west coast of, 331
Indian Ocean, 363
Indian Photo-Interpretation Institute, Dehra Dun, 402
Indonesia, 18, 189, 365
Indonesian archipelago, 357, 370; in stant shores, 364
Ingle, J.C., 343
Inman, D.L., 328, 337
Intensive linear erosion, 314
Interception, radiative aspect of, 30
Intermittent uplifts, age and cause of, 110ff
International Hydrographic Organisation, 387
Ipari Community, 289
Isaeva-Petrova, L.S., 94
Ishikawa, M., 216
Ivory coast, 356, 362, 363, 365, 367

Jack, R.L., 191
Jackson, I.J., 352
Jahn, A., 302
James, J.M., 224, 243
Japan, 73
Jawaharlal Nehru University, New Delhi, 169, 184
Jennings, J.N., 191, 209, 212, 226, 229, 232, 238, 243—246, 248, 368
Jessen, O., 154
Johnson, D.W., 65, 67, 71, 331, 334
Jones, L.J., 245
Jones, T.R., 128
Joppa ridge, 243
Julian, M., 212

Kabardo upland, 286, 287
Kala community, 285
Kangaroo Island, 129, 156, 195, 196
Kanker (lime-concretious), 173
Kapudjikh mountain, 282
Karroo, 73
Karst, 207, 208, 215, 225 229; basis, 272; belt, West Georgia, 257,

270; denudation, 211, 212, 219 development, geological substratum and geomorphological conditions, 256; processes, climatic conditions of, 256; spaces, 318
Kathiawar Peninsula, 356
Kavkasioni, mountainous, system, 83, 89, 90, 94, 96
Kazbegi mountain, 282, 285; region 287
Keli, volcanic plateau of, 279
Kemp, E.M., 152, 199
Kennedy, B.A., 3
Kent, L.E., 189, 191
Kentucky, 223
Kerala, 357
Keyes, C.R., 63
Khazaradze, R.D., 286, 292
Khevsuretis-Aragvi, river bed of, 295; valley of, 295, basin, 299
Khonu, coalescence of, 308
King, C.A.M., 4, 64, 302, 325
King, L.C., 63, 64, 71, 72, 130, 135, 137, 144
Kinetic energy, 30, 31, 33
Kipiani, Sh., 256
Kodori, basins of, 289, 294
Kolbezen, M., 212
Koppen, W., 353
Korea, 73
Kosak, H., 214
Koshtantan, summit of, 291
Kosov, B.F., 85
Kotarba, A., 159, 212
Krasnov, A., 285
Kruber, A., 256
Kudaro caves, complex of, 268 multi-tier cave system of, 274
Kudelin, B.I., 252
Kular ridge, 305
Kulla, J., 313
Kumar, Pramila, 5
Kunaver, J., 212
Kuznetsov, S., 85

LANDSAT data, 403; image, 404-406, 412, 416; interpretations of, 410
Lake Eyre, 192; basin, 191, 193
Landforms, shapping of, 26, 34
Landsberg, H.E., 354
Lang, S., 212
Langbein, L.B., 46, 47
Lankori pass, 89
Larsen, E.E., 159
Laterite, 187
Lateritised old plateau, 75
Laurasia peneplanes, 87
Le Bourdice, P., 363
Lechkhumi Ridge, 89
Lehmann, O., 64, 387
Leonov, L., 256
Lester, J.G., 130
Lichty, R.W., 38
Likhi ridge, 87
Lilien berg, D.A., 85
Lincoln uplands, 195
Lindsay, J.M., 149, 193
Linton, D.L., 132, 139, 154, 159
Lister, L.A., 137
Little, L.S., 63
Ljubin, V.P., 268
Loffler, E., 359, 366, 369
Logan, J.R., 75, 141
Lokhi range, mountains of, 293
Longitudinal profiles, study of, 108
Longyear, J., 224, 235
Louis, H., 130
Lovering, T.S., 191
Lower Gondwana sedimentary rocks, 125
Lower Hazaribagh Planation level, 121; plateau, 123, 124
Lower Proterozoic hatches creek group, 157
Lowry, D.C., 238, 248
Lozinski, Walery, 302
Ludbrook, N.H., 199

Mabbutt, J.A., 71, 75, 130, 153, 157, 198

Macar, P., 316
Macdonnell ranges, 161
Mackin, J.H., 69
Mahabar pahar, 118
Mahudi pahar, 118
Maignien, R., 187, 189
Maksimovic, G.A., 215, 218, 257, 273
Malaya, 368; coast of, 366
Malaysia, 189, 357, 359
Malott, C.A., 242; Vandose caves of, 242
Malwa plateau, 105
Mammoth cave ridge, 243
Mangroves, deltaic coasts of, 360ff; swamps, 370
Mann, A.W., 191, 192
Manohar, M., 338
Markovicz, Lohinowicz, 213, 214
Mars, natural stellites of, 14t
Martel, Edwar Afred, 256
Martian Volcanoe, 16ff
Maruashvili, L.I., 5, 93, 97, 209, 256-258 273, 279, 281, 286, 289, 296
Mass movement, 45
Mathes, F.E., 135
Maucci, W., 233
Maurin, V., 211
McArthur, J.A., 354
McIntire, N.G., 369
McGee, W.J., 72
Mediterranean, 94; climate, 41, 42; zone, 54
Meghvriki mountain, 293
Melnikov, P.I., 304
Mennell, F.P., 132
Mesozoic, 87, 88; peneplane, 89
Metasedimentary, Babahudan group of, 412
Mexico, 354
Micro-organisms, 32
Middendorf, A.F., 301
Middle Jurassic age, 198
Middle Khrami group, mountains of, 293
Mihijam pahar, 124
Miliankovich, M., 42
Milanivski, E.E., 296
Miller, R.P., 190
Milne Bay, 366
Milnes, A.R., 129, 150, 156, 189-191, 198
"Minor glacial epoch", 297
Miotke, F.D., 245
Misra, V.C., 112
Missouri, Cameron cave in, 235
Mongolia, 159
Monsoon lands, 54
Monsoons and cyclones, effects of, 369ff
Moore, G.W., 238
Morphogenetic regions, 56t
Morphogenic system, 29
Morphological types, caves of, 261
Morton, J.K., 368
Mount Boundry area, 128; lofty ranges, 195, 196; manypeaks camplex, 156; stirling, 134; Wudinna, 201
Mt. Owen plateau, New Zealand, 247
Multispectral Scanner, 402
Murgatona pahar, 124

Namibia, Kuiseb plain of, 74
Naseemuddin, M., 6
National Geographical Society, 387
National Remote Sensing Agency, Hyderabad, 402
Naval Hydrographic Office, Dehra Dun, 389
Neogene subtropical flora, 257
Neogenic, 85; planes, 89ff
Netarhat plateau, 102
Netterberg, F., 191
New Guinea, 365
Newsom, M.D., 207, 229
New South Wales, 47; Bungonia caves of, 242; Jenolon caves of, 237; Murray caves of, 230; River Cave Cooleman plain of, 229; Wyanbene caves of, 230
New York, Skull caves of, 249

New Zealand, 180, 248; Nairimu cave of, 244
Nicob, J., 212
Niger delta, 363; river basin, 32
Nordstrom, C.E., 328
Nossin, J.J., 360, 366, 368, 369
North America, 40, 223, 294
North Bihar, 111
North Chile, 30
North Europe, 294
North Queensland, 73; coast of, 361
North Westren Yugoslavia, 211
Norteast Queensland, 368
Northeastern Australia, 370
Northemphreatic, 239
Northern America, 211
Northern Australia, 195, 198
Northern Caucasus, 94; glaciers of, 288
Northern Israel, 214
Northern Sedimentary Alps, 211
Northwestern Dinaric Karst, 212
Nothephreas, 234ff
Nothephreatic, 237
Nubbins, 151ff

O' Reilly, P.M., 226
Oberlander, T., 130
Oberlander, T.M., 71
Obst, E., 135
Ocean, 381
Ocean Island, 358
Oceans, General bathymetric charts of, 387
Ochtinska Aragonite Cave, Slovakia, 237, 245
Offshore wind, 344
Okriba Argveti Ridge, 268; region, 89
Old planar surfaces, remnants of, 415
Ollier, C.D., 52, 130, 139
Olmstead, E.W., 63
Ongkosongo, O., 363
Onshore monsoon, 369; winds, 344
Opik, A.A., 74, 75, 191
Osetian plain, 286
Oshten mountains, 290
Ottmann, F., 365, 367

Padmaja, G., 105
Pafenholts, K.N., 286
Pakistan, 356
Paleo-glaciation, 281
Palaeoclimatic conditions, 46
Palgrave, W.G., 281
Palmer, A.N., 245, 249
Pantjukov, I., 256
Papua New Guinea, 357; Finim Tel caves of, 244
Paralana Fault zone, 193
Parker, F., 248
Parthsarthi tor, 180
Passarge, S., 63
Paton, I.R., 189
Pavoms Mons, 17t
Pediments and pedimentation, 68ff
Pediplains, 73ff
Peel, R.F., 74
Peela rock, 134
Pegmatites, 172
Peltier, B.C., 21, 52
Peltier, L.C., 21, 22, 45, 51, 56, 57
Penck, A., 52
Penck, W., 64
Pendleton, R.L., 187
Peneplains, 72, 87
Pereplanation planes, 64ff, 85
Pennsylvania, 50
Periglacial geomorphology, 280, 301; problems of, 302; processes, 52; zone, differentiation of, 303ff
Permafrost degradation, 306, 308
Perry, H.G., 76
Persati plateau, 293
Petit, M., 74
Peru, 30
Pewe, T.L., 313
Phobos, 4; exception of, 15; morphology of, 18
Photosynthesis, 28
Phreatic caves, enlargement of, 241
Picknett, R.G., 224, 225, 227
Pierce, J.W., 334

Pierre, St. Marten, 262
Planation surface, 115
Planet mars, volcanoes and satellites of, 12ff
Pleistocene fossiliferous sediments, 199; glaciation, 287, 291; characteristics of, 285ff; planation surface, 193
Plio-Pleistocene age, 199
Pliocene, 84
Pluvial erosion, 45
Podozerski, 284
Popov, I.W., 247
Post-cryogenic textures, 312, 313
Post-Gondwana erosion, 120
Post-pliocene age,111
Prasad, Nageshwar, 6
Pre-Cambrian granites: 102; rocks, 412; sedimentaries, 102
Pre-Caucasian plain, 281
Prescott, J.A., 187
Present day climatic range, 39ff
Price, W.A., 191
Prider, R., 188, 198
Priesnitz, K., 212, 214
Process and form (relationship in climato geomorphology), 50ff
Pugh, J.C., 71, 130, 141

Quaternary climates, 25; deposits, 172ff
Quartzitic tors, 179
Queensland, 75, 76, 235; great barrier reef of, 364; Humid tropica, 365; Ryan Imperial cave, 235

Rachna karst massif crest of, 268
Radiation, 28
Rahn, P.H., 69, 72
Rai, Raj Kumar, 5
Rajasthan, Aravallis of, 6
Ramsay, A.C., 63
Ranchi plateau, 102
Rao, K.R., 399
Rao, Prasada, 340
Rastvorova, V.A., 85
Regional erosion, rates of, 48t
Reinghard, A.L., 286, 292
Remote sensing, 402; data, 405: evolution of, 402ff; inter-pretation of, 406; nature of, 403
Remote Sensing Geographic Mapping, sequences of, 408
Remote Sensing techniques, 401; advantages of, 404; geomorphic mapping, 409
Renault, P., 229, 231, 245
Renault's, Paragenetic cave development in, 236
Resistant linear ridges, 184
Resurrected land surface, 75
Rewa plateau, 109
Reynolds Range, 152, 161
Rhodesia, 161
Richards, P.W., 351
Richthoffen, A., von, 21
Rione-kura, 83
Riss-Wriim glaciation, 296, 297
Ritchot, G., 154
Rittmann, A., 16, 18
River, Abasha, 273; Abudelauri, 295; Aldan, 306; Ardon, 292; Argun, 292; Assa, 292; Avad-khara, 290; Barakar, 119; Barsoti, 119; Bzibi, 261, 290, 291; Chambal, 104, Citarum, 361; Congo, 356, Ganga, 169, 329; Godavari, 340; Gvandra, 294; Indigirka, 305; Indus, 329; Kelasuri, 272; Kemp Welch, 369; Khakho, 122; Kodori, 287,20; Kusarchai, 293; Lena, 301, 306, 315; Marcha, 304; Mchista, 263; Meuse, 339; Mohana, 119; Mtkvari, 293, 294; Mzimta, 290; Niger, 329; Pre-Tergi, 287; Rhine, 339; Sakeni, 294; Samur, 293; Sassandra, 365; Shashe, 134; Sindh, 169; Subernarekha, 104; Shavi-Tskhali, 288; Teberda, 291; Tergi, 286, 292; Toba, 264; Umgeni, 137; Usri; 122; Yamuna 169, 173; Zaire, 329
Roberts, W.J..76

Robinson, A.H.W., 345
Rocklin Quarry, 128
Roglic, J., 218
Romani, Lucian, 4, 16
Rock, H., 213
Rosser, H., 190
Roy Choudhury, P.C., 119, 120
Russell, R.J., 340, 367
Russian plain, 281

Safronov, I.N., 85, 94
Saharan plains, 74
"Sakhinule" (ice caves), 264
Salt crystallisation, 231
Samsar mountain, 294
Sametskhvario mountain, 293
Sand-sized beach, 341
Sand waves, 345
Santal parganas, 124
Sapper, K., 358
Satya Prakash, 326
Saville, T., 343
Scarp zone, 118ff
Scarpfoot depressions, 141; zone, 160
Scherbakova, E.M., 85
Schneealp karst, 214
Scholz, D.L., 156
Schumm, S.A., 38, 46, 47, 50, 68, 146
Sea, 381; Arabian, 169; A sov, 281; Baltic, 330; Caspian, 281; Maikop, 83; Mediterranean, 330; Red, 356; coasts of, 329
Second World War, 301
Selby, M.J., 130
Semi-arid areas, 40; climates, 49, 51, 173
Selminum Tel, Papua New Guinea, 244
Semmel, A., 3
Sen, D., 174, 177
Sengupta, A.K., 102
Serebjany, L.R., 294, 297
Seurin, M., 140
'Shallow phreatic', 239
Shaori polje, 260
Sharma, H.S., 6, 105
Sharma, N.L., 124, 125
Sharp, R.P., 72
Shashe Castle Koppies, 161
Shepard, F.P., 331
Shephered, J.A., 198
Shergin, F., 301
Short, A.D., 336, 337
Siberia, 306, 313, 317
Siberian rivers, valleys of, 314
Siddiqi, J.A., 369
Sierra Nevada, 129
Silcrete, 189
Silvester, R., 335, 336
Singh, R.P., 102
Sinha, R.C., 121
Sivarajasingham, S., 187
Slope analysis, 181
Slope and drainage density (relationship), 51t
Slovakia, Low Tatra mountains of, 245
Smith, D.I., 208
Smith, D.M., 76, 141, 146, 152, 157, 193, 198, 199
Soen, Oen Ing, 135
Soil freezing, analysis of, 29
Solar radiation, 29
Sonar-Bearma basin, 101, 105ff, 111
South Africa, 134; cape fold belt of, 72
South America, 46, 367; Atlantic coast of, 354
South Australia, 76, 189, 191; flinders ranges of, 190; gulfs region of, 195; Murray Gorge, 149
South China, tower karst of, 216
South France, 212
South-Georgian upland, 90
South Poland, 21
Southeast, wheat belt of, 200
Southeast Asia, 367
Southeast England, 52
Southwest England, 50
Southwest Queensland, 192
Southern Africa, 144
Southern Australia, 154

Southern Europe, 214
Southern Florida, 353
Sourhern France, Sidobre of, 128
Southern-western and control Australia, stepped inselbergs, 145t
Speleological union, 215
Spitzberg, 27
Splash erosion, 33
Sri Lanka, 357, 369
Stenner, R.D., 224
Stereo-aerial photography, 402
Stereoscopic techniques, 408
Stoddart, D.R., 38, 346, 351, 367
Storm waves, 342; environment, 333
Stepped bornhardts, 141ff
Strahler, A.N., 46
Strakov, N.M., 48
Stride, A.H., 345
Sturt, C., 195
Submarine navigation, 379; terrain, 391; charts, 382; classification, 375ff; comparative analysis of the proportion of graphics in, 393t; evolution, 375; utility of, 379ff; graphics in, 382ff
Suess, E., 328
Surface denudation processes, 54
Surinam coast, 363
Svensson, H., 313
Swampy lowlands, 361
Swan, S.B., 369
Sweden, 313
Sweeting, M.M., 191, 212, 213, 215, 216, 232, 238, 246, 247
Swell waves, 336, 342
Swift, D.J.P., 334
Swinnerton, A.C., 239
Syngenetic permafrost, 309
System analysis, 32

Talkies and geomorphological significance, 314
Tasmania, 159, 189, 246; Baldocks cave of, 245
Tate, R., 76
Tator, B.A., 71
Tectonic geomorphology, 57
Tergi basin, 292
Ters, M., 74
Tertiary climate, 46
Thailand, 357
Thar-Aravalli barrier, 169
Thermo-erosion, 315
'Thermocirques', 306
Thom, B.G., 334, 337
Thomas, M.F., 139, 140, 180; 358
Thomasson, A.J., 52
Thomson, R.M., 198
Tintilozov, Z., 256, 257
Tolstikhin, N.I., 304
Tomirdiaro, S.V., 313
Torricelli ranges, 365
Tors, development of, 179
Trade, winds, 336
Transcaucasian highland, 293ff; upland, 84, 288
Trialeti ridge, 96
Tricart, J., 4, 5, 29, 57, 58, 130, 242, 243, 280, 301, 351, 358, 360, 361, 363, 365, 366
Trimmel, H., 215, 230
Triple scarps, 104
Tropical coasts, 353; zone, 55
Trudgill, S.T., 212, 326
Tsereteli, D.V., 298
Tsutskhvati multitier cave system, 268
Tuan, Y.F., 71
Tuddenham, W.G., 130
Turkey, 244
Twidale, C.R., 5-7, 67-69, 71-73, 75, 76, 127-130, 132, 135, 137-142, 144, 146, 149-157, 180, 181, 189-193, 195, 198, 199, 201
Two-stage model, 132

USSR, 280, 284, 302, 304, 313; Akhaliatoni cave, 262ff
Udaipur, 104
Udokan ridge, 304
Uganda, 75
United States, 30, 64, 256, 304, 402
Upper cretaceous limestones, 273
Upper Etch Plain, 415

Upper Hazaribagh planation, 118
Upper Miocene, 94
Urgonian limestone, 258
Utah, Big Creek cave, 249

Vakhushti Institute of Geography, 268
Valentin, H., 331, 333
Valley, Aoudelauri, 295: Ashburton, 193; Beda, 190; Calcretes, 191; Damodar, 111; Dolra, 295; Enguri, 289; Gobishuri, 295; Krokodiv river, 135; Lobster Rivulet, 246; Lower Chambal, 104, 105; Menskra, 295: Mole Creek, 246; Mulkhura, 295; Narmada, 111, Swakop, 129, 136, 152, 153; Teberda, 288; Tergi, 292; Thermokarst, 308; Yosemite, 135, 146
Van Straaten, L.M.J.U., 334-339
Van Zul, J.A., 72
Vann, J.H., 363
Vardoniants, L.A., 286, 292
Vardanjanz, L.A., 85, 86
Vegetal detritus, 33
Velichko, A.A., 313
Velikovskaia, I.M., 286
Vendose flow, 241
Venezuela, 211
Verma, P., 102
Verstappen, H.T., 361, 365, 369
Veverka, J., 12, 19
Vietnam, coast of, 357
Vindhyan range, 107
Volcanic upland, 282
Vysokoe Prilenskoe plateau, 315

Wadia, D.N., 111
Wahrhaftig, C., 72, 127, 144, 157, 180
Walker, H.J., 369
Waltham, A.C., 247-249
Washburn, A.L., 302
Water, percolation of, 28
Water table stream, 239
Waters, R.S. 74
Wave system, types of, 337
Waves Edge, 337
Wayland, E.J., 75
Weathered land surfaces, 74
Weathering and minor landforms, 159ff
Wellman, P.W., 198
West Africa, 55, 141, 363, 364, 368, 369; Ivory coast of, 356, 359, 368; Liberian coast of, 356; Sierra Leone, 356
West Australia, Dune limestone caves of, 237
West coast swell environments, 338
West Europe, 42
West European Alpine tradition, idea of, 279
West Georgia, 89, 255; karst belt, 256
Western Australia, 134, 140, 150, 156, 191, 198, 200; darling ranges of, 135; Meekatharra surface of, 74; pilbara of, 129, 136, 151, 157; ranges of, 149
Western Caucasus glaciers, 284
Western Georgia, 258
Western Ghana, 356
Western Ghat scarp, 412
Western Murray basin, 193
Western Nigeria, 362
Wet Mediterranean regions, 50; Neck cave, 244
Whitaker, C.R., 69
White, W.B., 224, 232, 235, 237, 239
White Wells Cave, Nullarbor plain, 235
Whitlow, J.R., 137
Wiegel, R.L., 341
Wilford, G.E., 248
Wilhelmy, H., 139
Williams, G.E., 75, 149
Williams, M.A.J., 189
Williams, M.A.J., 189
Williams, P.W., 225, 248
Willis, B., 127, 150
Wilson, L. 45-47
Wischmeier, soil loss equation of, 33

Wood, A., 64
Woodard, G.D., 76, 137
Wooldridge, S.W., 67
Wopfner, H., 150, 190, 192, 199
World War I, 21
Wright, L.D. 337

Yakoutia, 27
Yarwondutta, 160; rock, 142, 144
Yasso, W.E., 335
Yorke Peninsula, 193, 199
Yugoslavia, 218, 244, 256

Zangezuri range, 285
Zimbabwe, 160
Zoti, J., 211